VLSI Architectures for Modern Error-Correcting Codes

VLSI Architectures for Modern Error-Correcting Codes

Xinmiao Zhang

CRC Press
Taylor & Francis Group
Boca Raton London New York

CRC Press is an imprint of the
Taylor & Francis Group, an **informa** business

CRC Press
Taylor & Francis Group
6000 Broken Sound Parkway NW, Suite 300
Boca Raton, FL 33487-2742

First issued in paperback 2020

© 2016 by Taylor & Francis Group, LLC
CRC Press is an imprint of Taylor & Francis Group, an Informa business

No claim to original U.S. Government works

ISBN-13: 978-1-4822-2964-6 (hbk)
ISBN-13: 978-0-367-73803-7 (pbk)

Visit the Taylor & Francis Web site at
http://www.taylorandfrancis.com

and the CRC Press Web site at
http://www.crcpress.com

*To Amelia
and Ruoyu Chen*

Contents

Preface

Error-correcting codes are ubiquitous. They are adopted in almost every modern digital communication and storage system, such as wireless communications, optical communications, Flash memories, computer hard drives, sensor networks and deep-space probing. New-generation and emerging applications demand codes with better error-correcting capability. On the other hand, the design and implementation of those high-gain error-correcting codes pose many challenges.

It is difficult to follow the vast amount of literature to develop efficient very large scale integrated (VLSI) implementations of en/decoders for state-of-the-art error-correcting codes. The goal of this book is to provide a comprehensive and systematic review of available techniques and architectures, so that they can be easily followed by system and hardware designers to develop en/decoder implementations that meet error-correcting performance and cost requirements. This book can also be used as a reference for graduate courses on VLSI design and error-correcting coding. Particularly, the emphases are given to soft-decision Reed-Solomon (RS) and Bose-Chaudhuri-Hocquenghem (BCH) codes, binary and non-binary low-density parity-check (LDPC) codes. These codes are among the best candidates for modern and emerging applications, due to their good error-correcting performance and lower implementation complexity compared to other codes. To help explain the computations and en/decoder architectures, many examples and case studies are included. More importantly, discussions are provided on the advantages and drawbacks of different implementation approaches and architectures.

Although many books have been published on both coding theory and circuit design, no book explains the VLSI architecture design of state-of-the-art error-correcting codes in great detail. High-performance error-correcting codes usually involve complex mathematical computations. Mapping them directly to hardware often leads to very high complexity. This book serves as a bridge connecting advancements in coding theory to practical hardware implementations. Instead of circuit-level design techniques, focus is given to integrated algorithmic and architectural transformations that lead to great improvements on throughput, silicon area requirement, and/or power consumption in the hardware implementation.

The first part of this book (Chapters 1 and 2) introduces some fundamentals needed for the design and implementation of error-correcting codes. Specifically, Chapter 1 focuses on the implementation of finite field arithmetic, which is the building block of many error-correcting coding and cryptographic schemes. Chapter 2 briefly introduces commonly-used techniques for achieving speed, area, and/or power consumption tradeoffs in hardware designs. The second part (Chapters 3-7) of this book is devoted to the implementation of RS and BCH codes. Simplification techniques and VLSI architectures are presented in Chapter 3 for root computation of polynomials over finite fields, which is a major functional block in both hard- and soft-decision RS and BCH decoders. Chapter 4 gives the encoding and hard-decision decoding algorithms of RS codes and their implementation architectures. Transformations for lowering the complexity of algebraic soft-decision (ASD) RS decoding algorithms and their implementations are discussed thoroughly in Chapter 5. Chapter 6 focuses on the interpolation-based Chase decoder, which is one special case of ASD decoders, and it achieves a good performance-complexity tradeoff. BCH encoder and decoders, including both hard-decision and soft-decision Chase decoders, are presented in Chapter 7. The third part of the book (Chapters 8 and 9) addresses the implementation of LDPC decoders. Decoding algorithms and VLSI architectures for binary and non-binary LDPC codes are presented, compared, and discussed in Chapters 8 and 9, respectively.

Chapter 1 of this book reviews finite field arithmetic and its implementation architectures. Besides error-correcting codes, such as RS, BCH, and LDPC codes, cryptographic schemes, including the Advanced Encryption Standard (AES) and elliptic curve cryptography, are also defined over finite fields. Definitions and properties of finite fields are first introduced. Then the implementation architectures of finite field operations using different representations of field elements, such as standard basis, normal basis, dual basis, composite field, and power presentation, are detailed and compared. The conversions among different representations are also explained. To assist the understanding of the en/decoder designs presented in later chapters, some fundamental concepts used in VLSI architecture design are introduced in Chapter 2. Brief discussions are given to pipelining, retiming, parallel processing, and folding, which are techniques that can be used to manipulate circuits to trade off speed, silicon area, and power consumption. More detailed discussions on these techniques are available in [21].

Chapter 3 discusses the implementation of root computation for polynomials over finite fields, which is a major block in hard-decision decoders and the factorization step of ASD decoders of RS and BCH codes. This is separated from the other Chapters on RS and BCH decoders to make the discussion more focused. Although exhaustive Chien search is necessary to find the roots of a general polynomial over finite field, the roots of an affine polynomial can be computed directly. Additionally, when the polynomial degree is at most three, the root computation is greatly simplified by converting the polynomial to a special format and making use of normal basis representation of finite field elements.

RS codes are among the most extensively used error-correcting codes because of their good error-correcting capability and flexibility on the codeword length and code rate. Besides traditional systems, including optical and magnetic recording, wireless communications, and deep-space probing, RS codes are finding their ways in emerging applications, such as sensor networks and biomedical implants. Traditionally, hard-decision RS decoders are adopted in many systems, since they can achieve very high throughput with relatively low complexity. Chapter 4 first reviews the construction of RS codes, the encoding algorithms, and encoder architectures. Then the focus is given to popular hard-decision decoding algorithms, such as the Peterson's and Berlekamp-Massey algorithms, their complexity-reducing modifications, and corresponding VLSI implementation architectures.

Chapters 5 and 6 present VLSI architectures for ASD decoders of RS codes. Compared to hard-decision decoders, soft-decision decoders correct more errors by making use of the channel reliability information. Over the past decade, significant advancement has been made on soft-decision RS decoding. In particular, by incorporating the reliability information from the channel into an interpolation process, ASD decoders achieve a better performance-complexity tradeoff than other soft-decision decoders. Nevertheless, the computations involved in ASD decoders are fundamentally different from those in hard-decision decoders, and they would lead to very high hardware complexity if implemented directly. Many algorithmic and architectural optimization techniques have been developed to reduce the complexity of ASD decoders to practical level. These techniques are reviewed in Chapters 5 and 6. Chapter 5 discusses general ASD decoders, and Chapter 6 focuses on the Chase decoder, which can be interpreted as an ASD algorithm with test vectors consisting of flipped bits. The major steps of ASD decoders are the interpolation and factorization. Chapter 5 covers the re-encoding and coordinate transformation techniques that allow substantial complexity reduction on these two steps, the implementations of two interpolation algorithms, namely the Kötter's and Lee-O'Sullivan algorithms, as well as prediction-based and partial-parallel factorization architectures. When the multiplicities of all interpolation points are one in the Chase algorithm, additional modifications are enabled to simplify the interpolation-based decoding process as presented in Chapter 6. Examples are systematic re-encoding, backward-forward interpolation, eliminated factorization, and low-power Chien-search-based codeword recovery. Moreover, the interpolation-based approach also leads to substantial complexity reduction if applied to the generalized minimum distance (GMD) decoding. Chapter 6 also introduces a generalized backward interpolation that can eliminate arbitrary points of any multiplicity.

Chapter 7 is dedicated to binary BCH codes. Similar to RS codes, BCH codes enjoy the flexibility on the codeword length and code rate. Nevertheless, compared to RS codes, BCH codes are more suitable for systems that have random bit errors and require low-complexity encoders and decoders. Example applications of BCH codes include optical communications, digital video broadcasting, and Flash memories. By making use of the binary property, the encoders and hard-decision decoders of RS codes discussed in Chapter 4 are simplified for BCH codes. Moreover, special treatment is given to 3-error-correcting BCH decoders, which are able to achieve very high throughput and are being considered for 100G optical transport network. A binary BCH code can be interpreted as a subcode of a RS code. Accordingly, the Chase ASD decoder architectures presented in Chapter 6 can be used to implement interpolation-based Chase BCH decoding. Nevertheless, the binary property allows additional transformations to be made to reduce the hardware complexity of these decoders.

Chapter 8 discusses the implementation of binary LDPC codes. Although LDPC codes have higher implementation costs than RS and BCH codes and need long codeword length to achieve good error-correcting capability, they are gaining popularity due to their capacity-approaching performance. Binary LDPC codes are adopted in many applications and standards, such as 10GBase-T Ethernet, WiMAX wireless communications, digital video broadcasting, and magnetic and solid-state drives. The emphasis of this chapter is given to the VLSI architectures of the Min-sum decoders for quasi-cyclic (QC)-LDPC codes, which are usually employed in practical systems. The most popular decoding scheduling schemes, including the flooding, sliced message-passing, layered and shuffled schemes, are reviewed and compared. Then the details of the computation units are presented. Moreover, techniques for reducing the power consumption of LDPC decoders are briefly discussed at the end of this chapter. They are applicable to any LDPC decoding algorithm.

Compared to binary LDPC codes, non-binary LDPC codes can achieve better performance when the code length is moderate, and are more suitable for systems that have bursts of errors. As a result, they are being considered for emerging applications. Chapter 9 deals with the VLSI architectures for non-binary LDPC decoders. The major hardware bottlenecks of these decoders are complex check node processing and large memory requirement. The focus of this chapter has been given to the Min-max algorithm, which allows efficient tradeoff between the error-correcting performance and hardware complexity. Reviews are provided for various implementation approaches of the Min-max check node processing, such as the forward-backward, trellis-based path construction, syndrome-based, and basis-construction methods. Their relative complexity, advantages, drawbacks, and impacts on the memory requirement of the overall decoder are also discussed. Besides Min-max decoders, architectures are also presented in this chapter for another two types of popular decoders: the extended Min-sum and iterative majority-logic decoders.

The preparation of this book was part of the task to be accomplished from the author's National Science Foundation (NSF) Faculty Early Career Development (CAREER) Award. It was originally intended to be used as a reference book for graduate classes on error-correcting en/decoder hardware implementation. Nevertheless, it would also serve as an introduction for hardware engineers, system designers, and researchers interested in developing error-correcting en/decoders for various systems. The materials on finite field arithmetic and hardware design fundamentals in Chapters 1 and 2, respectively, are needed to understand the following chapters. Chapters 3 to 7 for BCH and RS codes and Chapters 8, 9 for LDPC codes can be read independently. However, the chapters in these two groups need to be followed in the order they are presented. Although the architectures, transformation techniques, and simplification methods are presented for BCH, RS and LDPC codes in this book, they may be extended for other algorithms and systems. For example, an interpolation-based approach is also used in the fuzzy vault biometric encryption scheme, and message-passing algorithms can be employed to reconstruct compressively sensed signals.

Most of the architectures presented in this book are from the author's research results and papers. The author is grateful to her colleagues, friends, and students. Without them, this book would not have been possible. The author would like to thank Prof. Keshab Parhi for his encouragement and support of exploiting various research topics of interest. She is also grateful to Profs. Shu Lin and Daniel Costello for all their guidance, advise and kind support over the years. The author is thankful to Prof. Alexander Vardy for his encouragement for undertaking the research on ASD decoder design. Thanks are also due to Profs. and Drs. David Declercq, Fuyun Ling, Krishna Narayanan, Michael O'Sullivan, William Ryan, Gerald Sobelman, Myung Hoon Sunwoo, Bane Vasic, Zhongfeng Wang, and Yingquan Wu. It has been a real pleasure working and being friends with them.

The author's research included in this book has been supported by the National Science Foundation, and the Air Force Office of Scientific Research. Many thanks are due to Profs. Scott Midkiff and Zhi Tian for their support and encouragement.

Special thanks to Jennifer Ahringer and Nora Konopka at Taylor & Francis for their interest in this topic and their great help in the production of this book.

List of Figures

List of Tables

1

Finite field arithmetic

CONTENTS

Many error-correcting codes, such as Reed-Solomon (RS), Bose-Chaudhuri-Hocquenghem (BCH) and low-density parity-check (LDPC) codes, as well as cryptographic schemes, including the Advanced Encryption Standard (AES) and elliptic curve cryptography, are defined over finite fields. Finite fields are also referred to as *Galois Fields*, which are named after a nineteenth-century French mathematician, Evariste Galois. Thorough discussions of finite field theory can be found in [1, 2, 3]. Moreover, numerous coding books, such as [4, 5, 6], have chapters introducing the basics of finite fields. The implementation of finite field arithmetic largely affects the hardware complexities of the error-correcting encoders and decoders. The same computation can be implemented by different architectures using different representations of finite field elements, and the properties of finite fields can be exploited to simplify the implementation architectures. In this chapter, definitions and relevant properties of finite fields are introduced first. Then implementation architectures of finite field arithmetic are explained in detail.

1.1   Definitions, properties, and element representations

Definition 1 (Group) *A group is a set of elements G on which an operation $\cdot$ is defined such that the following three properties are satisfied:*

- *The $\cdot$ operation is associative, i.e., for any $a, b, c \in G$,*

$$a \cdot (b \cdot c) = (a \cdot b) \cdot c.$$

- *G contains an identity element e, such that for all $a \in G$,*

$$a \cdot e = e \cdot a = a.$$

- *For each $a \in G$, there exists an inverse element $a^{-1} \in G$, such that*

$$a \cdot a^{-1} = a^{-1} \cdot a = e.$$

If the $\cdot$ operation is commutative, i.e., for all $a, b \in G$, $a \cdot b = b \cdot a$, then the group is called commutative or abelian.

Example 1 *The set of all integers forms a commutative group with the $\cdot$ operation defined as integer addition. In this group, '0' is the identity element. The associativity is obvious and the inverse of an integer a is $-a$.*

Definition 2 (Field) *A field is a set F, together with two operations: the additive operation denoted by $+$, and the multiplicative operation denoted by $\cdot$, such that the following conditions hold:*

- *The elements in F form a commutative group with respect to $+$. The identity element with respect to $+$ is called the zero of the field, and is labeled as 0.*

- *All nonzero elements of F form a commutative group with respect to $\cdot$. The identity element with respect to $\cdot$ is called the identity of the field, and is labeled as 1.*

- *The multiplicative operation is distributive over the additive operation: $a \cdot (b + c) = a \cdot b + a \cdot c$ for all $a, b, c \in F$.*

Example 2 *The set of integers $F = \{0, 1, 2, 3, 4, 5, 6\}$ together with $+$ defined as integer addition modulo 7 and $\cdot$ defined as integer multiplication modulo 7 form a field. '0' and '1' are the zero and identity, respectively, of the field.*

In the case that a field has a finite number of elements as in the above example, it is called a *finite field*. The number of elements in a finite field can be either a prime or a power of prime. The field with p^q elements is denoted by $GF(p^q)$. The number of elements in a field is also called the order of the field. $GF(2)$ is the finite field with the least number of elements. In this field, the additive operation is logic XOR and the multiplicative operation is logic AND.

Some definitions and properties of finite fields are given below.

Definition 3 *The characteristic of a finite field is the smallest positive integer r such that adding up r copies of the identity elements leads to zero.*

Definition 4 *The order of $\alpha \in GF(p^q)$ is the smallest positive integer r, such that α^r equals the identity element of $GF(p^q)$.*

The order of $\alpha \in GF(p^q)$ always divides $p^q - 1$. Hence α^{p^q-1} always equals the identity. If t divides $p^q - 1$, then there are $\Phi(t)$ elements of order t in $GF(p^q)$, where $\Phi(\cdot)$ is the Euler's totient function. It is the number of integers in the set $\{1, 2, \cdots, t-1\}$ that are relatively prime to t. $\Phi(1)$ is defined to be 1. For $t > 1$, $\Phi(t)$ is computed by

$$\Phi(t) = t \prod_{s|t} \left(1 - \frac{1}{s}\right).$$

The product in the above equation is taken over all distinct positive prime integers s, such that s divides t. For example, $\Phi(24) = 24(1 - \frac{1}{3})(1 - \frac{1}{2}) = 8$.

Definition 5 (Primitive element) *When the order of an element $\alpha \in GF(p^q)$ is $p^q - 1$, α is called a primitive element of $GF(p^q)$.*

A polynomial whose coefficients are elements of $GF(p^q)$ is referred to as a polynomial over $GF(p^q)$.

Definition 6 (Irreducible polynomial) *A polynomial over $GF(p^q)$ is irreducible if it does not have any non-trivial factor over the same field.*

Definition 7 (Primitive polynomial) *An irreducible polynomial $f(x)$ over $GF(p)$ of degree q is said to be a primitive polynomial if the smallest integer s, for which $f(x)$ divides $x^s - 1$, is $p^q - 1$.*

It should be noted that any root of a primitive polynomial of degree q over $GF(p)$ is a primitive element of $GF(p^q)$.

Definition 8 (Minimal polynomial) *Let $\alpha \in GF(p^q)$. The minimal polynomial of α with respect to $GF(p^q)$ is the non-zero polynomial $m(x)$ of the lowest degree over $GF(p)$ such that $m(\alpha) = 0$.*

For each element $\alpha \in GF(p^q)$, there exists a unique monic (the coefficient of the highest degree term is identity) minimal polynomial over $GF(p)$. Minimal polynomials are irreducible.

TABLE 1.1
Conjugacy classes, cyclotomic cosets, and minimal
polynomials for $GF(2^4)$

Conjugacy class	cyclotomic coset	minimal polynomial
$\{1\}$	$\{0\}$	$x + 1$
$\{\alpha, \alpha^2, \alpha^4, \alpha^8\}$	$\{1,2,4,8\}$	$x^4 + x + 1$
$\{\alpha^3, \alpha^6, \alpha^{12}, \alpha^9\}$	$\{3,6,12,9\}$	$x^4 + x^3 + x^2 + x + 1$
$\{\alpha^5, \alpha^{10}\}$	$\{5,10\}$	$x^2 + x + 1$
$\{\alpha^7, \alpha^{14}, \alpha^{13}, \alpha^{11}\}$	$\{7,14,13,11\}$	$x^4 + x^3 + 1$

Definition 9 (Conjugates of field elements) *Let $\alpha \in GF(p^q)$. The conjugates of α with respect to $GF(p)$ are $\alpha, \alpha^p, \alpha^{p^2}, \cdots$.*

All the conjugates of α form a set called the *conjugacy class* of α. The cardinality of a conjugacy class divides q.

Example 3 *Assume that α is a primitive element of $GF(2^4)$. The conjugates of α with respect to $GF(2)$ are $\alpha, \alpha^2, \alpha^4, \alpha^8, \alpha^{16}, \ldots$. Since the order of α is 15, $\alpha^{16} = \alpha$. Thus, the conjugacy class of α with respect to $GF(2)$ is $\{\alpha, \alpha^2, \alpha^4, \alpha^8\}$.*

Definition 10 (Cyclotomic coset) *The cyclotomic cosets modulo n with respect to $GF(p)$ divide the integers $0, 1, \ldots, n-1$ into sets in the format of*

$$\{a, ap, ap^2, ap^3, \ldots\}$$

It can be easily seen that each cyclotomic coset modulo p^q corresponds to a conjugacy class of $GF(p^q)$. The elements in a conjugacy class can be expressed as powers of a primitive element. The elements in a cyclotomic coset are the exponents of those powers in the corresponding conjugacy class.

Theorem 1 *Assume that $\alpha \in GF(p^q)$, and $m(x)$ is the minimal polynomial of α with respect to $GF(p)$. Then the roots of $m(x)$ in $GF(p^q)$ are exactly the conjugates of α.*

The proof of this theorem can be found in [5].

Example 4 *Let α be a root of the primitive polynomial $x^4 + x + 1$ over $GF(2)$. Then α is a primitive element of $GF(2^4)$, and the order of α is $2^4 - 1 = 15$. Table 1.1 lists all conjugacy classes of the non-zero elements in $GF(2^4)$, as well as their corresponding cyclotomic cosets and minimal polynomials.*

Definition 11 *The trace of $\alpha \in GF(p^q)$ with respect to $GF(p)$ is*

$$Tr(\alpha) = \sum_{i=0}^{q-1} \alpha^{p^i}.$$

For $\alpha, \beta \in GF(p^q)$, $c \in GF(p)$, the *trace* has the following properties:

- $Tr(\alpha + \beta) = Tr(\alpha) + Tr(\beta)$

- $Tr(c\alpha) = cTr(\alpha)$

The *trace* function of $GF(p^q)$ over $GF(p)$ is a linear transformation from $GF(p^q)$ to $GF(p)$. For any $\alpha \in GF(p^q)$, $Tr(\alpha) \in GF(p)$.

Extension fields $GF(2^q)(q \in \mathbb{Z}^+)$ are usually used in digital communication and storage systems, since each field element can be represented by a binary vector. The remainder of this chapter focuses on the element representations, properties, and implementations of the arithmetic over these fields.

Assume that α is a root of a degree-q primitive polynomial $p(x) = x^q + a_{q-1}x^{q-1} + \cdots + a_1 x + a_0$ $(a_i \in GF(2))$. Then α is a primitive element of $GF(2^q)$ and all the elements of $GF(2^q)$ can be represented as $\{0, 1, \alpha, \alpha^2, \ldots, \alpha^{2^q - 2}\}$. This is called the power representation of finite field elements. Define the multiplication of two nonzero elements α^i, α^j as $\alpha^i \cdot \alpha^j = \alpha^{(i+j) \mod (2^q - 1)}$. Apparently, all the nonzero elements form a commutative group with respect to this multiplication, and 1 is the identity element. Each element can be also expressed as a polynomial in terms of $1, \alpha, \alpha^2, \ldots, \alpha^{q-1}$ by making use of the property that $p(\alpha) = \alpha^q + a_{q-1}\alpha^{q-1} + \cdots + a_1\alpha + a_0 = 0$. All the elements form a commutative group under polynomial addition over $GF(2)$, and 0 is the zero element.

Example 5 *Let α be a root of the degree-4 primitive polynomial $p(x) = x^4 + x + 1$ over $GF(2)$. Then the elements of $GF(2^4)$ can be expressed as $\{0, 1, \alpha, \alpha^2, \ldots, \alpha^{14}\}$. The polynomial representation of each element is listed in Table 1.2. They are derived iteratively by using the property that $\alpha^4 + \alpha + 1 = 0$. In finite fields of characteristic two, additions and subtractions of field elements are the same, and hence $\alpha^4 = \alpha + 1$. Then $\alpha^5 = \alpha\alpha^4 = \alpha(\alpha + 1) = \alpha^2 + \alpha$, $\alpha^6 = \alpha\alpha^5 = \alpha^3 + \alpha^2$, and $\alpha^7 = \alpha\alpha^6 = \alpha(\alpha^3 + \alpha^2) = \alpha^4 + \alpha^3 = \alpha^3 + \alpha + 1$. The other polynomial representations can be derived similarly.*

$GF(2^q)$ can be also viewed as a q-dimensional vector space over $GF(2)$ spanned by a basis. Assume that $\{\alpha_0, \alpha_1, \alpha_2, \cdots, \alpha_{q-1}\}$ form a basis $(\alpha_i \in GF(2^q))$. Then an element $a \in GF(2^q)$ can be expressed as $a = a_0\alpha_0 + a_1\alpha_1 + \cdots + a_{q-1}\alpha_{q-1}$, where $a_0, a_1, \cdots, a_{q-1} \in GF(2)$. Alternatively, a can be represented by the q-bit binary tuple $a_0 a_1 \ldots a_{q-1}$. These polynomial and vector representations are used interchangeably throughout this book. The three most commonly used bases are the standard basis (also called the canonical or polynomial basis), the normal basis, and the dual basis.

Definition 12 (Standard basis) *The set $\{1, \alpha, \alpha^2, \cdots, \alpha^{q-1}\}$, where α is a root of an irreducible polynomial of degree q over $GF(2)$, is a standard basis of $GF(2^q)$ over $GF(2)$.*

TABLE 1.2

Power representation vs. polynomial representation for elements of $GF(2^4)$ constructed using $p(x) = x^4 + x + 1$

Power representation	Polynomial representation	Power representation	Polynomial representation
0	0	α^7	$\alpha^3 + \alpha + 1$
1	1	α^8	$\alpha^2 + 1$
α	α	α^9	$\alpha^3 + \alpha$
α^2	α^2	α^{10}	$\alpha^2 + \alpha + 1$
α^3	α^3	α^{11}	$\alpha^3 + \alpha^2 + \alpha$
α^4	$\alpha + 1$	α^{12}	$\alpha^3 + \alpha^2 + \alpha + 1$
α^5	$\alpha^2 + \alpha$	α^{13}	$\alpha^3 + \alpha^2 + 1$
α^6	$\alpha^3 + \alpha^2$	α^{14}	$\alpha^3 + 1$

$\{1, \alpha, \alpha^2, \alpha^3\}$, where α is a root of the primitive polynomial $x^4 + x + 1$ used in Example 5, is actually a standard basis of $GF(2^4)$. It should be noted that a non-primitive irreducible polynomial can be also used to construct a standard basis. Using a standard basis, an element of $GF(2^q)$ is represented as a degree-$q - 1$ polynomial of α over $GF(2)$. The field addition is polynomial addition over $GF(2)$, which is bit-wise XOR of the coefficients. The field multiplication can be also done using basis representations. In standard basis, it is a polynomial multiplication modulo the irreducible polynomial.

Example 6 *Let α be a root of the irreducible polynomial $p(x) = x^4 + x^3 + x^2 + x + 1$. Then $\{1, \alpha, \alpha^2, \alpha^3\}$ is a standard basis of $GF(2^4)$. $a(\alpha) = 1 + \alpha + \alpha^3$ and $b(\alpha) = \alpha^2 + \alpha^3$ are two elements of $GF(2^4)$. $a(\alpha) + b(\alpha) = 1 + \alpha + \alpha^2$, and $a(\alpha)b(\alpha) = (1 + \alpha + \alpha^3)(\alpha^2 + \alpha^3) \mod p(\alpha) = \alpha^3$.*

Definition 13 (Normal basis) $\{\omega, \omega^2, \omega^{2^2}, \cdots, \omega^{2^{(q-1)}}\}$, *where $\omega \in GF(2^q)$, is a normal basis of $GF(2^q)$ over $GF(2)$ if the q elements in this set are linearly independent.*

Definition 14 (Dual basis) *Let $A = \{\alpha_0, \alpha_1, \cdots, \alpha_{q-1}\}$ be a basis. The basis $B = \{\beta_0, \beta_1, \cdots, \beta_{q-1}\}$ is a dual basis of A if:*

$$Tr(\alpha_i \beta_j) = \begin{cases} 1 & if \quad i = j \\ 0 & if \quad i \neq j. \end{cases}$$

It has been shown that for a given basis, there always exists a unique dual basis [3]. For any basis representation, the addition operation is bit-wise XOR. The multiplication using normal and dual bases will be detailed in the next section.

Higher-order finite fields can be also constructed iteratively from lower-order finite fields. A basis for a composite field $GF((2^n)^m)$,

$\{\alpha_0, \alpha_1, \ldots, \alpha_{m-1}\}$, has m elements and each element of $GF((2^n)^m)$ is represented by $a_0\alpha_0 + a_1\alpha_1 + \cdots + a_{m-1}\alpha_{m-1}$, where $a_i \in GF(2^n)$. A composite field $GF((2^n)^m)$ is isomorphic to $GF(2^q)$ for $q = nm$. This means if $\alpha, \beta, \gamma, \delta \in GF(2^q)$ are mapped to $\alpha', \beta', \gamma', \delta' \in GF((2^n)^m)$, respectively. Then $\gamma = \alpha + \beta$ iff $\gamma' = \alpha' + \beta'$ and $\delta = \alpha\beta$ iff $\delta' = \alpha'\beta'$. This isomorphism also holds for different representations of elements for the same finite field.

1.2 Finite field arithmetic

As explained previously, finite fields can be constructed differently, and the field elements can be represented as linear combinations of the elements in a basis, as well as powers of a primitive element. Although the finite fields of the same order are isomorphic and there exists one-to-one mapping among the element representations, the hardware complexities of finite field operations are heavily dependent on the element representations. This section discusses the hardware architectures and complexities of finite field operations using different element representations. Proper representations can be chosen based on the computations involved in a given application. Moreover, a single system can employ different representations in different units to minimize the overall hardware complexity.

Power representation is beneficial to finite field multiplications, exponentiations and multiplicative inversions. Multiplication in $GF(2^q)$ using power representation is done as

$$\alpha^i \alpha^j = \alpha^{(i+j) \mod (2^q-1)}.$$

Hence, only an integer addition modulo $2^q - 1$ is needed. Similarly, exponentiation in power representation is carried out by

$$(\alpha^i)^j = \alpha^{ij \mod (2^q-1)}.$$

For $\alpha^i \in GF(2^q)$, its multiplicative inverse is α^{-i} intuitively because $\alpha^i \alpha^{-i} = \alpha^0 = 1$. However, for $0 \leq i \leq 2^q - 2$, $-i$ is negative and α^{-i} is not a legal power representation. Instead, $\alpha^i \alpha^{2^q-1-i} = \alpha^{2^q-1} = 1$. Therefore, the inverse of α^i is α^{2^q-1-i}. Since $2^q - 1$ is a q-bit string of '1's, $2^q - 1 - i$ can be derived by simply flipping each bit in the q-bit binary representation of i.

Example 7 *Assume $\alpha \in GF(2^8)$. The inverse of α^{71} is computed as α^{2^8-1-71}. The 8-bit binary representation of 71 is '01000111'. Flipping each bit leads to '10111000', which is the binary representation of 184. Hence, $(\alpha^7)^{-1} = \alpha^{184}$.*

Although power representation facilitates finite field multiplications, exponentiations and inversions, finite field additions can not be done directly using power representation. Power representation has to be converted to a basis representation before the additions are done. An example for such a conversion in $GF(2^4)$ is shown in Table 1.2. After the addition is completed in a basis representation, a reverse mapping is needed to get the power representation of the sum. For $GF(2^q)$, the conversion or mapping table has 2^q lines and each line has q bits. The implementation of this table becomes quite hardware-consuming when q is not small. When finite field additions are needed in a system, the overhead of the mapping tables makes power representation less attractive.

Using any of the basis representations, additions over $GF(2^q)$ are simply bit-wise XOR operations. Assume that $\{\alpha_0, \alpha_1, \ldots, \alpha_{q-1}\}$ is a basis of $GF(2^q)$. If two elements $a, b \in GF(2^q)$ are expressed as $a_0\alpha_0 + a_1\alpha_1 + \cdots + a_{q-1}\alpha_{q-1}$ and $b_0\alpha_0 + b_1\alpha_1 + \cdots + b_{q-1}\alpha_{q-1}$, respectively, where $a_i, b_i \in GF(2)$, then $a + b = (a_0 \oplus b_0)\alpha_0 + (a_1 \oplus b_1)\alpha_1 + \cdots + (a_{q-1} \oplus b_{q-1})\alpha_{q-1}$. Here $\oplus$ denotes the XOR operation.

Using basis representations, the other computations over finite fields are more complicated than those based on power representation. Next, the multiplication and inversion using different basis representations are discussed in detail.

1.2.1 Multiplication using basis representations

1.2.1.1 Standard basis multiplier

Standard basis multipliers are the most commonly-used basis multipliers. Assume that a degree-q irreducible polynomial $p(x)$ is used to construct $GF(2^q)$, and α is a root of $p(x)$. Then $\{1, \alpha, \alpha^2, \ldots, \alpha^{q-1}\}$ form a standard basis of $GF(2^q)$. Let $a(\alpha) = a_0 + a_1\alpha + \ldots, +a_{q-1}\alpha^{q-1}$ and $b(\alpha) = b_0\alpha + b_1\alpha + \cdots + b_{q-1}\alpha^{q-1}$ ($a_i, b_i \in GF(2)$) be two elements of $GF(2^q)$. Their product is $c(\alpha) = \alpha(\alpha)\beta(\alpha) \mod p(\alpha)$, and can be rewritten as

$$c(\alpha) = a_0 b(\alpha) \mathrm{mod}\ p(\alpha) + a_1(b(\alpha)\alpha)\mathrm{mod}\ p(\alpha) + \cdots + a_{q-1}(b(\alpha)\alpha^{q-1})\mathrm{mod}\ p(\alpha). \tag{1.1}$$

$b(\alpha)\alpha^i \mod p(\alpha)$ can be iteratively computed as $(b(\alpha)\alpha^{i-1} \mod p(\alpha))\alpha \mod p(\alpha)$. Accordingly, a standard basis multiplier can be implemented according to the architecture in Fig. 1.1. The block labeled 'mult-mod' multiplies the input by α and then modulo-reduces the product by $p(\alpha)$. This block can be implemented by $w - 2$ XOR gates if w is the number of nonzero terms in the irreducible polynomial $p(x)$.

Example 8 *Assume that the irreducible polynomial* $p(x) = x^8 + x^4 + x^3 + x + 1$

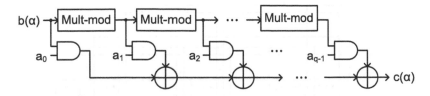

FIGURE 1.1
Standard basis multiplier architecture (Modified from [7])

is used to construct $GF(2^8)$.

$$b(\alpha)\alpha \mod p(\alpha) = b_7\alpha^8 + b_6\alpha^7 + b_5\alpha^6 + b_4\alpha^5 + b_3\alpha^4 + b_2\alpha^3 + b_1\alpha^2 + b_0\alpha$$
$$= b_7(\alpha^4 + \alpha^3 + \alpha + 1) + b_6\alpha^7 + b_5\alpha^6 + b_4\alpha^5 + b_3\alpha^4 + b_2\alpha^3$$
$$+ b_1\alpha^2 + b_0\alpha$$
$$= b_6\alpha^7 + b_5\alpha^6 + b_4\alpha^5 + (b_3 \oplus b_7)\alpha^4 + (b_2 \oplus b_7)\alpha^3 + b_1\alpha^2$$
$$+ (b_0 \oplus b_7)\alpha + b_7$$

Therefore, the 'mult-mod' block is implemented by three XOR gates, which compute $b_3 \oplus b_7$, $b_2 \oplus b_7$ and $b_0 \oplus b_7$.

Many studies have addressed the hardware implementation of standard basis multiplication according to (1.1) [8, 9, 10, 11]. Alternatively, (1.1) can be converted to a matrix multiplication format. Such matrix multiplication-based multipliers are also referred to as Mastrovito multipliers [12]. Simplification schemes for Mastrovito multipliers can be found in [13, 14, 15].

1.2.1.2 Normal basis multiplier

Normal basis multipliers are also referred to as Massey-Omura multipliers. Assume that $\{\omega, \omega^2, \omega^{2^2}, \cdots, \omega^{2^{(q-1)}}\}$ is a normal basis of $GF(2^q)$. Then $a \in GF(2^q)$ can be expressed in normal basis as:

$$a = a_0\omega + a_1\omega^2 + \cdots + a_{q-1}\omega^{2^{(q-1)}}, \tag{1.2}$$

where $a_i \in GF(2)\,(0 \le i < q)$. Since $a_i^2 = a_i$ and $sa_i = 0$ for even s, the square of a becomes

$$a^2 = a_0\omega^2 + a_1\omega^{2^2} + \cdots + a_{q-1}\omega^{2^q}.$$

For $\omega \in GF(2^q)$, $\omega^{2^q} = \omega^{2^q-1} \cdot \omega = \omega$. Therefore

$$a^2 = a_{q-1}\omega + a_0\omega^2 + \cdots + a_{q-2}\omega^{2^{(q-1)}}. \tag{1.3}$$

Comparing (1.2) and (1.3), it can be observed that the square (square root) of an element in normal basis representation can be implemented by cyclically

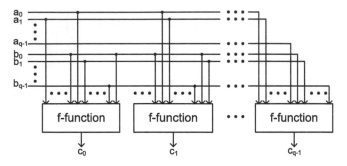

FIGURE 1.2
Normal basis multiplier architecture

shifting all the bits one position to the direction of the most significant bit (MSB) (least significant bit (LSB)). Similarly, a^{2^e} $(a^{1/2^e})$, where e is a non-negative integer, can be derived by cyclically shifting all the bits in the normal basis representation of a by e positions to the MSB (LSB).

Assume another field element b is expressed in normal basis as

$$b = b_0\omega + b_1\omega^2 + b_2\omega^{2^2} + \cdots + b_{q-1}\omega^{2^{(q-1)}},$$

where $b_i \in GF(2)(0 \leq i < q)$, and the product of a and b is

$$c = ab = c_0\omega + c_1\omega^2 + c_2\omega^{2^2} + \cdots + c_{q-1}\omega^{2^{(q-1)}}. \tag{1.4}$$

Then the highest coefficient c_{q-1} can be expressed as a bilinear function of the two sets of coefficients a_i and b_i:

$$c_{q-1} = f(a_0, a_1, \cdots, a_{q-1}; b_0, b_1, \cdots, b_{q-1}).$$

This bilinear function is called the f-function. If both sides of (1.4) are squared,

$$c^2 = a^2b^2 = c_{q-1}\omega + c_0\omega^2 + c_1\omega^{2^2} + \cdots + c_{q-2}\omega^{2^{(q-1)}}. \tag{1.5}$$

Comparing (1.4) and (1.5), it can be derived that

$$c_{q-2} = f(a_{q-1}, a_0, \cdots, a_{q-2}; b_{q-1}, b_0, \cdots, b_{q-2}).$$

Similarly, other coefficients in the product, c, are obtained by using the f-function with the two sets of input coefficients cyclically shifted. Hence, a parallel normal basis multiplier over $GF(2^q)$ can be implemented by the architecture illustrated in Fig. 1.2. This architecture consists of q f-function blocks, each of which computes one bit in the product c.

Example 9 *The finite field considered is $GF(2^4)$ constructed by using $p(x) = x^4 + x^3 + 1$. Taking ω as a root of $p(x)$, it can be verified that ω, ω^2, ω^4 and ω^8 are linearly independent, and hence form a normal basis. Express two elements $a, b \in GF(2^4)$ as $a = a_0\omega + a_1\omega^2 + a_2\omega^4 + a_3\omega^8$ and $b = b_0\omega + b_1\omega^2 + b_2\omega^4 + b_3\omega^8$. Then the product of a and b is*

$$
\begin{aligned}
c =& c_0\omega + c_1\omega^2 + c_2\omega^4 + c_3\omega^8 \\
=& ab = (a_0\omega + a_1\omega^2 + a_2\omega^4 + a_3\omega^8)(b_0\omega + b_1\omega^2 + b_2\omega^4 + b_3\omega^8) \\
=& \omega(a_3b_3) + \omega^2(a_0b_0) + \omega^3(a_0b_1 \oplus a_1b_0) + \omega^4(a_1b_1) + \omega^5(a_0b_2 \oplus a_2b_0) \\
& + \omega^6(a_1b_2 \oplus a_2b_1) + \omega^8(a_2b_2) + \omega^9(a_0b_3 \oplus a_3b_0) + \omega^{10}(a_1b_3 \oplus a_3b_1) \\
& + \omega^{12}(a_2b_3 \oplus a_3b_2).
\end{aligned}
$$

In the product c, the terms $\omega^3, \omega^5, \omega^6, \omega^9, \omega^{10}$ and ω^{12} do not belong to the normal basis. They need to be expressed in terms of $\omega, \omega^2, \omega^4$ and ω^8 in order to make c in normal basis representation. Since ω is a root of $p(x)$, $\omega^4 + \omega^3 + 1 = 0$. Using this property, it can be derived that

$$
\begin{aligned}
\omega^{12} &= \omega^8 + \omega^4 + \omega^2 \\
\omega^{10} &= \omega^8 + \omega^2 \\
\omega^9 &= \omega^8 + \omega^4 + \omega \\
\omega^6 &= \omega^4 + \omega^2 + \omega \\
\omega^5 &= \omega^4 + \omega \\
\omega^3 &= \omega^8 + \omega^2 + \omega.
\end{aligned}
$$

Collecting all coefficients for ω^8, c_3 is expressed in terms of $a_i, b_i (0 \leq i < 4)$ as

$$
\begin{aligned}
c_3 &= f(a_0, a_1, a_2, a_3; b_0, b_1, b_2, b_3) \\
&= a_2b_3 \oplus a_3b_2 \oplus a_1b_3 \oplus a_3b_1 \oplus a_3b_0 \oplus a_0b_3 \oplus a_2b_2 \oplus a_0b_1 \oplus a_1b_0.
\end{aligned}
$$

The above equation defines the f-function. c_2, c_1, c_0 can be computed by this function with inputs cyclically shifted.

Compared to standard basis representation, normal basis representation has the advantage that the square and square root operations are cyclical shifts. Various architectures have been proposed to reduce the complexity of normal basis multipliers [16, 17, 18, 19, 20]. For $q = 2, 4, 10, 12, 18, 28, 36, 52, \ldots$, there exists a degree-$q$ all-one irreducible polynomial over $GF(2)$ whose roots form a normal basis of $GF(2^q)$. Such a normal basis is called an optimal normal basis. In this case, the normal basis multiplier can be greatly simplified [17], and its complexity is q^2 AND gates and $q^2 - 1$ XOR gates. This gate count is the same as that of the standard basis multipliers in Fig. 1.1 when an irreducible trinomial is employed. Simplification schemes have been proposed in [19] for general normal basis multipliers over finite fields of any order. Nevertheless, normal basis multipliers usually have higher complexity than standard basis multipliers, except when optimal normal bases are used.

1.2.1.3 Dual basis multiplier

In dual basis multiplications, one of the operands is represented in standard basis, while the other operand and product are in dual basis. Let $B_s = \{1, \alpha, \alpha^2, \cdots, \alpha^{q-1}\}$ be a standard basis of $GF(2^q)$, and $B_d = \{\beta_0, \beta_1, \cdots, \beta_{q-1}\}$ be a dual basis of B_s. Assume that operand a is represented in standard basis as

$$a = a_0 + a_1\alpha + \cdots + a_{q-1}\alpha^{q-1},$$

and operand b is represented in dual basis as

$$b = b_0\beta_0 + b_1\beta_1 + \cdots + b_{q-1}\beta_{q-1},$$

where $a_i, b_i \in GF(2)(0 \leq i < q)$. According to the properties of the *trace* function and the definition of dual basis, it can be derived that

$$\begin{aligned} Tr(\alpha^j b) &= Tr(\alpha^j b_0\beta_0 + \alpha^j b_1\beta_1 + \cdots + \alpha^j b_{q-1}\beta_{q-1}) \\ &= b_0 Tr(\alpha^j \beta_0) + b_1 Tr(\alpha^j \beta_1) + \cdots + b_{q-1} Tr(\alpha^j \beta_{q-1}) \qquad (1.6) \\ &= b_j \end{aligned}$$

for $j = 0, 1, \cdots, q - 1$. Hence, if the jth coefficient of ab is denoted by $(ab)_j$, then

$$(ab)_j = Tr(\alpha^j(ab)) = Tr(\alpha^{j+1}b) = \begin{cases} b_{j+1} & for \ j = 0, 1, \cdots, q-2 \\ Tr(\alpha^q b) & for \ j = q-1 \end{cases}$$

$$(1.7)$$

Therefore, all the coefficients of ab, except the highest one, can be obtained by shifting the coefficients of b. Assume that α is a root of the irreducible polynomial $p(x) = 1 + p_1 x + p_2 x^2 + \cdots + x^q$. Then the coefficient $(ab)_{q-1}$ can be computed as

$$\begin{aligned} (ab)_{q-1} = Tr(\alpha^q b) &= Tr((1 + p_1\alpha + p_2\alpha^2 + \cdots + p_{q-1}\alpha^{q-1})b) \\ &= b_0 + b_1 p_1 + \cdots + b_{q-1} p_{q-1} = b \circ p, \end{aligned}$$

where $\circ$ stands for inner product. If there are w nonzero terms in $p(x)$, then the inner product requires $w - 2$ XOR gates to compute.

Now consider the computation of the product $c = ab$. From (1.6),

$$c_j = Tr(\alpha^j c) = Tr(\alpha^j ab) = Tr((\alpha^j b)a).$$

Therefore,

$$
\begin{aligned}
c_0 = Tr(ba) &= Tr(a_0 b) + Tr(a_1 \alpha b) + \cdots + Tr(a_{q-1} \alpha^{q-1} b) \\
&= a_0 Tr(b) + a_1 Tr(\alpha b) + \cdots + a_{q-1} Tr(\alpha^{q-1} b) \\
&= a_0 b_0 + a_1 b_1 + \cdots + a_{q-1} b_{q-1} = a \circ b.
\end{aligned}
$$

Similarly, $c_1 = Tr((\alpha b)a) = a \circ (\alpha b)$. The coefficients in αb can be computed by (1.7), and the rest of the coefficients of c are computed iteratively as

$$
c_2 = Tr((\alpha^2 b)a) = a \circ (\alpha(\alpha b)),
$$
$$
c_3 = Tr((\alpha^3 b)a) = a \circ (\alpha(\alpha(\alpha b))),
$$
$$
\vdots
$$

In matrix form, the dual basis multiplication can be expressed by

$$
\begin{bmatrix} c_0 \\ c_1 \\ c_2 \\ \vdots \\ c_{q-1} \end{bmatrix}
=
\begin{bmatrix}
b_0 & b_1 & \cdots & b_{q-2} & b_{q-1} \\
b_1 & b_2 & \cdots & b_{q-1} & b \circ p \\
b_2 & b_3 & \cdots & b \circ p & (\alpha b) \circ p \\
\vdots & \vdots & \ddots & \vdots & \vdots \\
b_{q-1} & b \circ p & \cdots & (\alpha^{q-3} b) \circ p & (\alpha^{q-2} b) \circ p
\end{bmatrix}
\begin{bmatrix} a_0 \\ a_1 \\ a_2 \\ \vdots \\ a_{q-1} \end{bmatrix}.
$$

1.2.1.4 Multiplication in composite field

Assume that $\alpha_0, \alpha_1, \ldots, \alpha_{m-1} \in GF((2^n)^m)$ form a basis of $GF((2^n)^m)$. Then each element $a \in GF((2^n)^m)$ can be expressed as $a_0 \alpha_0 + a_1 \alpha_1 + \cdots + a_{m-1} \alpha_{m-1}$, where $a_i \in GF(2^n)$. Although the basis used to construct $GF(2^n)$ can be a different type of basis (standard or normal) from the basis $\{\alpha_0, \alpha_1, \ldots, \alpha_{m-1}\}$, this book focuses on the cases where standard bases are used in the constructions of both $GF(2^n)$ and from $GF(2^n)$ to $GF((2^n)^m)$.

Let $p(x)$ be a degree-n irreducible polynomial over $GF(2)$, and $g(x)$ be a degree-m irreducible polynomial over $GF(2^n)$. Irreducible polynomials over $GF(2)$ are listed in many coding books, such as [5]. When n is not very large, a simple way to find out whether a polynomial over $GF(2^n)$ is irreducible is to try each element of $GF(2^n)$ to see if it is a root. If α and δ are roots of $p(x)$ and $g(x)$, respectively, then $\{1, \alpha, \alpha^2, \ldots, \alpha^{n-1}\}$ is a standard basis of $GF(2^n)$ and $\{1, \delta, \delta^2, \ldots, \delta^{m-1}\}$ is a standard basis of $GF((2^n)^m)$. Let a and b be two elements of $GF((2^n)^m)$, and $a(\alpha) = a_0 + a_1 \alpha + \cdots + a_{m-1} \alpha^{m-1}$ and $b(\alpha) = b_0 + b_1 \alpha + \cdots + b_{m-1} \alpha^{m-1}$ ($a_i, b_i \in GF(2^n)$). $a(\alpha) + b(\alpha) = (a_0 + b_0) + (a_1 + b_1)\alpha + \cdots + (a_{m-1} + b_{m-1})\alpha^{m-1}$, and the addition of a_i and b_i ($0 \le i < m$) is still done as bit-wise XOR. The multiplication of $a(\alpha)$ and $b(\alpha)$ over $GF((2^n)^m)$ is computed as $a(\alpha)b(\alpha) \mod g(\alpha)$. During the polynomial multiplication and modulo reduction, the computations over the coefficients a_i, b_i follow the computations defined for the subfield $GF(2^n)$.

When $q = 2^t$, where t is a positive integer, a composite field of $GF(2^q)$

can be built iteratively from $GF(2)$ as $GF(((((2)^2)^2)^{\cdots 2})$, where the extension is repeated for t times. Such a composite field is also referred to as a tower field.

Example 10 *$GF(2^4)$ can be constructed using $p(x) = x^4 + x + 1$, which is irreducible over $GF(2)$. Let β be a root of $p(x)$. Then $\{1, \beta, \beta^2, \beta^3\}$ is a standard basis of $GF(2^4)$. It can be found that $g(x) = x^2 + x + \beta^3$ is a degree-2 irreducible polynomial over $GF(2^4)$. β^3 is also represented in vector format as '0001'. Let α be a root of $g(x)$, then $\{1, \alpha\}$ is a standard basis of $GF((2^4)^2)$. Let $a = $'00010110' and $b = $'10111001' be two elements of $GF((2^4)^2)$. They are also interpreted as $a(\alpha) = a_0 + a_1\alpha$ with $a_0 = $'0001', $a_1 = $'0110' and $b(\alpha) = b_0 + b_1\alpha$ with $b_0 = $'1011', $b_1 = $'1001'. Their product is*

$$a(\alpha)b(\alpha) \mod g(\alpha) = a_0b_0 + (a_1b_0 + a_0b_1)\alpha + a_1b_1\alpha^2 \mod g(\alpha)$$
$$= (a_0b_0 + a_1b_1\beta^3) + (a_1b_0 + a_0b_1 + a_1b_1)\alpha$$

In polynomial format, $a_1(\beta) = \beta + \beta^2$ and $b_0(\beta) = 1 + \beta^2 + \beta^3$. Hence, their product is $a_1(\beta)b_0(\beta) \mod p(\beta) = \beta^5 + \beta^3 + \beta^2 + \beta \mod p(\beta) = \beta^3$. Accordingly, $a_1b_0 = $'0001' in vector representation. The multiplications between the other coefficients a_i and b_i can be computed in a similar way.

When the extension degree is small, such as two as in Example 10, simplifications can be done on the composite field multiplications. Consider the field $GF(p^2)$ constructed using an irreducible polynomial $g(x) = x^2 + x + \phi$ over $GF(p)$, and α is a root of $g(x)$. The multiplication of two elements $a(\alpha) = a_0 + a_1\alpha$ and $b(\alpha) = b_0 + b_1\alpha$ $(a_0, a_1, b_0, b_1 \in GF(p))$ can be performed as:

$$c(\alpha) = a(\alpha)b(\alpha) \mod g(\alpha)$$
$$= (a_0b_0 + \phi a_1b_1) + (a_1b_1 + a_1b_0 + a_0b_1)\alpha$$
$$= (a_0b_0 + \phi a_1b_1) + ((a_1 + a_0)(b_1 + b_0) + a_0b_0)\alpha.$$

The second row of the above equation has four multiplications between the a_i and b_i coefficients, while the third row needs three. Here the multiplication by ϕ is not counted since ϕ is known and a constant multiplier is much simpler than a general multiplier. Although the third row needs one more addition, its overall complexity is lower since general multiplications over $GF(p)$ are more expensive than additions over $GF(p)$ to implement. When the extension degree is not two, the corresponding formula can be also manipulated to reduce the number of multiplications at the cost of more less-expensive additions to lower the overall complexity. However, it becomes more difficult to find sharable terms as the extension degree becomes higher. A similar technique is also adopted in many other applications, such as fast filtering, convolution, and complex number multiplications [21].

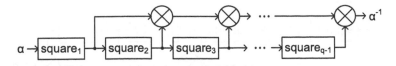

FIGURE 1.3
Implementation architecture of inversion over $GF(2^q)$ (Scheme A)

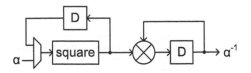

FIGURE 1.4
Implementation architecture of inversion over $GF(2^q)$ (Scheme A, iterative approach)

1.2.2 Inversion using basis representations

Finite field inversions have broad applications in cryptography and error-correcting coding systems, such as the AES algorithm [22] and RS codes [23]. The inverse of each field element can be pre-computed and stored in a look-up table (LUT). For $GF(2^q)$, the size of such a LUT is $2^q \times q$, and it would take significant amount of silicon area to implement if q is not small. Next, methods for mathematically computing the inverse using basis representations are presented.

1.2.2.1 Inversion in original field

For $\alpha \in GF(2^q)$, $\alpha^{-1} = \alpha^{2^q-2}$. The complexity of α^{2^q-2} computation can be reduced by sharing common terms.

- Scheme A
$$\alpha^{-1} = \alpha^{a^q-2} = \alpha^2 \cdot \alpha^{2^2} \cdots \alpha^{2^{(q-1)}}.$$

Hence, inversions can be implemented by repeated squaring and multiplying as illustrated in Fig. 1.3. The inverter over $GF(2^q)$ in this figure consists of $q-1$ square operators and $q-2$ multipliers over $GF(2^q)$. Its critical path has $q-1$ square operators and one multiplier. Pipelining can be achieved to achieve higher clock frequency at the cost of longer latency. If $q-1$ clock cycles can be spent, then the inverse can be calculated iteratively by the architecture shown in Fig. 1.4.

Terms other than α^2 can be utilized to simplify the inversion. As an example, another two schemes for computing the inverse over $GF(2^8)$ by using common terms α^3 and α^7 are exploited below [24].

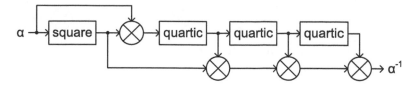

FIGURE 1.5
Implementation architecture of inversion over $GF(2^8)$ (Scheme B)

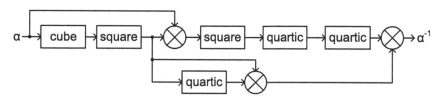

FIGURE 1.6
Implementation architecture of inversion over $GF(2^8)$ (Scheme C)

- Scheme B

$$\alpha^{-1} = \alpha^{254} = (\alpha^3)^{4^3} \cdot (\alpha^3)^{4^2} \cdot (\alpha^3)^4 \cdot \alpha^2.$$

Accordingly, α^3 can be computed first and re-used in the inverse computation. The corresponding architecture is shown in Fig. 1.5.

- Scheme C

$$\alpha^{-1} = \alpha^{254} = (((\alpha^7)^2)^4)^4 \cdot ((\alpha^3)^2)^4 \cdot (\alpha^3)^2.$$

Hence, the inversion over $GF(2^8)$ can be also implemented by the architecture in Fig. 1.6.

Compared to a general multiplier, the square, cube, and quartic operators are simpler. It has been shown in [24] that among the three non-iterative architectures shown in Figs. 1.3, 1.5, and 1.6, the one according to Scheme A has the highest complexity for the inversion over $GF(2^8)$ despite its more regular architecture. The one corresponding to Scheme C requires the least hardware and has the shortest critical path. Nevertheless, as the order of the field increases, the complexities of these exponentiation-based inversions become overwhelming. Alternatively, composite field arithmetic can be employed to simplify the inversions.

1.2.2.2 Inversion in composite field

For $q = mn$, $GF(2^q)$ is isomorphic to $GF((2^n)^m)$. The following theorem can be utilized to map the inversion in $GF(2^q)$ to that over subfield $GF(2^n)$.

Theorem 2 *If* $\alpha \in GF(p^q)$, *then* $\alpha^{\frac{p^q-1}{p-1}}$ *is an element of the subfield* $GF(p)$.

The proof of this theorem can be found in [1]. The inverse of an element $\alpha \in GF((2^n)^m)$ can be written as

$$\alpha^{-1} = (\alpha^r)^{-1}\alpha^{r-1}. \tag{1.8}$$

Let $r = \frac{2^{nm}-1}{2^n-1}$, then α^r is an element of $GF(2^n)$ from Theorem 2. (1.8) implies that the inverse over $GF(2^q)$ can be computed in four steps [25]:

1. Compute α^{r-1} (exponentiation in $GF((2^n)^m)$)
2. Compute $\alpha^{r-1}\alpha = \alpha^r$ (multiplication in $GF((2^n)^m)$, and the product is an element of $GF(2^n)$)
3. Compute $(\alpha^r)^{-1} = \alpha^{-r}$ (inversion in $GF(2^n)$)
4. Compute $\alpha^{-r}\alpha^{r-1} = \alpha^{-1}$ (multiplication of an element from $GF(2^n)$ with an element from $GF((2^n)^m)$)

Since $n > 1$, r is at least around three times smaller than 2^{mn}. The inversion of $\alpha \in GF(2^n)$ can be computed as α^{2^n-2}. Its complexity is much lower than that of computing $\alpha^{2^{nm}-2}$, which is needed if the inversion is done over the original field. In addition, if the product or an operand is a subfield element, the complexity of the composite field multiplication can be further reduced. Hence, employing composite field arithmetic leads to significant reduction in the inversion complexity.

In the case of $m = 2$, inversions can be implemented by even simpler architectures. An element of $GF((2^n)^2)$ can be expressed as $s_h\alpha + s_l$, where $s_h, s_l \in GF(2^n)$ and α is a root of the irreducible polynomial, $p(x)$, used in the construction of $GF((2^n)^2)$ from $GF(2^n)$. $p(x)$ should be in the form of $x^2 + x + \lambda$, where λ is an element of $GF(2^n)$. Using the Extended Euclidean Algorithm, the inverse of $s_h\alpha + s_l$ modulo $p(\alpha)$ can be computed as

$$(s_h\alpha + s_l)^{-1} = s_h\Theta\alpha + (s_h + s_l)\Theta, \tag{1.9}$$

where $\Theta = (s_h^2\lambda + s_hs_l + s_l^2)^{-1}$.

(1.9) is derived as follows. The problem of finding the inverse of $s(\alpha) = s_h\alpha + s_l$ modulo $p(\alpha) = \alpha^2 + \alpha + \lambda$ is equivalent to finding polynomials $a(\alpha)$ and $b(\alpha)$ satisfying the following equation:

$$a(\alpha)p(\alpha) + b(\alpha)s(\alpha) = 1. \tag{1.10}$$

Then $b(\alpha)$ is the inverse of $s(\alpha)$ modulo $p(\alpha)$. Such $a(\alpha)$ and $b(\alpha)$ can be found by using the Extended Euclidean Algorithm for one iteration. First, $p(x)$ is rewritten as

$$p(\alpha) = q(\alpha)s(\alpha) + r(\alpha), \tag{1.11}$$

where $q(\alpha)$ and $r(\alpha)$ are the quotient and the remainder polynomials of dividing $p(\alpha)$ by $s(\alpha)$, respectively. By long division, it can be derived that

$$q(\alpha) = s_h^{-1}\alpha + (1 + s_h^{-1}s_l)s_h^{-1}, \quad r(\alpha) = \lambda + (1 + s_h^{-1}s_l)s_h^{-1}s_l. \tag{1.12}$$

Substitute (1.12) into (1.11) and multiply s_h^2 to both sides of the equation,

$$s_h^2 p(\alpha) = (s_h \alpha + (s_h + s_l)) s(\alpha) + (s_h^2 \lambda + s_h s_l + s_l^2). \qquad (1.13)$$

Multiplying $\Theta = (s_h^2 \lambda + s_h s_l + s_l^2)^{-1}$ to both sides of (1.13), it can be derived that

$$\Theta s_h^2 p(\alpha) = \Theta (s_h \alpha + (s_h + s_l)) s(\alpha) + 1. \qquad (1.14)$$

Since addition and subtraction are the same in extension fields of $GF(2)$, the first term on the right-hand side of (1.14) can be moved to the left-hand side. Comparing (1.10) and (1.14), (1.9) is derived.

All the computations in (1.9) are over subfield $GF(2^n)$, and only a few multiplications, additions and an inverse are needed. Hence, in the case of $m = 2$, computing the inverse according to (1.9) has lower complexity than the four-step process.

1.3   Mapping between finite field element representations

The types of finite field operations involved in the given application decide the most suitable element representations to use. Different representations may be adopted in different units to minimize the overall complexity. The overheads of these mappings also need to be considered when making decisions on which representations to use. In the following, the mappings between some commonly used representations are discussed. The key for deriving the mappings is that the isomorphism needs to be maintained. In other words, if $\alpha, \beta, \gamma, \delta$ in one representation are mapped to $\alpha', \beta', \gamma', \delta'$, respectively, in another representation, then $\gamma = \alpha + \beta$ iff $\gamma' = \alpha' + \beta'$ and $\delta = \alpha \beta$ iff $\delta' = \alpha' \beta'$.

1.3.1   Mapping between standard basis and composite field representations

The mapping between standard basis and composite field representations can be done as a constant matrix multiplication. Such a mapping matrix can be found using the approach proposed in [26]. Although the method in [26] assumes that the roots of a primitive polynomial are used to construct the standard basis, it can be also extended to the case that an irreducible but non-primitive polynomial is used. Consider the mapping from the representation in $GF(2^q)$ using standard basis $\{1, \alpha, \alpha^2, \ldots, \alpha^{q-1}\}$ to the representation in composite field $GF((2^n)^m)$, where $q = nm$. Assume that α is a root of a degree-q irreducible polynomial $p(x)$ over $GF(2)$. Clearly, the "1" element in the standard basis must be mapped to the "1" element in the composite field. To maintain the isomorphism, α must be mapped to an element $\beta \in$

$GF((2^n)^m)$ such that $p(\beta) = 0$. If an element $\omega \in GF((2^n)^m)$ is not a root of $p(x)$, then none of the other elements in the same conjugacy class as ω is a root of $p(x)$. This helps to increase the search speed. Once the $\beta \in GF((2^n)^m)$ such that $p(\beta) = 0$ is found, the other elements $\alpha^2, \alpha^3, \ldots, \alpha^{q-1}$ are mapped to $\beta^2, \beta^3, \ldots, \beta^{q-1}$, respectively. The exponentiation of β is done by following the computation rules over composite field. Then the vector representations of $1, \beta, \beta^2, \ldots, \beta^{q-1}$ are collected as the columns of the mapping matrix.

Example 11 *Assume that α is a root of the irreducible polynomial $p(x) = x^8 + x^7 + x^6 + x + 1$. Then $\{1, \alpha, \alpha^2, \ldots, \alpha^7\}$ is a standard basis of $GF(2^8)$. The composite field $GF(((2^2)^2)^2)$ can be built iteratively from $GF(2)$ using the following irreducible polynomials:*

$$\begin{cases} GF(2) \Rightarrow GF(2^2) & : p_0(x) = x^2 + x + 1 \\ GF(2^2) \Rightarrow GF((2^2)^2) & : p_1(x) = x^2 + x + \phi \\ GF((2^2)^2) \Rightarrow GF(((2^2)^2)^2) & : p_2(x) = x^2 + x + \lambda \end{cases}$$

where $\phi = $ '10' and $\lambda = $ '1100'. The mapping of $a = a_7\alpha^7 + a_6\alpha^6 + \cdots + a_1\alpha + a_0 \in GF(2^8)$ in standard basis, which is also written in vector format as $a = $ '$a_7a_6a_5a_4a_3a_2a_1a_0$', to $b = $ '$b_7b_6b_5b_4b_3b_2b_1b_0$' $\in GF(((2^2)^2)^2)$ can be done as

$$[b_7, b_6, \ldots, b_0]^T = \begin{bmatrix} 0 & 0 & 1 & 1 & 0 & 1 & 1 & 0 \\ 0 & 1 & 1 & 1 & 1 & 0 & 1 & 0 \\ 1 & 1 & 1 & 0 & 0 & 0 & 1 & 0 \\ 1 & 1 & 1 & 0 & 1 & 1 & 1 & 0 \\ 1 & 0 & 1 & 1 & 0 & 0 & 1 & 0 \\ 1 & 0 & 1 & 1 & 1 & 1 & 0 & 0 \\ 1 & 0 & 0 & 1 & 1 & 1 & 0 & 0 \\ 0 & 0 & 0 & 0 & 0 & 0 & 1 & 1 \end{bmatrix} \begin{bmatrix} a_7 \\ a_6 \\ a_5 \\ a_4 \\ a_3 \\ a_2 \\ a_1 \\ a_0 \end{bmatrix}.$$

Clearly, the last column of this matrix is the vector representation of "1". From an exhaustive search, $\beta = $ '11111001' $\in GF(((2^2)^2)^2)$ satisfies $p(\beta) = 0$, and hence becomes the second to the last column of the matrix. The other columns are derived by computing $\beta^2, \beta^3, \ldots$ over $GF(((2^2)^2)^2)$.

The mapping matrix can be pre-computed, and the implementation of a constant binary matrix multiplication is simple. In addition, substructure sharing can be applied to further reduce the gate count. For example, $b_5 = (a_7 \oplus a_6 \oplus a_5 \oplus a_1)$ and $b_4 = (a_7 \oplus a_6 \oplus a_5 \oplus a_1) \oplus a_3 \oplus a_2$. $(a_7 \oplus a_6 \oplus a_5 \oplus a_1)$ is calculated once and shared in these two computations.

1.3.2 Mapping between power and standard basis representations

In standard basis $\{1, \alpha, \alpha^2, \ldots, \alpha^{q-1}\}$ of $GF(2^q)$, α is a root of a degree-q irreducible polynomial $p(x)$ over $GF(2)$. If $p(x)$ is also primitive, then

$\{0, 1, \alpha, \alpha^2, \ldots, \alpha^{2^q-2}\}$ is a power representation of all the elements in $GF(2^q)$. In this case, the power representation can be converted to standard basis representation by making use of the property that $p(\alpha) = 0$. An example has been given in Table 1.2.

In the case that $p(x)$ is not primitive, the order of α divides $2^q - 1$ but is less than $2^q - 1$. Then the powers of α do not consist of all nonzero elements, and they do not form a complete power representation of the elements in $GF(2^q)$. Let β be a primitive element, and the power representations of all the elements are $\{0, 1, \beta, \beta^2, \ldots, \beta^{2^q-2}\}$. Inspired by the method in [26], the mapping between power presentation and standard basis derived from a non-primitive irreducible polynomial can be done as follows. The "0" and "1" in power representation should be mapped to the "0" and "1", respectively, in standard basis representation. The mapping should not change the order of the elements. If the order of α is r, then α should be mapped to an element β^t whose order is also r. However, there are multiple elements of the same order. To maintain the isomorphism, β^t should be a root of $p(x)$. Once the mapping of β^t is decided, the mapping of the other elements in the same conjugacy class can be derived accordingly. $\beta^{2t}, \beta^{4t}, \ldots$ are mapped to $\alpha^2, \alpha^4, \ldots$, respectively. Here the powers of α need to be written in terms of the standard basis $\{1, \alpha, \alpha^2, \ldots, \alpha^{q-1}\}$. The mapping of the rest of the elements is completed by exploiting the isomorphism. Among the powers of β that have already been mapped, pick two elements randomly and calculate their sum or product, as well as the sum or product of their images in standard basis representation. If the sum or product is not one of the elements that have been mapped, then the mapping of a new element is defined. Similarly, the mapping of the other elements in the same conjugacy class follows. This process is repeated until the mappings for all elements are found.

Example 12 $p(x) = x^4 + x^3 + x^2 + x + 1$ *is an irreducible but non-primitive polynomial over* $GF(2)$. *If* α *is a root of* $p(x)$, $\{1, \alpha, \alpha^2, \alpha^3\}$ *is a standard basis of* $GF(2^4)$. *Taking* β *as a root of the primitive polynomial* $p_{prime}(x) = x^4 + x + 1$, *then* $\{0, 1, \beta, \beta^2, \ldots, \beta^{14}\}$ *is a power representation for* $GF(2^4)$. *The order of* α *is 5.* β^3 *also has order 5 and is a root of* $p(x)$. *Hence,* α *can be mapped to* β^3. *Moreover, the other element in the same conjugacy class,* β^6, β^{12}, *and* β^9, *are mapped to* α^2, $\alpha^4 = \alpha^3 + \alpha^2 + \alpha + 1$ *and* $(\alpha^4)^2 = \alpha^3$, *respectively. Now pick* β^3 *and* β^6. *Their sum in power representation is* β^2. *This can be derived by using the property that* $p_{prime}(\beta) = \beta^4 + \beta + 1 = 0$. *Adding the standard basis representations of* β^3 *and* β^6, β^2 *should be mapped to* $\alpha + \alpha^2$. *Similarly, the other elements in the same conjugacy class of* β^2, *which are* β^4, β^8, β, *are mapped to* $(\alpha + \alpha^2)^2 = \alpha^3 + \alpha + 1$, $(\alpha^3 + \alpha + 1)^2 = \alpha^2 + \alpha + 1$, *and* $(\alpha^2 + \alpha + 1)^2 = \alpha^3 + \alpha$, *respectively. The mappings of all elements are listed in Table 1.3.*

The above method can be also used to derive the mapping between power representation and standard basis representation of composite field, if the computations are done with regard to the rules of the composite field. For a

TABLE 1.3

Mapping between power representation and standard basis representation using non-primitive irreducible polynomial $x^4 + x^3 + x^2 + x + 1$ for elements of $GF(2^4)$

Power representation	Standard basis representation	Power representation	Standard basis representation
0	0	β^7	$\alpha^3 + 1$
1	1	β^8	$\alpha^2 + \alpha + 1$
β	$\alpha^3 + \alpha$	β^9	α^3
β^2	$\alpha^2 + \alpha$	β^{10}	$\alpha^3 + \alpha^2 + 1$
β^3	α	β^{11}	$\alpha^3 + \alpha^2 + \alpha$
β^4	$\alpha^3 + \alpha + 1$	β^{12}	$\alpha^3 + \alpha^2 + \alpha + 1$
β^5	$\alpha^3 + \alpha^2$	β^{13}	$\alpha^2 + 1$
β^6	α^2	β^{14}	$\alpha + 1$

standard basis multiplier, the complexity is lower if there are fewer nonzero terms in the corresponding irreducible polynomial. Usually there are degree-q binary primitive polynomials that have the same or fewer nonzero terms than degree-q binary non-primitive irreducible polynomials, and they can be used to construct standard bases. However, this is not necessarily true for irreducible and primitive polynomials over extension fields. On the other hand, some computations, such as finite Fourier transform, are defined using power representation. For finite Fourier transform over subfields, the power representation may need to be mapped to a standard basis constructed with a non-primitive irreducible polynomial over subfields in order to reduce the multiplication complexity.

As mentioned before, the mapping between power representation and basis representation of $GF(2^q)$ elements is implemented by a LUT, whose size is $2^q \times q$ bits. When the order of the finite field is not low, the LUT becomes large and results in significant hardware overhead.

1.3.3 Mapping between standard and normal basis representations

Although normal basis multipliers are usually more complicated than standard basis multipliers, square or square root computation only takes a cyclical shift in normal basis. When a system requires these computations as well as quite a few multiplications, a converter between standard and normal basis representations may be needed.

The mapping from standard to normal basis can be also done as a $q \times q$ constant matrix multiplication. Assume that $\{1, \alpha, \alpha^2, \ldots, \alpha^{q-1}\}$ is a standard basis of $GF(2^q)$, and $\{\omega, \omega^2, \omega^4, \ldots, \omega^{2^{q-1}}\}$ is a normal basis. The columns of the mapping matrix are derived by expressing each element in $\{1, \alpha, \alpha^2, \ldots, \alpha^{q-1}\}$ in terms of $\omega, \omega^2, \omega^4, \ldots, \omega^{2^{q-1}}$.

Example 13 *Let α be a root of the irreducible polynomial $p(x) = x^8 + x^7 + x^6 + x + 1$. Then $\{1, \alpha, \alpha^2, \ldots, \alpha^7\}$ is a standard basis of $GF(2^8)$. The eight elements $\alpha, \alpha^2, \ldots, \alpha^{2^7}$ are linearly independent, and hence form a normal basis. The mapping of an element $a = a_0 + a_1\alpha + \cdots + a_7\alpha^7$ to its normal basis representation $b_0\alpha + b_1\alpha^2 + \cdots + b_7\alpha^{2^7}$ can be done as follows*

$$
[b_0, b_1, \ldots, b_7]^T =
\begin{bmatrix}
1 & 1 & 0 & 0 & 0 & 1 & 1 & 1 \\
1 & 0 & 1 & 0 & 0 & 1 & 0 & 1 \\
1 & 0 & 0 & 0 & 1 & 1 & 0 & 1 \\
1 & 0 & 0 & 1 & 0 & 0 & 0 & 0 \\
1 & 0 & 0 & 0 & 0 & 0 & 1 & 0 \\
1 & 0 & 0 & 1 & 0 & 1 & 0 & 1 \\
1 & 0 & 0 & 1 & 0 & 1 & 1 & 0 \\
1 & 0 & 0 & 1 & 0 & 0 & 1 & 0
\end{bmatrix}
[a_0, a_1, \ldots, a_7]^T.
$$

It can be derived that $(\alpha + \alpha^2 + \cdots + \alpha^{2^7})^2 = \alpha^2 + \alpha^4 + \cdots + \alpha^{2^7} + \alpha^{2^8} = \alpha^2 + \alpha^4 + \cdots + \alpha^{2^7} + \alpha$, The only element whose square equals itself is the identity element. Therefore, $1 = \alpha + \alpha^2 + \cdots + \alpha^{2^7}$, and the first column of the mapping matrix is '11111111'. The normal basis representation of the other elements in the standard basis can be derived by using this property as well as $p(\alpha) = 0$. For example, $\alpha^5 = \alpha + \alpha^2 + \alpha^4 + \alpha^{32} + \alpha^{64}$, and hence the 6th column of the mapping matrix is '11100110'.

2

VLSI architecture design fundamentals

CONTENTS

Some fundamental concepts used in VLSI architecture design are first introduced in this chapter. Then brief discussions are given to pipelining, retiming, parallel processing, and folding, which are techniques that can be used to manipulate circuits to tradeoff speed, silicon area, and power consumption. These definitions and techniques are necessary to understand the error-correcting decoder designs introduced in later chapters. For more detailed discussions on these techniques, the interested reader is referred to [21].

2.1   Definitions and graph representation

To measure the achievable performance of a digital circuit, several definitions are given below.

Definition 15 (Critical path) *For a combinational logic circuit, the critical path is the data path with the longest computation time between inputs and outputs. For a sequential logic circuit, the critical path is defined as the longest path in terms of computation time between any pair of delay elements or registers.*

The minimum achievable clock period, T_{min}, of a circuit can be estimated as $T_{critical} + t_{setup} + t_{hold}$, where $T_{critical}$ is the computation time of the critical path, and t_{setup} and t_{hold} are the setup and hold time of the registers, respectively. Usually, $T_{critical}$ is much longer than $t_{setup} + t_{hold}$, and hence, the minimum achievable clock period is largely decided by the critical path.

Definition 16 (Throughput) *The throughput of a sequential circuit is defined as the average number of input bits processed per second.*

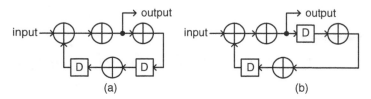

FIGURE 2.1
Circuits with feedback loops

For a sequential circuit, the maximum achievable throughput can be estimated as

$$Throughput = \frac{\# \text{ of bits processed each time}}{\# \text{ of clock cycles needed} \times T_{min}}$$

Definition 17 (Latency) *Latency is the difference between the time that an output is generated and the time at which the corresponding input is available.*

The minimum achievable latency of a sequential circuit is computed as the number of clock cycles between when the first output is generated and when the first input is available, multiplied by the minimum achievable clock period.

For a circuit without any feedback loop, the critical path can be changed by adding registers to or deleting registers from the data paths using the pipelining technique. Pipelining can not be applied to circuits with feedback loops. Nevertheless, a technique called retiming can be adopted to move registers around in feedback loops. Both the pipelining and retiming will be discussed later in this chapter.

Example 14 *The circuit shown in Fig. 2.1(a) has a feedback loop. It can be observed that the critical path of this circuit has three adders. The register in the bottom right corner of this circuit can be moved to before the adder in the top right corner to derive the circuit in Fig. 2.1(b) without changing the function. However, the circuit in Fig. 2.1(b) has two adders in the critical path, and hence can achieve higher clock frequency.*

To characterize the minimum achievable clock period of circuits with feedback loops, the iteration bound needs to be used.

Definition 18 (Loop bound) *Loop bound is defined as the sum of the computation time of all the units in the loop divided by the number of delay elements in the loop.*

Definition 19 (Critical loop) *Critical loop is the loop with the maximum loop bound.*

Definition 20 (Iteration bound) *Iteration bound, denoted by T_∞, equals the loop bound of the critical loop.*

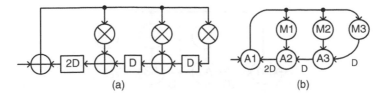

FIGURE 2.2
Example of DFG. (a) block diagram of a filter; (b) corresponding DFG

For simple circuits, all the loops can be found by observation and the iteration bound is computed accordingly. The iteration bound can also be computed by using the longest path matrix algorithm [27] or the minimum cycle mean algorithm [28]. Both of them can be written into computer programs, which facilitates the computation of the iteration bound for large complex circuits. Note that compound loops going through a loop multiple times or formed by concatenating different loops do not need to be considered in the iteration bound computation, since they do not lead to longer loop bounds.

For an architecture, G, the minimum achievable clock period is decided by $\max(\lceil T_\infty \rceil, \max_{i \in G} T_i)$, where T_i is the computation time of each unit in the architecture that can not be divided into smaller parts. Here the ceiling function is taken in case T_∞ is a fraction of the unit computation time. The critical path of an architecture with feedback loops can be reduced to meet this minimum clock period by the retiming technique [21].

Example 15 *In the architecture of Fig. 2.1 (a), there is only one loop. Denote the computation time of an adder by T_A. Then the loop bound is $4T_A/2 = 2T_A$. Therefore, the iteration bound, and hence the minimum achievable clock period, of this architecture is $2T_A$. On the other hand, the critical path of this architecture has three adders. Applying re-timing, the architecture in Fig. 2.1(a) can be transferred to that in Fig. 2.1(b), which has two adders in the critical path and achieves the minimum clock period.*

Transformation techniques, such as pipelining, retiming, folding and unfolding are more easily explained using the data flow graph (DFG) representation of circuits. For a given circuit, the computation units are represented by nodes in the corresponding DFG, and the data paths connecting computation units are denoted by directed edges. Each edge is also associated with the same number of delay elements as in the corresponding data path. Fig. 2.2(a) shows the block diagram of a filter, and the corresponding DFG is illustrated in Fig. 2.2(b).

Definition 21 (Cutset) *A cutset is a set of edges, such that the graph becomes two disjoint parts when they are removed.*

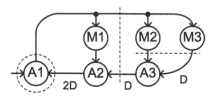

FIGURE 2.3
Cutset examples

Definition 22 (Feed-forward cutset) *If the edges in a cutset go in the same direction, then the cutset is called a feed-forward cutset.*

Example 16 *In the DFG shown in Fig. 2.3, the vertical dashed line shows a cutset. If the two edges in this cutset are removed, the parts of the graph consisting of nodes A1, A2 and M1 will become disjoint from that consisting of nodes A3, M2 and M3. This cutset is not a feed-forward cutset since the top and bottom edges go in opposite directions. The dashed circle around node A1 is also a cutset. It covers the three edges going into and out of node A1. If they are removed, the node A1 will become separated from the rest of the graph. Of course, it is not a feed-forward cutset either, since two of the edges go into node A1, while the other comes out of A1. The horizontal dashed line is not a cutset. Even if the two edges covered by this dashed line are removed, all the nodes in the graph are still connected.*

2.2   Pipelining and retiming

Pipelining is a technique that is applied to feed-forward cutsets to change the critical path of the circuit. It adds or removes the same number of delay elements from each edge in a feed-forward cutset. Although the overall function of the circuit remains the same after pipelining, the latency may be changed.

Example 17 *Fig. 2.4 (a) shows a finite impulse response (FIR) filter. Assume that the computation time of an adder is T_a, and that of a multiplier is $T_m = 2T_a$. Then both the critical path and latency of this architecture are $T_m + 2T_a = 4T_a$. This architecture can be pipelined by adding one register to each of the edges in the feed-forward cutset denoted by the horizontal dashed line shown in Fig. 2.4(b). In this pipelined architecture, the critical path is reduced to $\max\{T_m, 2T_a\} = 2T_a$. Hence, it can achieve twice the throughput if the setup and hold time of the registers are ignored. It can be easily checked*

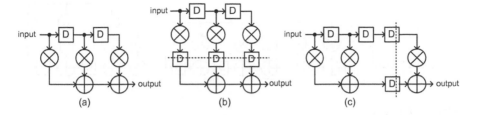

FIGURE 2.4
Pipelining example. (a) an FIR filter; (b)(c) pipelined filters

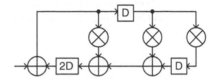

FIGURE 2.5
A retimed version of the filter

that the outputs of this pipelined architecture are the same as those of the original architecture, except that every output sample appears one clock cycle later. The latency of this pipelined architecture is $2 \times 2T_a = 4T_a$, the same as the original architecture in Fig. 2.4(a) if the register setup and hold time are ignored. Pipelining can be also done on the vertical cutset shown in Fig.2.4(c). The critical path of this second pipelined architecture is $T_m + T_a = 3T_a$, and the latency is $2 \times 3T_a = 6T_a$, which is longer than that of the original architecture.

Retiming is a technique that is used to change the locations of delay elements in a circuit without changing the overall function. It can be applied to any cutset, and is not limited to feed-forward cutsets. Pipelining can be considered as a special case of retiming. Let G_1 and G_2 be the two disjointed subgraphs resulted from removing the edges in a cutset. In a retiming solution, k delay elements can be removed from each edge going from G_1 to G_2 if k delay elements are added to each edge from G_2 to G_1. It should be noted that the number of delay elements on any edge should be non-negative before and after retiming.

Example 18 *Now consider applying retiming to the filter in Fig. 2.2(a) on the vertical cutset shown in Fig. 2.3. The top edge in this cutset goes to the right, while the edge on the bottom goes to the left. Applying retiming, one delay element is removed from the bottom edge and one delay element is added to the top edge. As a result, the architecture in Fig. 2.5 is derived. It has the same function as that in Fig. 2.2(a).*

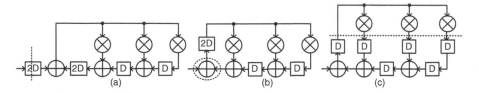

FIGURE 2.6
An example of filter retiming for achieving iteration bound

    While pipelining is usually adopted to reduce the critical path of circuits without any feedback loop, re-timing can help to achieve iteration bound in circuits with feedback loops.

Example 19 *For the architecture in Fig. 2.2(a), there are three loops. Assume that the computation time of an adder is T_a, and that of a multiplier is $T_m = 2T_a$. Then the iteration bound of this architecture is $\max\{(T_m + 2T_a)/2, (T_m + 3T_a)/3, (T_m + 3T_a)/4\} = 2T_a$. Pipelining alone can not make the critical path of this architecture equal to the iteration bound. Fig. 2.6 shows one solution to achieve the iteration bound by retiming and pipelining. The vertical cutset in Fig. 2.6(a) has only one edge, and it is a feed-forward cutset. Hence, retiming (also pipelining in this case) can be applied to add two delay elements to this edge. This enables the application of retiming to the cutset circling the first adder on the left. Two delay elements have been removed from each edge going into the circle, and two delays have been added to each edge going out of the circle to derive the architecture in Fig. 2.6(b). After that, retiming is applied again to a cutset that covers all four vertical edges. One register is eliminated from the edge on the left going to the top, and one register is added to each of the other three edges going to the bottom. As a result, the architecture in Fig. 2.6(c) is derived. The critical path of this architecture is $T_m = 2T_a$, and the iteration bound is achieved.*

    Although the retiming solution in the above example is derived by observation, it can be also described quantitatively and solved by systematic algorithms. A retiming solution of a DFG can be decided by setting a retiming variable to each node in the graph. Assume that node U is connected to node V by edge e with $w(e)$ delays in the original DFG. Let $r(U)$ and $r(V)$ be the retiming variables for node U and V, respectively. Then the number of delays on edge e after retiming is defined as $w_r(e) = w(e) + r(V) + r(U)$. Following the rule that the number of delays on any edge must be non-negative, a set of inequalities is written. Solving the system of inequalities would lead to a set of retiming variables, from which the retimed graph is derived. The shortest path algorithms, such as the Bellman-Ford and Floyd-Warshall algorithms, find the solutions of the inequalities if they are mapped to a constraint graph first. These algorithms are described in [21]. Building on this method for

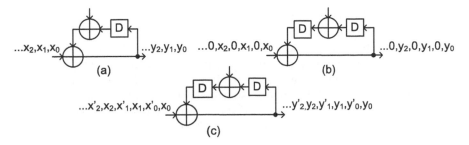

FIGURE 2.7
Example of applying slow-down. (a) a feedback loop; (b) 2-slow-downed and retimed architecture; (c) slow-downed architecture with interleaved inputs

finding retiming solutions, techniques for minimizing clock period or number of registers through retiming have also been developed. The interested reader is referred to [21] for more details.

The minimum achievable clock period of an architecture consisting of feedback loops is limited by the iteration bound. Although retiming may reduce the critical path, it can not go beyond the iteration bound. If higher clock frequency is desired, a technique referred to as slow-down can be adopted. The basic idea of slow-down is to replace each delay element by N ($N \in Z^+, N > 1$) delay elements. Accordingly, the iteration bound, and hence the minimum achievable clock period, is reduced by almost N times if retiming is applied to place the registers evenly around each feedback loop. On the other hand, to maintain the same circuit function, $N - 1$ zeros need to be inserted between each input sample. Of course, there will be $N - 1$ zeros between each valid output sample.

Example 20 *Fig. 2.7 (a) shows a feedback loop whose critical path has two adders. Adopting 2-slow-down, the register in this loop is replaced by two registers. After retiming is applied around the adder on the top branch to move one register at its input to its output, the architecture in Fig. 2.7(b) is derived. The critical path of the 2-slow-downed and retimed architecture is one adder. However, one zero needs to be inserted between each input sample and output sample. Hence, it takes twice the number of clock cycles to process an input.*

In a slow-downed architecture, the overall throughput is not increased despite the larger area caused by the extra delay elements, and much computation is wasted on the zeros. However, for an N-slow-downed architecture. if N independent input sequences are available, then they can be interleaved, and the samples in the second through Nth sequences can take the places of the zeros. For example, two independent input sequences x and x' are interleaved for the 2-slow-downed architecture in Fig. 2.7(c). When N sequences are interleaved, the samples in these sequences are processed in about the same amount of time as that needed for processing one single sequence in the original architecture, and the area overhead of the slow-downed architecture only includes the extra registers. Considering these, the slow-down is actually a method that can be used to reduce the hardware requirement when multiple independent input sequences are available. It should be noted that as N increases, the data path is cut into shorter segments after retiming, and the setup and hold time of registers become more prominent. Hence, N should not be a large number in order to be cost-effective.

2.3 Parallel processing and unfolding

Different from pipelining and retiming, which increase the processing speed by enabling higher clock frequency, parallel processing improves the throughput by duplicating hardware units and processing multiple inputs simultaneously. The hardware overhead of retiming and pipelining is extra registers. However, the area requirement of a parallel architecture increases proportionally with the number of duplicated copies. Hence, to achieve higher throughput, pipelining and retiming should be considered before parallel processing if higher clock frequency is allowed. It should also be noted that duality exists between pipelining and parallel processing. If multiple independent sequences are processed in an interleaved manner by a pipelined architecture, they can be also computed by duplicated units in a parallel architecture.

If the output sequence of a circuit, $y(n)$ $(n = 0, 1, 2, \ldots)$, can be expressed as a mathematical formula in terms of the input sequence $x(n)$, then parallel processing can be achieved through rewriting the formula in terms of parallel inputs and outputs. To obtain a J-parallel architecture that processes J inputs and generates J outputs in each clock cycle, the inputs and outputs need to be divided into J groups $x(Jk), x(Jk+1), \ldots, x(Jk+J-1)$ and $y(Jk), y(Jk+1), \ldots, y(Jk+J-1)$ $(k = 0, 1, 2, \ldots)$. Once the formulas for expressing the parallel outputs in terms of parallel inputs are derived, the parallel architecture can be drawn accordingly.

Example 21 *Let $x(n)$ and $y(n)$ $(n = 0, 1, 2, \ldots)$ be the input and output sequences, respectively, of the 3-tap filter shown in Fig. 2.8(a). The function of this filter can be described by*

$$y(n) = a_0 x(n) + a_1 x(n-1) + a_2 x(n-2). \qquad (2.1)$$

Here $x(n-1)$ is the sequence $x(n)$ delayed by one clock cycle. To achieve 3-parallel processing, the inputs are divided into three sequences $x(3k), x(3k+1)$ and $x(3k+2)$. Similarly, the outputs are divided into $y(3k), y(3k+1)$ and

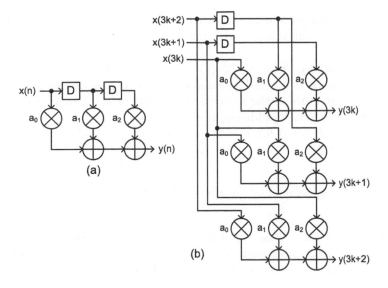

FIGURE 2.8
Parallel processing of 3-tap filter. (a) serial architecture; (b) 3-parallel architecture

$y(3k+2)$. *Replacing n in* (2.1) *by* $3k$, $3k+1$ *and* $3k+2$, *it can be derived that*

$$\begin{cases} y(3k) = a_0 x(3k) + a_1 x(3k-1) + a_2 x(3k-2) \\ y(3k+1) = a_0 x(3k+1) + a_1 x(3k) + a_2 x(3k-1) \\ y(3k+2) = a_0 x(3k+2) + a_1 x(3k+1) + a_2 x(3k) \end{cases} \qquad (2.2)$$

In a J-parallel architecture, since the indexes of the samples in each sequence are J apart, each register delays its input by J samples. Hence, $x(3k-1)$ is obtained by delaying $x(3k+2)$ with one register. Similarly, delaying $x(3k+1)$ by one register would yield $x(3k-2)$. As a result, a 3-parallel architecture of the filter in Fig. 2.8(a) is derived as shown in Fig. 2.8(b) according to (2.2).

In many occasions, the function of a circuit may not be easily described by mathematical formulas or the formulas are complicated. In this case, the unfolding [21] technique can be applied to the DFG of the circuit to achieve parallel processing. In a DFG, the computation units are represented by nodes, and no detail is required on the computations carried out. Unfolding is a graphical method for deriving parallel architectures.

To achieve J-unfolding and hence J-parallel processing, first make J copies of every node in the DFG. Assume that the output of node U is sent to node V after w delay elements in the original architecture. Let U_i and V_j ($0 \le i, j < J$) be the copies of the nodes in the unfolded architecture. Then the output of U_i should be sent to $V_{(i+w) \mod J}$ after $\lfloor (i+w)/J \rfloor$ delay elements [21]. Similarly, the inputs and outputs of a J-unfolded architecture should be divided into J groups collecting the samples with indexes $Jk, Jk + 1, \ldots, Jk + J - 1$ ($k = 0, 1, 2, \ldots$).

It can be derived that $\sum_{i=0}^{J-1} \lfloor (i + w)/J \rfloor = w$. Hence the total number of delay elements on the edges connecting the copies of the nodes in the unfolded architecture is the same as that between the pair of nodes in the original architecture. In addition, if $w \geq J$, then $\lfloor (i + w)/J \rfloor \geq 1$. If $w < J$, then $\lfloor (i + w)/J \rfloor = 0$ for $i = 0, 1, \ldots, J - w - 1$, and $\lfloor (i + w)/J \rfloor = 1$ for $i = J - w, J - w + 1, \ldots, J - 1$. Therefore, unfolding has the following properties.

Property 1 *Unfolding does not change the total number of delay elements in the architecture.*

Property 2 *If the number of delay elements between two units is greater than or equal to J, then each edge connecting the copies of the units in the J-unfolded architecture has at least one delay element.*

Property 3 *If the number of delay elements, w, between two units is less than J, then in the J-unfolded architecture, there are w edges connecting the copies of the nodes with exactly one delay element and the rest $J - w$ edges do not have any delay element.*

Properties 2 and 3 help to identify the critical path in the unfolded architecture. If an edge in the original architecture has $w \geq J$ delays, then the critical path will not be increased as a result of unfolding the edge. Otherwise, some copies of the computation units in the unfolded architecture are no longer connected through delay elements, and the corresponding data paths become longer.

Example 22 *Fig. 2.9 shows an example of unfolding. The architecture in part (a) is a linear feedback shift register, and the adders are labeled as A, B, and C. Now apply 3-unfolding. First, three copies are made for each of the adders as illustrated in the 3-unfolded architecture of Fig. 2.9 (b), and they are labeled A_i, B_i and C_i ($i = 0, 1, 2$). Take the edge from unit B to unit A in the original architecture as an example. It has $w = 2$ delay elements. Hence, in the $J = 3$ unfolded architecture, the output of B_1 should be sent to $A_{(1+2)} \bmod 3 = A_0$ after $\lfloor (1+2)/3 \rfloor = 1$ delay element. Similarly, the output of B_2 should be sent to $A_{(2+2)} \bmod 3 = A_1$ after $\lfloor (2+2)/3 \rfloor = 1$ delay element. The output of B_0 is used by $A_{(0+2)} \bmod 3 = A_2$ after $\lfloor (0+2)/3 \rfloor = 0$ delay element.*

The total number of delay elements in Fig. 2.9 (b) is four, which is the same as in the original architecture. Since the output of node B is sent to node A after $w = 2 < J = 3$ delay elements in the original architecture, there are $w = 2$ edges connecting copies of node A and B in the 3-unfolded architecture with one delay element, and the other $J - w = 1$ edge does not have any delay.

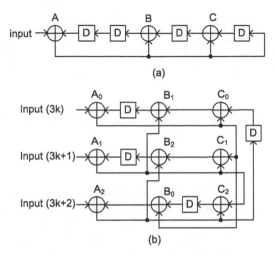

FIGURE 2.9
Unfolding example. (a) original architecture; (b) 3-unfolded architecture

The following lemmas are also useful in understanding the structure and performance of the unfolded architecture without actually carrying out unfolding. The proofs can be found in [21].

Lemma 1 *Applying J-unfolding to a loop with w delay elements leads to* $\gcd(w, J)$ *loops in the unfolded architecture, and each loop has* $w/\gcd(w, J)$ *delay elements.*

Here gcd denotes the greatest common divisor.

Lemma 2 *If the iteration bound of an architecture is* T_∞, *then the iteration bound of the J-unfolded architecture is* JT_∞.

Example 23 *In the architecture shown in Fig. 2.9(a), there is a loop that goes through nodes A, B and has two delay elements. Now examine the loops formed by nodes* A_i, *and* B_i *($i = 0, 1, 2$) in the 3-unfolded architecture shown in Fig. 2.9(b). It can be seen that* $A_0 \to B_0 \to A_2 \to B_2 \to A_1 \to B_1 \to A_0$ *form one single loop. This agrees with* $\gcd(w, J) = \gcd(2, 3) = 1$. *Also this loop has* $w/\gcd(w, J) = 2/1 = 2$ *delay elements.*

For the architecture in Fig. 2.9(a), there are three loops, and the iteration bound is $T_\infty = \max\{2T_a/2, 3T_a/3, 3T_a/4\} = T_a$, *where* T_a *is the computation time of an adder. In the 3-unfolded architecture, there are four non-compound loops besides the one consisting of the three copies of A and B. They are* $A_0 \to C_0 \to B_1 \to A_0$, $A_1 \to C_1 \to B_2 \to A_1$, $A_2 \to C_2 \to B_0 \to A_2$, *and* $A_0 \to C_1 \to B_2 \to A_1 \to C_2 \to B_0 \to A_2 \to C_0 \to B_1 \to A_0$. *Accordingly, the iteration bound of the 3-unfolded architecture is* $\max\{6T_a/2, 3T_a/1, 3T_a/1, 3T_a/1, 9T_a/4\} = 3T_a = 3T_\infty$.

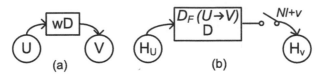

FIGURE 2.10
Folding of an edge. (a) an edge in the original architecture; (b) corresponding data path in N-folded architecture

For an architecture without any feedback loop, the same clock frequency can be achieved in its parallel or unfolded architecture by applying pipelining. For an architecture with feedback loops, the iteration bond in the J unfolded architecture is J times higher, although J samples are processed simultaneously. However, this does not mean that the parallel processing of an architecture with feedback loops can not achieve any speedup despite the almost J times larger area. This is because the critical path is decided by $\max(\lceil T_\infty \rceil, \max_{i \in G} T_i)$ as previously mentioned, and it can not always be made equal to the iteration bound. One such case is when the iteration bound is a fraction of the computation time of the units that can not be divided into smaller parts. For example, the iteration bound of the original architecture may be $T_\infty = 1/3$, and hence the minimum achievable clock period is $\lceil T_\infty \rceil = 1$ unit of time. In the 3-unfolded architecture, the iteration bound becomes $3T_\infty = 1$, and the minimum achievable clock period is still 1 unit of time. As a result, the 3-unfolded architecture can achieve three times the throughput.

2.4   Folding

Folding is the reverse of the unfolding technique. It is useful when the fully-parallel architecture is easier to design, but the application demands smaller area and can tolerate lower speed. Hardware units can be shared to carry out multiple operations in a time-multiplexed manner. The folding technique systematically decides the switching of the inputs to the hardware units in order to carry out each mapped operation.

In an N-folded architecture, N operations are implemented by a single hardware unit in a time-multiplexed way. Therefore, the computations done in a single clock cycle in the original architecture would take N clock cycles in the N-folded architecture. Assume that node U is originally connected to node V by w delay elements as shown in Fig. 2.10 (a). Consider that the executions of nodes U and V for the lth iteration in the N-folded architecture are done at clock cycles $Nl + u$ and $Nl + v$, respectively. Here u and v are called the folding orders, and $0 \leq u, v < N$. The folding orders decide at which time instance the computations in the corresponding nodes are executed. The nodes that are executed by the same unit in the folded architecture can be collected and written in a set according to their folding order. Such a set is called a folding set. Let H_U and H_V be the hardware units in the folded architecture executing the functions of nodes U and V, respectively. If H_U has P_U pipelining stages itself, then the output of node U in the lth iteration is ready at clock cycle $Nl + u + P_U$. Each delay in the original architecture is translated to an iteration of N clock cycles in the N-folded architecture. Since there are w delays on the edge connecting nodes U and V, the output generated by node U in the lth iteration should be consumed by node V in the $(l + w)$th iteration. Considering the folding order of node V, it is actually consumed at clock cycle $N(l+w)+v$. Accordingly, the output of node U needs to be stored for

$$D_F(U \to V) = (N(l + w) + v) - (Nl + u + P_U) = Nw - P_U + v - u$$

clock cycles. Therefore, the edge connecting nodes U and V with w delay elements in the original architecture shown in Fig. 2.10 (a) should be mapped to the data path in Fig. 2.10 (b) in the N-folded architecture. The entire folded architecture can be derived by carrying out similar computations to decide the data path corresponding to each edge in the original architecture. It is possible that the value of $D_F(U \to V)$ is negative. In this case, pipelining and/or retiming can be applied to adjust the number of delays on each edge in order to make every $D_F(U \to V)$ non-negative. In addition, assigning different folding orders also affects the value of $D_F(U \to V)$.

Example 24 *Consider applying 2-folding to the 4-tap FIR filter shown in Fig. 2.11(a). The 2-folded architecture has two multipliers M_1 and M_2 and two adders A_1 and A_2. The mapping and the folding order of each node are labeled in Fig. 2.11(a). For example, node 1 is executed by multiplier M_1 in clock cycle $2l + 0$, and node 7 is executed by adder A_2 in clock cycle $2l + 1$. The folding set for M_1 is $\{1, 3\}$ and that for A_2 is $\{6, 7\}$. Assume that each multiplier has one pipelining stage inside, i.e. $P_U = 1$. The mapping of each edge for applying 2-folding is decided by the $D_F(U \to V)$ values computed as follows*

$$D_F(1 \to 5) = 2 \times 1 - 1 + 0 - 0 = 1$$
$$D_F(2 \to 5) = 2 \times 0 - 1 + 0 - 0 = -1$$
$$D_F(3 \to 6) = 2 \times 0 - 1 + 0 - 1 = -2$$
$$D_F(4 \to 7) = 2 \times 0 - 1 + 1 - 1 = -1$$
$$D_F(5 \to 6) = 2 \times 1 - 0 + 0 - 0 = 2$$
$$D_F(6 \to 7) = 2 \times 1 - 0 + 1 - 0 = 3$$

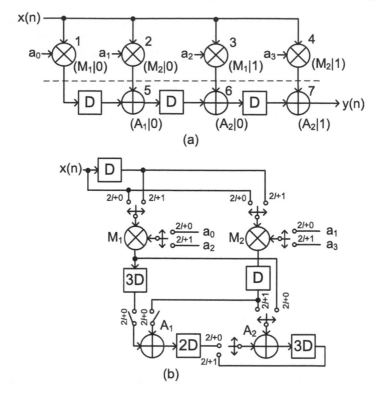

FIGURE 2.11
Folding example. (a) A 4-tap FIR filter; (b) 2-folded filter

From the above computations, the mapping of the edges $2 \to 5$, $3 \to 6$ and $4 \to 7$ leads to negative delay elements, which are not implementable. To solve this problem, delay elements need to be added to these paths. A systematic method was given in [21] to eliminate negative delays in the mapped edges through retiming and pipelining. However, for simple architectures like the FIR filter shown in Fig. 2.11(a), this can be done by direct observation. Pipelining is applied to the cutset shown by the horizontal dashed line. One delay element is added to each edge in the cutset. After that, the numbers of delays on the *mapped edges are updated as*

$$D_F(1 \to 5) = 2 \times 2 - 1 + 0 - 0 = 3$$
$$D_F(2 \to 5) = 2 \times 1 - 1 + 0 - 0 = 1$$
$$D_F(3 \to 6) = 2 \times 1 - 1 + 0 - 1 = 0$$
$$D_F(4 \to 7) = 2 \times 1 - 1 + 1 - 1 = 1$$
$$D_F(5 \to 6) = 2 \times 1 - 0 + 0 - 0 = 2$$
$$D_F(6 \to 7) = 2 \times 1 - 0 + 1 - 0 = 3$$

(2.3)

The equations in (2.3) tell the number of delay on each edge in the folded architecture. The folding sets decide at which time instances the hardware units in the folded architecture take certain inputs. Combining this information, the 2-folded architecture of the 4-tap FIR filter in Fig. 2.11 (a) is derived as shown in Fig. 2.11(b). For example, the adder A2 in the folded architecture takes care of the addition in nodes 6 and 7 of Fig. 2.11 (a) at time instances $2l + 0$ and $2l + 1$, respectively. The inputs of node 6 come from node 5 and 3. Node 5 is implemented by the adder A1 in the folded architecture. From (2.3), the output of node 5 needs to be stored for 2 clock cycles before it is consumed by node 6. Therefore, the output of unit A1 in the folded architecture is taken as the input of A2 after two delay elements at time instance $2l + 0$. Similarly, node 3 is implemented by multiplier M1 in the folded architecture. Its result needs to be delayed for 0 clock cycle before being used in the computation in node 6. Accordingly, the output of M1 becomes the other input of A2 at time instance $2l + 0$ in the folded architecture.

Unlike unfolding, folding does not preserve the number of delay elements. The FIR filter in Fig. 2.11 (a) would have 7 delay elements after pipelining is applied to the cutset shown by the dashed line. However, the folded architecture in Fig. 2.11 (b) has 10 delays. The folding sets affect the number of delays needed in the folded architecture since they participate in the computations of the edge delays. For given folding sets, the minimum number of registers required in the folded architecture can be found by making use of a linear lifetime chart, which records the signals that need to be stored in each clock cycle. Then an allocation table is utilized to re-organize the storage of the signals so that they can fit into the minimum number of registers at the cost of more complex switching. The interested reader is referred to [21] for the details of this register minimization technique.

3

Root computations for polynomials over finite fields

CONTENTS

Root computations for polynomials over finite fields are needed in Reed-Solomon and Bose-Chaudhuri-Hocquenghem (BCH) decoding algorithms. This chapter presents methods and implementation architectures for such root computations.

3.1   Root computation for general polynomials

The roots of a general polynomial over finite fields are computed by exhaustive Chien search, which tests each finite field element to see if it is a root. Efficient Chien search architectures can be found in many papers, such as [29].

Fig. 3.1 shows a J-parallel Chien search architecture for computing the roots of $f(x) = f_0 + f_1 x + \cdots + f_n x^n$ over $GF(2^q)$. Let α be a primitive element of $GF(2^q)$. This architecture tests J field elements, $\alpha^{Ji}, \alpha^{Ji+1}, \ldots, \alpha^{Ji+J-1}$, in clock cycle i ($i = 0, 1, 2, \ldots$). $f(\alpha^{Ji+j}) = f_0 + f_1 \alpha^{Ji} \alpha^j + f_2 \alpha^{2Ji} \alpha^{2j} + \cdots + f_n \alpha^{nJi} \alpha^{nj}$. $f_1 \alpha^{Ji}, f_2 \alpha^{2Ji}, \ldots f_n \alpha^{nJi}$ are generated iteratively by the feedback loops in the first column of Fig. 3.1, and they are multiplied by the same powers of α in the other multipliers in each clock cycle. The n rows and J columns of multipliers in Fig. 3.1 are constant multipliers. A constant multiplier over $GF(2^q)$ can be implemented as a $q \times q$ constant binary matrix multiplication, and hence the bits in the product are written as XOR functions over the input bits. Compared to general finite field multipliers, constant multipliers take much fewer logic gates to implement. In addition, common terms can be shared among the XOR functions for the J constant multipliers in the same row of Fig. 3.1 since their inputs are the same [30].

To implement serial Chien search, only the column of the feedback loops

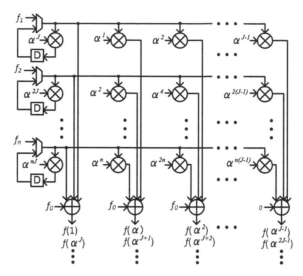

FIGURE 3.1
J-parallel Chien search architecture

and the first $n + 1$-input XOR gate in Fig. 3.1 need to be kept. The design in Fig. 3.1 can be also extended to a fully-parallel Chien search engine by having $2^q - 2$ columns of constant multipliers, and $2^q - 1$ $n + 1$-input XOR gates. In this case, the column of the feedback loops in Fig. 3.1 is not needed.

3.2　Root computation for linearized and affine polynomials

Since the Chien search needs to test each finite field element, it has very high complexity when the order of the finite field is high. Alternatively, if the polynomial is in an affine format, the root computation can be done in an easier way without Chien search [4].

Definition 23 (Linearized polynomial) *A polynomial $f(x)$ over $GF(2^q)$ is called a linearized polynomial iff it is in the format of $f(x) = \sum_i f_i x^{2^i}$.*

If $\{1, \alpha, \alpha^2, \cdots, \alpha^{q-1}\}$ is a standard basis of $GF(2^q)$ and $a_i \in GF(2)$, then for any integer k

$$(\sum_{i=0}^{q-1} a_i \alpha^i)^{2^k} = \sum_{i=0}^{q-1} (a_i^{2^k} (\alpha^i)^{2^k}) = \sum_{i=0}^{q-1} (a_i (\alpha^i)^{2^k}).$$

This serves as the proof of the following theorem.

Theorem 3 *If $f(x)$ is a linearized polynomial, and an element $a \in GF(2^q)$ is represented in standard basis as $a = \sum_{i=0}^{q-1} a_i \alpha^i$ ($a_i \in GF(2)$), then $f(a) = \sum_{i=0}^{q-1} a_i f(\alpha^i)$.*

Therefore, if each $f(\alpha^i)$ for $0 \le i < q$ is pre-computed, and the coefficients in their standard basis representations are collected as the rows of a $q \times q$ binary matrix M, then $f(a)$ can be derived by multiplying the row vector $[a_0, a_1, \ldots, a_{q-1}]$ with M.

Example 25 *Consider the polynomial $f(x) = x^4 + \alpha^5 x^2 + \alpha^7 x$ over $GF(2^4)$. α is a root of the primitive irreducible polynomial $x^4 + x + 1$ and $\{1, \alpha, \alpha^2, \alpha^3\}$ is a standard basis of $GF(2^4)$. It can be computed that*

$$f(1) = \alpha^2 + \alpha^3$$
$$f(\alpha) = 1 + \alpha^2 + \alpha^3$$
$$f(\alpha^2) = 1 + \alpha^2$$
$$f(\alpha^3) = \alpha + \alpha^2$$

Accordingly,

$$M = \begin{bmatrix} 0 & 0 & 1 & 1 \\ 1 & 0 & 1 & 1 \\ 1 & 0 & 1 & 0 \\ 0 & 1 & 1 & 0 \end{bmatrix}$$

As an example, let us compute $f(\alpha^{12})$. In standard basis, $\alpha^{12} = 1 + \alpha + \alpha^2 + \alpha^3$. Hence, $f(\alpha^{12}) = [1\,1\,1\,1]M = [0\,1\,0\,0]$, which is the vector representation of α.

Definition 24 (Affine polynomial) *A polynomial $f(x)$ over $GF(2^q)$ is an affine polynomial iff it is in the format of $f(x) = l(x) + u$, where u is a constant in $GF(2^q)$ and $l(x)$ is a linearized polynomial.*

Accordingly, finding a root $a = a_0 + a_1\alpha + \cdots + a_{q-1}\alpha^{q-1}$ of $f(x)$ in $GF(2^q)$ is equivalent to solving the following set of q simultaneous binary linear equations for $a_0, a_1, \ldots, a_{q-1}$.

$$[a_0, a_1, \ldots, a_{q-1}]M = [u_0, u_1, \ldots, u_{q-1}]$$

Here $[u_0, u_1, \ldots, u_{q-1}]$ is the vector representation of u in standard basis and M is the $q \times q$ binary matrix whose rows are the vector representations of $l(1), l(\alpha), \ldots, l(\alpha^{q-1})$ in standard basis.

The set of simultaneous binary linear equations can be solved by using the method in [4]. It reduces M to a triangular idempotent form through an iterative process that has long latency. The corresponding implementation architecture can be found in [31]. In addition, for an arbitrary polynomial with degree higher than four, the conversion to affine format incurs much hardware overhead [4]. Therefore, finding the roots through converting the polynomial to affine format and solving simultaneous linear equations becomes less practical when the degree of the polynomial is higher than four.

When the polynomial degree is two or three, the root computation can be substantially simplified as presented next.

3.3   Root computation for polynomials of degree two or three

A degree-2 polynomial over $GF(2^q)$ can be written as $f(x) = f_2x^2 + f_1x + f_0$, where $f_2 \neq 0$. If $f_1 = 0$, $f_2x^2 + f_1x + f_0 = 0$ can be reduced to $x^2 = f_0f_2^{-1}$. Using normal basis representation, the square root is found by cyclically shifting each bit in the vector representation of $f_0f_2^{-1}$ by one position. When $f_1 \neq 0$, $f_2x^2 + f_1x + f_0 = 0$ can be transformed to

$$y^2 + y = f_2f_0(f_1^{-1})^2 \tag{3.1}$$

by substituting x with $f_1f_2^{-1}y$.

Equation (3.1) is a special case of affine polynomial. The corresponding root computation does not need any simultaneous linear equation solving. A polynomial in the format of $y^2 + y = u$, where $u \in GF(2^q)$, only has solutions in $GF(2^q)$ iff $Tr(u)=0$. Let α be a primitive element of $GF(2^q)$, and α^z is an element of $GF(2^q)$ such that $Tr(\alpha^z) = 1$. Note that the *trace* of a $GF(2^q)$ element is an element in $GF(2)$, *i.e.* either 0 or 1. From an exhaustive search, a set of $y(i) \in GF(2^q)$ $(i = 0, 1, \ldots, q-1)$ satisfying the following conditions can be found.

$$(y(i))^2 + y(i) = \begin{cases} \alpha^i & \text{if } Tr(\alpha^i) = 0 \\ \alpha^i + \alpha^z & \text{if } Tr(\alpha^i) \neq 0 \end{cases} \tag{3.2}$$

Write u in standard basis as $\sum_{i=0}^{q-1} u_i\alpha^i$, where $u_i \in GF(2)$. Then $y =$

$\sum_{i=0}^{q-1} u_i y(i)$ is a root of $y^2 + y = u$ [4]. This is because

$$
\begin{aligned}
y^2 + y &= \sum_{i=0}^{q-1} u_i (y(i)^2 + y(i)) \\
&= \sum_{i=0}^{q-1} u_i (\alpha^i + Tr(\alpha^i)\alpha^z) \\
&= u + \sum_{i=0}^{q-1} u_i Tr(\alpha^i)\alpha^z \qquad (3.3) \\
&= u + Tr(\sum_{i=0}^{q-1} u_i \alpha^i)\alpha^z \\
&= u + Tr(u)\alpha^z \\
&= u
\end{aligned}
$$

The other root is $y' = \sum_{i=0}^{q-1} u_i y(i) + 1$, since $y'^2 + y' = (\sum_{i=0}^{q-1} u_i y(i) + 1)^2 + \sum_{i=0}^{q-1} u_i y(i) + 1 = (\sum_{i=0}^{q-1} u_i y(i))^2 + \sum_{i=0}^{q-1} u_i y(i) = y^2 + y$. $y(i)$ can be precomputed, and their vector representations are collected to form a $q \times q$ binary matrix. Accordingly, the roots of $y^2 + y = u$ are computed by multiplying the vector representation of u with this pre-computed binary matrix.

Example 26 *Let α be a root of $p(x) = x^8 + x^7 + x^6 + x + 1$. It is a primitive polynomial over $GF(2)$. $\{1, \alpha, \alpha^2, \ldots, \alpha^7\}$ is a standard basis of $GF(2^8)$. It can be computed that $Tr(1) = Tr(\alpha^3) = Tr(\alpha^6) = Tr(\alpha^7) = 0$ and $Tr(\alpha) = Tr(\alpha^2) = Tr(\alpha^4) = Tr(\alpha^5) = 1$. Take $z = 1$, then the set of $y(i)$ satisfying (3.2) are found through exhaustive search as*

$$
\begin{cases}
y(0) = \alpha^{85} = \text{`}00110101\text{'} & y(1) = 0 = \text{`}00000000\text{'} \\
y(2) = \alpha = \text{`}01000000\text{'} & y(3) = \alpha^{18} = \text{`}00111110\text{'} \\
y(4) = \alpha^{151} = \text{`}11100000\text{'} & y(5) = \alpha^{51} = \text{`}11001001\text{'} \\
y(6) = \alpha^{36} = \text{`}00101110\text{'} & y(7) = \alpha^{96} = \text{`}01110011\text{'}.
\end{cases}
\qquad (3.4)
$$

The binary tuples in the above equations are the vector representations in standard basis. Accordingly, one root of $y^2 + y = u$ is

$$
[u_0, u_1, \ldots, u_7]
\begin{bmatrix}
0 & 0 & 1 & 1 & 0 & 1 & 0 & 1 \\
0 & 0 & 0 & 0 & 0 & 0 & 0 & 0 \\
0 & 1 & 0 & 0 & 0 & 0 & 0 & 0 \\
0 & 0 & 1 & 1 & 1 & 1 & 1 & 0 \\
1 & 1 & 1 & 0 & 0 & 0 & 0 & 0 \\
1 & 1 & 0 & 0 & 1 & 0 & 0 & 1 \\
0 & 0 & 1 & 0 & 1 & 1 & 1 & 0 \\
0 & 1 & 1 & 1 & 0 & 0 & 1 & 1
\end{bmatrix},
$$

and the other is only different in the first bit.

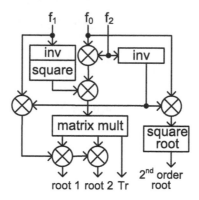

FIGURE 3.2
Root computation architecture for degree-2 polynomials (modified from [7])

To check if the roots computed by the matrix multiplication procedure are real roots, $Tr(u)$ also needs to be computed. According to the linearity of trace, $Tr(u) = u_0 Tr(1) + u_1 Tr(\alpha) + u_2 Tr(\alpha^2) + \cdots u_7 Tr(\alpha^7)$. Since $Tr(1) = Tr(\alpha^3) = Tr(\alpha^6) = Tr(\alpha^7) = 0$ and $Tr(\alpha) = Tr(\alpha^2) = Tr(\alpha^4) = Tr(\alpha^5) = 1$, $Tr(u)$ is computed as $u_1 \oplus u_2 \oplus u_4 \oplus u_5$.

The architecture for computing the roots of a degree-2 polynomial $f_2 x^2 + f_1 x + f_0$ is shown in Fig. 3.2. The matrix multiplication block multiplies the vector representation of $u = f_2 f_0 (f_1^{-1})^2$ by the pre-computed binary matrix. The two multipliers after are for reversing the coordinate transformation. When $f_1 = 0$, the second-order root derived from the square root unit is the real root. If $f_1 \neq 0$ and the trace value is zero. Then 'root 1' and 'root 2' are the two real roots. Otherwise, the polynomial does not have any root.

Standard basis representation is adopted in the above process for computing the roots of a polynomial in the format of $y^2 + y + u$. On the other hand, the square root computation unit as shown in Fig. 3.2 needs to use a normal basis. Hence, converters to and from the normal basis need to be adopted. Both of them are implemented as constant binary matrix multiplications. In addition, the two matrix multiplications and the cyclical shift in the normal basis can be combined to reduce the complexity.

The root computation for $y^2 + y + u$ can be also done in normal basis representation [32]. The roots are derived in a similar way by multiplying the vector representation of u in normal basis with a pre-computed binary matrix. Nevertheless, the matrix is derived by collecting the vector representations of $y(i)$ satisfying the following conditions instead.

$$y(i)^2 + y(i) = \begin{cases} \omega^{2^i}, & if\ Tr(\omega^{2^i}) = 0 \\ \omega^{2^i} + \omega^z, & if\ Tr(\omega^{2^i}) \neq 0 \end{cases}$$

Here $\{\omega, \omega^2, \ldots \omega^{2^{q-1}}\}$ is a normal basis, and ω^z is an element of $GF(2^q)$ such that $Tr(\omega^z) \neq 0$. If u is written in normal basis as $y = \sum_{i=0}^{q-1} u_i \omega^{2^i}$, then the two roots are given by $y = \sum_{i=0}^{q-1} u_i y(i)$ and $y' = \sum_{i=0}^{q-1} u_i y(i) + 1$ in normal basis representation. This can be proved by derivations similar to those in (3.3). Note that in normal basis, the identity element is represented as $\omega + \omega^2 + \cdots + \omega^{2^{(q-1)}}$.

Any degree-3 polynomial can be scaled so that it becomes a monic polynomial $f(x) = x^3 + f_2 x^2 + f_1 x + f_0$. Let $a = f_2^2 + f_1$, $b = f_0 + f_1 f_2$ and $c = ba^{-3/2}$. In the case that $a \neq 0$, let $y = x + f_2$ and $z = ya^{-1/2}$. $f(x)$ can be transformed to the following format [33]

$$z^3 + z + c = 0.$$

c is the only variable in the above equation. Hence, the roots corresponding to each possible value of c can be pre-computed and stored in a look-up table (LUT). Then the root computation becomes table lookups followed by inverse coordinate transformations.

The LUT-based root computation can achieve very fast speed. However, for $c \in GF(2^q)$, the size of the LUT is $2^q \times 3q$ bits. It occupies a large silicon area for high-order finite fields. For example, the size of the LUT is $2^{10} \times 30$ when the finite field involved is $GF(2^{10})$. One way to reduce the LUT size is to store only one root and compute the other two by solving a degree-2 polynomial. Specifically, let β be a root. Then

$$(z^3 + z + c)/(z + \beta) = z^2 + \beta z + (1 + \beta^2).$$

The other two roots are computed by solving $z^2 + \beta z + (1 + \beta^2) = 0$ using the approach described previously. At the cost of extra computation for solving a degree-2 polynomial, this method reduces the size of the LUT to 1/3.

The LUT size can be further reduced by the method in [32]. Assume that $z^3 + z + c$ has a root β. In other words, $\beta^3 + \beta + c = 0$. Then $(\beta^3 + \beta + c)^2 = 0$, and hence $\beta^6 + \beta^2 + c^2 = 0$. Therefore, if β is a root of $z^3 + z + c$, then β^2 is a root of $z^3 + z + c^2$. As a result, the roots of $z^3 + z + c^2$ do not need to be stored. They can be computed from those of $z^3 + z + c$ by square operations. The roots of $z^3 + z + c^{2^i}$ can be also derived from those of $z^3 + z + c^{2^{i-1}}$ by square operations. In the ideal case, for all the finite field elements in a conjugacy class, *i.e.* $\{c, c^2, c^4, \ldots, c^{2^{q-1}}\}$, only one root corresponding to c is stored, and the roots corresponding to the conjugates are derived by square operations. In this case, the LUT for root computation is reduced by a factor of q. Nevertheless, the address generation for such a LUT is very difficult.

For the uncompressed LUT that stores the root for each possible value of c, c is used as the address to read out the root. When the LUT is compressed, considerations need to be given to not only the achievable compression ratio, but also the way the access address is generated. The roots for different c should be stored in consecutive addresses in order to avoid wasting memory. Due to the irregularity of the finite field elements in a conjugacy class, it is very difficult to label each conjugacy class by a single address that can be easily mapped to the elements in the class and at the same time make the addresses for all conjugacy classes consecutive. Hence, the roots for multiple elements in each conjugcay class may need to be stored to trade off the LUT compression ratio with address generation complexity.

To simplify the address generation, normal basis representation of finite field elements is employed. Using normal basis, the square or square root computation is done as cyclically shifting the bits in the vector representation by one position. For a given c in normal basis representation, if the root corresponding to a cyclically shifted version of c is stored in the LUT, then this root is read out first. After that, reversely shifting the root would yield the root corresponding to c. Assume that $c_0 c_1 \ldots c_{q-1}$ is the binary vector representation of c in normal basis. The zero element is an all '0' vector and the identity element is an all '1' vector. Except these two elements, each of the other elements has both '0's and '1's in the corresponding vector representation. Such a vector can always be cyclically shifted to become a vector of '01' followed by $q - 2$ bits. Considering these, the roots for all possible values of c whose two most significant bits (MSBs) are '01' are stored in the LUT, while the roots corresponding to zero and identity elements are taken care of separately. As a result, the roots are stored in 2^{q-2} consecutive lines, and the LUT size is reduced by a factor of four. For a given c, a priority encoder is used to locate the first '01' pattern in its normal basis binary vector representation. Using the output of the priority encoder, the vector is shifted so that the MSBs become '01'. The other $q - 2$ bits are used as the address to read the root from the LUT. After that, the root corresponding to c is derived by reversely shifting the root read from the LUT. Using this method, multiple roots may be stored for a conjugacy class.

Example 27 *For $GF(2^7)$, the elements '1100101', '1110010', '0111001', '1011100', '0101110', '0010111', and '1001011' in normal basis representation form a conjugacy class. The roots corresponding to '0101110' and '0111001' are stored in the addresses '01110' and '11001', respectively, in the LUT. Given the vector representation of an element c that is in this conjugacy class, it is shifted to one of these two vectors depending on the location of the first '01' pattern, and the other 5 bits are used as the address to access the LUT. For example, if c='1110010', then it is cyclically shifted to the left by four positions to become '0101110', and '01110' is used as the address to read the root from the LUT. The root is cyclically shifted to the right by four positions to derive the actual root corresponding to c='1110010'. If c='1011100', then it is cyclically shifted to the left by one position to become '0111001', and the root stored at the address '11001' is read out and cyclically shifted back.*

Equivalently, the LUT can be compressed by storing the roots for all finite field elements whose MSBs are '10'. Except the zero and identity elements, the given c can be shifted into a vector starting with '10' for LUT access. The target MSBs for shifting can also be other patterns, such as '11' or '00'.

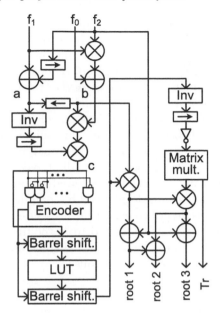

FIGURE 3.3
Root computation architecture for degree-3 polynomials (modified from [32])

Nevertheless, besides the zero and identity elements, there are other field elements that can not be shifted into a vector starting with these patterns. For example, in $GF(2^{10})$, there are '1010101010' and '0101010101'. No matter how they are shifted, the vector will not have '11' or '00' as the MSBs. The roots corresponding to these outliers are not stored in the LUT and are handled separately. If a pattern of p-bit is used as the target MSBs for shifting, then the size of the LUT is reduced by a factor of 2^p. On the other hand, there will be many more elements that can not be shifted into a vector starting with the target MSBs as p increases. These special cases need to be handled separately outside of the LUT. As a result, more logic is required, and the data path for root computation may be increased.

Fig. 3.3 shows the architecture for computing the roots of degree-3 polynomials, $f(x) = x^3 + f_2 x^2 + f_1 x + f_0$, using compressed LUT when $a = f_2^2 + f_1 \neq 0$. The variables a, b, and c needed for transforming a degree-3 polynomial into the format of $z^3 + z + c$ are calculated by the units in the top left part of Fig. 3.3. The blocks with right and left arrows denote the square and square root computations, respectively. The priority encoder discovers the first '10' pattern in the normal basis representation of c. The result of the priority encoder decides how many positions the bits in c need to be cyclically shifted so that the MSBs become '10'. The remaining $q - 2$ bits of c are used as the address to read a root out of the LUT. If the polynomial does not have any root, zero is stored in the corresponding address of the LUT to indicate that the root is invalid. After all, zero can only be a root of $z^3 + z + c$ if $c = 0$, but the root corresponding to $c = 0$ is not handled by the LUT. The root read from the LUT is reversely shifted to get β, a root of $z^3 + z + c = 0$. After the coordinate transformations are reversed, $\beta a^{1/2} + f_2$ is a root of the monic degree-3 polynomial $f(x)$.

The other two roots are computed through solving $(z^3 + z + c)/(z + \beta) = z^2 + \beta z + (1 + \beta^2) = 0$. Letting $z = \beta w$, this polynomial is transformed to $w^2 + w + (1 + 1/\beta^2) = 0$, which is an affine polynomial. In normal basis representation, the identity element is an all '1' vector. Hence $1 + 1/\beta^2$ is implemented by cyclically shifting $1/\beta$ followed by bit-wise NOT. The roots of a degree-2 polynomial in affine format can be computed by a binary constant matrix multiplication as described earlier in this section. Similarly, if the trace value is nonzero, the two roots computed from the matrix multiplication are invalid. The multipliers and adders in the bottom right part of Fig. 3.3 are for reversing the coordinate transformations.

When $a = f_2^2 + f_1 = 0$, $f(x) = x^3 + f_2 x^2 + f_1 x + f_0$ is transformed to $y^3 + b = 0$ by letting $y = x + f_2$. In unsigned binary representation, $2^q - 1$ is a string of q '1's. For example, $2^5 - 1 = 31 = $'11111'. Since the binary representation of 3 is '11', and a string of even number of '1's is divisible by '11', $3 \nmid (2^q - 1)$ if q is odd, and $3 \mid (2^q - 1)$ if q is even. When q is odd, $y^3 = b$ has a unique cube root b^s, where s is an integer satisfying $b^{3s} = b$. In other words, s is an integer such that $3s = 1 \mod (2^q - 1)$. Since $q + 1$ is even when q is odd, $(2^{q+1} - 1) = 2(2^q - 1) + 1$ is divisible by 3. Therefore, $s = (2(2^q - 1) + 1)/3$ for odd q. In unsigned binary format, such an s has q bits and $s = $' $1010\ldots1$'. For example, the cube root of $b \in GF(2^7)$ is b^{85}, and $85 = $'1010101' in binary format. Accordingly, from b, b^s is computed by iteratively taking the $4th$ power and multiplying for $(q - 1)/2$ times. If normal basis representation is employed, the $4th$ power is implemented as cyclically shifting the bits in the vector representation by two positions.

When q is even, $3 \mid (2^q - 1)$. Let $s = (2^q - 1)/3$, and α be a primitive root of $GF(2^q)$. Then α^s is a primitive cube root of unity. Assume that $b = \alpha^r$. If r is not a multiple of 3, then $y^3 + b$ does not have any root. Otherwise, it has three distinct roots $\alpha^{r/3 + si}$ $(i = 0, 1, 2)$. Let $2^q - 1 = 3^t u$, where $3 \nmid u$. When $t = 1$, the following lemma was given in [34] to find a cube root of b.

Lemma 3 *Assume that $2^q - 1 = 3u$ with $3 \nmid u$, and b is a cube in $GF(2^q)$. Let $m \equiv 3^{-1} \mod u$. Then b^m is a cube root of b.*

There are three possible values for m (modulo $2^q - 1$), and their difference is $(2^q - 1)/3$. The one whose binary representation has the least number of '1's can be chosen to compute b^m in order to reduce the complexity. When normal basis is not adopted, and hence the square operation costs logic gates, it also needs to be considered that having '1's at higher bit positions requires more square operators. When normal basis is used, the square root operation is also a cyclical shift. Hence the digit '-1' can be also included to represent m in order to further simplify the computation of b^m. For example, for $GF(2^{10})$, $2^{10} - 1 = 1023 = 3 \cdot 341$. Possible values of m are 114, 455, and 796. In binary representation, these integers are '0001110010', '0111000111', and '1100011100'. $m = 114$ can be selected since it has the least number of '1's in its binary representation. In this case, computing b^m only needs three multiplications after b^2, b^{16}, b^{32} and b^{64} are made available. Additionally, adopting canonical signed digit representation [21], '0001110010' can be also expressed as '001001̄0010'. Each bit in this representation has the same weight as in the traditional binary representation, except that the digit $\bar{1}$ has negative weight. For example, the decimal value of '001001̄0010' is $128 - 16 + 2$, which is also 114. Using the canonical signed digit representation, b^m only needs two multiplications to compute after b^{128}, b^{-16} and b^2 are derived. After b^m is calculated, the other two roots are derived by multiplying $\alpha^{(2^q-1)/3}$ and $\alpha^{2(2^q-1)/3}$, which are pre-computed.

When $t \neq 1$, the lemma below can be used to compute the cube root [34]. Actually this lemma also applies for the cases with $t = 1$.

Lemma 4 *Assume that $2^q - 1 = 3^t u$ and $3 \nmid u$, and b is a cube in $GF(2^q)$. Let m be an integer divisible by 3^{t-2} and $m = 3^{-1} \mod u$. Let k be an integer divisible by u and $k = 1 \mod 3^t$. Then $b^{m+k/3}$ is a cube root of b.*

From this lemma, b^k is a 3^{t-1}th root of unity of $GF(2^q)$. Hence, only a very limited number of field elements can be b^k, and their cube roots can be pre-computed. For example, for $GF(2^{12})$, $2^{12} - 1 = 3^2 \times 455$ and hence $t = 2$. There are three 3rd roots of unity in $GF(2^{12})$, and they are 1, α^{1365} and α^{2730}. Their cube roots are 1, α^{455}, and α^{910}, respectively. There may be multiple choices of m and k satisfying the conditions in Lemma 4. Similarly, the ones whose binary or canonical signed digit representations have fewer nonzero digits can be chosen to reduce the number of multiplications needed. For example, 910=1024+16-128-2 is the only choice for k in $GF(2^{12})$. However, m has nine possible values, and taking $m = 152 = 128 + 16 + 8$ leads to the simplest design. In this case, a cube root of $b \in GF(2^{12})$ is computed as

1. compute $b^m = b^{152}$ as $b^{128}b^{16}b^8$.

2. compute $b^k = b^{910}$ as $(b^{128}b^2)^8/(b^{128}b^2)$.

3. select the pre-computed value for $b^{k/3}$.

4. $b = b^m b^{k/3}$.

4

Reed-Solomon encoder & hard-decision and erasure decoder architectures

CONTENTS

In this information age, the reliability of digital communication and storage becomes paramount in an unprecedented broad range of applications. Along with the ever increasing craving for lower error rates, this has created the demand for error-correcting codes that are powerful yet versatile.

Reed-Solomon (RS) codes [35] are the answer to this demand, due to their good burst-error correcting capability. In addition, compared to their competitors, such as low-density parity-check (LDPC) codes and Turbo codes, RS codes enjoy the luxury of having a much wider range of codeword length and code rate. Besides traditional applications, such as magnetic and optical recording, digital televisions, cable modems, frequency-hopping wireless communications, and deep-space probing, the forever young RS codes are also finding their way into emerging applications. To name a few, they are employed as forward error-correcting codes in medical implants. Carefully designed RS encoders and decoders can be employed to significantly reduce the power consumption of wireless sensor nodes, and hence increase the lifetime of sensor networks [36].

This chapter first introduces the construction of RS codes. Then the encoder architecture design is detailed. There exist various decoding algorithms for RS codes. The emphasis of this chapter is given to traditional hard-decision decoding algorithms, such as the Peterson-Gorenstein-Zierler algorithm and

Berlekamp-Massey algorithm (BMA) [4], and their implementation architectures. These decoders are currently employed in practical applications due to the existence of efficient and high-speed hardware implementations.

4.1   Reed-Solomon codes

RS codes are linear block codes. An (n, k) RS code encodes k message symbols into n codeword symbols. The minimum distance of RS codes, which is the minimum Hamming distance between any pair of codewords, is $n-k+1$. Hence, RS codes reach the Singleton bound, and are maximum distance separable (MDS) codes. Since the distance between any codewords is at least $n - k + 1$, hard-decision decoding algorithms of RS codes are capable of correcting $t = (n - k)/2$ errors.

For an (n, k) RS code constructed over $GF(2^q)$, n does not exceed $2^q - 1$. When $n = 2^q - 1$, it is called a primitive RS code. Consider the message symbols $m_0, m_1, \ldots, m_{k-1}$ as the coefficients of a message polynomial $m(x)$ whose degree is $k - 1$, and the codeword symbols $c_0, c_1, \ldots, c_{n-1}$ are the coefficients of a codeword polynomial $c(x)$. The encoding of RS codes can be done as

$$c(x) = m(x)g(x), \tag{4.1}$$

where $g(x)$ is the generator polynomial. Let α be a primitive element of $GF(2^q)$. Then the generator polynomial for a t-error-correcting RS code can be constructed as

$$g(x) = \prod_{i=0}^{2t-1} (x + \alpha^{b+i}), \tag{4.2}$$

whose degree is $2t = n - k$. When $b = 1$, it is called a narrow-sense RS code.

Example 28 *Let α be a primitive element of $GF(2^4)$. Then the generator polynomial of a narrow-sense $(15, 11)$ 2-error-correcting RS code is constructed as*

$$g(x) = (x + \alpha)(x + \alpha^2)(x + \alpha^3)(x + \alpha^4).$$

Although the generator polynomial is defined using power representation of finite field elements, the powers need to be converted to basis representations to carry out additions. Assume that α is a root of the primitive polynomial $p(x) = x^4 + x + 1$. Then $\{1, \alpha, \alpha^2, \alpha^3\}$ is a standard basis of $GF(2^4)$. The conversion between the power representation and the representation using this standard basis is listed in Table 1.2. From this table, it can be computed that

$$g(x) = x^4 + \alpha^{13} x^3 + \alpha^6 x^2 + \alpha^3 x + \alpha^{10}.$$

Assume that $g(x) = g_0 + g_1 x + \cdots + g_{n-k} x^{n-k}$. Equivalently, the codeword vector $c = [c_0, c_1, \ldots, c_{n-1}]$ can be derived by multiplying the message vector $m = [m_0, m_1, \ldots, m_{k-1}]$ with a generator matrix G

$$c = mG = m \begin{bmatrix} g_0 & g_1 & \cdots & g_{n-k} & & & & \mathbf{0} \\ & g_0 & g_1 & \cdots & g_{n-k} & & & \\ & & \ddots & \ddots & \ddots & \ddots & & \\ & & & g_0 & g_1 & \cdots & g_{n-k} & \\ \mathbf{0} & & & & g_0 & g_1 & \cdots & g_{n-k} \end{bmatrix} \qquad (4.3)$$

The above G matrix has k rows and n columns, and each row is the previous row shifted by one position.

Although the encoding according to (4.1) or (4.3) is straightforward, systematic encoding is preferred in practical systems since the message symbols can be directly read from the codeword symbols. Systematic encoding can be done as

$$c(x) = m(x)x^{n-k} + (m(x)x^{n-k})_{g(x)}, \qquad (4.4)$$

where $(\cdot)_{g(x)}$ denotes the reminder polynomial from the division by $g(x)$. Accordingly, the k message symbols become the highest k symbols in the codeword, and the $n-k$ parity symbols in the codeword are derived through taking the remainder of dividing $m(x)x^{n-k}$ by $g(x)$. It can be easily proved that the $c(x)$ computed from (4.4) is divisible by $g(x)$, and hence is a valid codeword polynomial. Similarly, linear combinations can be carried out on the rows of the generator polynomial G, so that the first k columns become an identity matrix. Then systematic encoding can be also done equivalently in matrix multiplication format.

Besides multiplying the generator polynomial or matrix, there are other RS encoding methods, such as evaluation map encoding. In this method, the codeword symbols are the evaluation values of the message polynomial at n distinct nonzero elements of $GF(2^p)$. Choosing the fixed-ordered set $\{\alpha_0, \alpha_1, \cdots, \alpha_{n-1}\}$ as the n distinct evaluation elements, the codeword symbols corresponding to message polynomial $m(x) = m_0 + m_1 x + \cdots + m_{k-1}x^{k-1}$ are $[m(\alpha_0), m(\alpha_1), \cdots, m(\alpha_{n-1})]$. There exists one-to-one mapping between the codewords generated from different encoding methods.

4.2 Reed-Solomon encoder architectures

Although the evaluation map encoding method may facilitate the interpretation of interpolation-based RS decoding algorithms, which will be discussed in Chapter 5, systematic encoding is usually adopted in practical systems to save the trouble of recovering message symbols from codeword symbols after the decoding is done. Hence, this section focuses on the implementation architectures of systematic RS encoders.

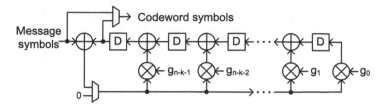

FIGURE 4.1
Serial systematic Reed-Solomon encoder architecture

The systematic encoding according to (4.4) can be implemented by a linear feedback shift register (LFSR) architecture. Note that the highest coefficient, g_{n-k}, is always one in the generator polynomial, $g(x)$, constructed according to (4.2). A serial RS encoder based on LFSR is shown in Fig. 4.1. In this architecture, there are $n - k$ taps of constant multipliers and adders, and the coefficients of $g(x)$ are the constants for the multipliers. In the first k clock cycles, the k message symbols are input to this architecture serially, starting with m_{k-1}. After k clock cycles, the remainder $(m(x)x^{n-k})_{g(x)}$ is located at the registers. Then the multiplexor in the bottom left corner passes zero, and the reminder coefficients are serially shifted out to become the parity symbols in the codeword. Using such an architecture, the encoding is completed in n clock cycles. If higher speed is required, the unfolding technique described in Chapter 2 can be applied to the serial LFSR encoder architecture to achieve partial-parallel designs. For ultra-high speed applications, the systematic encoding can be done in a fully parallel way as multiplying the message symbols to the systematic generator matrix.

4.3   Hard-decision Reed-Solomon decoders

Many decoding algorithms have been developed for RS codes. For a hard-decision decoder, each received symbol takes one of the possible values. These decoders have lower complexity but can not correct as many errors as soft-decision decoders, in which the probability information from the channel is used to decide the likely values for each received symbol. The most popular hard-decision RS decoding algorithms are the Peterson-Gorenstein-Zierler algorithm, BMA, the Euclidean algorithm [37], and the Berlekamp-Welch algorithm [38]. This section focuses on the Peterson-Gorenstein-Zierler algorithm and the BMA, as well as their implementation architectures.

To simplify the notations, narrow-sense RS codes are considered. For a t-error-correcting RS code constructed over $GF(2^q)$, the generator polynomial,

$g(x)$, has roots over $2t$ consecutive elements $\alpha, \alpha^2, \ldots, \alpha^{2t}$. From (4.1), a codeword polynomial $c(x)$ is defined as the product of $g(x)$ and the message polynomial $m(x)$. Hence, $c(x)$ is a codeword if and only if it has the same $2t$ consecutive elements as the roots. Let the received symbols be $r = [r_0, r_1, \ldots, r_{n-1}]$, and the corresponding polynomial be $r(x)$. $r(x)$ can be written as $c(x) + e(x)$, where $e(x) = e_0 + e_1 x + \cdots + e_{n-1} x^{n-1}$ is the error polynomial. Syndromes are defined as the evaluation values of $r(x)$ over the $2t$ roots of $g(x)$ as:

$$S_j = r(\alpha^j) = c(\alpha^j) + e(\alpha^j) = e(\alpha^j) = \sum_{i=0}^{n-1} e_i(\alpha^j)^i, \quad 1 \leq j \leq 2t.$$

Assume that r has v errors in positions $i_1, i_2, \ldots, i_v$, and let $X_l = \alpha^{i_l}$. Then

$$S_j = \sum_{l=1}^{v} e_{i_l}(\alpha^j)^{i_l} = \sum_{l=1}^{v} e_{i_l} X_l^j, \quad 1 \leq j \leq 2t. \tag{4.5}$$

Since X_l indicates the locations of the errors, they are called the error locators. From (4.5), a set of $2t$ equations with v unknown variables are written.

$$
\begin{aligned}
S_1 &= e_{i_1} X_1 + e_{i_2} X_2 + \cdots + e_{i_v} X_v \\
S_2 &= e_{i_1} X_1^2 + e_{i_2} X_2^2 + \cdots + e_{i_v} X_v^2 \\
&\vdots \\
S_{2t} &= e_{i_1} X_1^{2t} + e_{i_2} X_2^{2t} + \cdots + e_{i_v} X_v^{2t}
\end{aligned}
\tag{4.6}
$$

It was shown in [39] that these equations can be translated into a set of linear functions that are easier to solve by making use of an error locator polynomial defined as

$$\Lambda(x) = \prod_{l=1}^{v} (1 - X_l x) = \Lambda_0 + \Lambda_1 x + \cdots + \Lambda_v x^v.$$

From the above definition, $\Lambda_0 = 1$ and $\Lambda(X_l^{-1}) = 0$. Hence

$$e_{i_l} X_l^j \Lambda(X_l^{-1}) = e_{i_l}(\Lambda_0 X_l^j + \Lambda_1 X_l^{j-1} + \cdots + \Lambda_v X_l^{j-v}) = 0. \tag{4.7}$$

Taking the sum of (4.7) over $l = 1, \ldots, v$, it can be derived that

$$\sum_{l=1}^{v} e_{i_l}(\Lambda_0 X_l^j + \Lambda_1 X_l^{j-1} + \cdots + \Lambda_v X_l^{j-v})$$

$$= \Lambda_0 \sum_{l=1}^{v} e_{i_l} X_l^j + \Lambda_1 \sum_{l=1}^{v} e_{i_l} X_l^{j-1} + \cdots + \Lambda_v \sum_{l=1}^{v} e_{i_l} X_l^{j-v}$$

$$= \Lambda_0 S_j + \Lambda_1 S_{j-1} + \cdots + \Lambda_v S_{j-v} = 0$$

Since $\Lambda_0 = 1$, RS codes can be decoded through finding the error locator polynomial that satisfies

$$S_j = \Lambda_1 S_{j-1} + \cdots + \Lambda_v S_{j-v}. \tag{4.8}$$

4.3.1   Peterson-Gorenstein-Zierler algorithm

Peterson first proposed a method to solve (4.8) for binary Bose-Chaudhuri-Hocquenghem (BCH) codes [39]. Then the method was generalized by Gorenstein and Zierler for RS decoding in [40].

In the case of $v = t$, (4.8) can be rewritten as the following matrix multiplication format.

$$A\Lambda = \begin{bmatrix} S_1 & S_2 & S_3 & \cdots & S_{t-1} & S_t \\ S_2 & S_3 & S_4 & \cdots & S_t & S_{t+1} \\ S_3 & S_4 & S_5 & \cdots & S_{t+1} & S_{t+2} \\ \vdots & \vdots & \vdots & \ddots & \vdots & \vdots \\ S_{t-1} & S_t & S_{t+1} & \cdots & S_{2t-3} & S_{2t-2} \\ S_t & S_{t+1} & S_{t+2} & \cdots & S_{2t-2} & S_{2t-1} \end{bmatrix} \begin{bmatrix} \Lambda_t \\ \Lambda_{t-1} \\ \Lambda_{t-2} \\ \vdots \\ \Lambda_2 \\ \Lambda_1 \end{bmatrix} = \begin{bmatrix} S_{t+1} \\ S_{t+2} \\ S_{t+3} \\ \vdots \\ S_{2t-1} \\ S_{2t} \end{bmatrix} \quad (4.9)$$

The matrix A is nonsingular if and only if there are exactly t errors. When the number of errors is less than t, the rightmost column and bottom row of A is removed to see if a nonsingular matrix results. This process is repeated until the reduced matrix becomes nonsingular. Then Λ_i for $1 \leq i \leq v$ are found by solving the system of linear equations. After that, the methods described in Chapter 3 can be used to find the roots of $\Lambda(x)$. If there is a root α^{-i}, then the *ith* received symbol is erroneous. If the number of errors is $v \leq t$, then $\deg(\Lambda(x)) = v$ and $\Lambda(x)$ has exactly v distinct roots in $GF(2^q)$. If $v > t$, an error locator polynomial may still be computed from (4.9). However, the number of its distinct roots in $GF(2^q)$ does not equal its degree.

After the error locators are found, they are substituted back into (4.6), which becomes a set of equations with e_{i_l} as unknowns. The X_l^j coefficients in the first v syndromes as listed in (4.10) form a Vandermonde matrix, which is nonsingular. Therefore, the v magnitudes e_{i_l} can be computed by solving (4.10).

$$\begin{bmatrix} X_1 & X_2 & \cdots & X_v \\ X_1^2 & X_2^2 & \cdots & X_v^2 \\ \vdots & \vdots & \ddots & \vdots \\ X_1^v & X_2^v & \cdots & X_v^v \end{bmatrix} \begin{bmatrix} e_{i_1} \\ e_{i_2} \\ \vdots \\ e_{i_v} \end{bmatrix} = \begin{bmatrix} S_1 \\ S_2 \\ \vdots \\ S_v \end{bmatrix} \quad (4.10)$$

After the error magnitudes are derived, they are added to the received symbols to derive the codeword. In summary, the Peterson-Gorenstein-Zierler decoding algorithm for RS codes is carried out according to Algorithm 1.

TABLE 4.1

Power representation vs. standard basis representation with $p(x) = x^3 + x + 1$ for elements of $GF(2^3)$

Power representation	Standard basis representation	Power representation	Standard basis representation
0	0	α^3	$\alpha + 1$
1	1	α^4	$\alpha^2 + \alpha$
α	α	α^5	$\alpha^2 + \alpha + 1$
α^2	α^2	α^6	$\alpha^2 + 1$

Algorithm 1 Peterson-Gorenstein-Zierler Algorithm

inputs: S_j $(1 \le j \le 2t)$

1. *Construct the syndrome matrix A using the syndromes.*
2. *Compute the determinant of A.*
3. *If the determinant is nonzero, go to step 5.*
4. *Reduce A by deleting the rightmost column and bottom row;*
   Go back to step 2.
5. *Solve for Λ_i $(0 < i \le v)$.*
6. *Compute the roots of $\Lambda(x)$.*
7. *If the number of distinct roots of $\Lambda(x)$ in $GF(2^q)$ does not equal*
   $\deg(\Lambda(x))$, declare decoding failure and stop.
8. *Substitute the error locators into (4.10);*
   Solve for error magnitudes.
9. *Add error magnitudes to the received symbols to recover the codeword.*

Example 29 *Consider a (7,3) narrow-sense 2-error-correcting RS code over $GF(2^3)$. $GF(2^3)$ is constructed from $GF(2)$ using the primitive polynomial $p(x) = x^3 + x + 1$. A root, α, of $p(x)$ is a primitive element of $GF(2^3)$. The generator polynomial of this RS code is*

$$g(x) = (x + \alpha)(a + \alpha^2)(x + \alpha^3)(x + \alpha^4).$$

$\{1, \alpha, \alpha^2\}$ is a standard basis of $GF(2^3)$, and the mapping between the power and standard basis representation is listed in Table 4.1. From this table, it can be computed that

$$g(x) = x^4 + \alpha^3 x^3 + x^2 + \alpha x + \alpha^3.$$

Assume that the received polynomial is $r(x) = 1 + \alpha^4 x + \alpha^2 x^2 + \alpha^2 x^3 + \alpha^4 x^4 + \alpha x^5$. Then the syndromes are $S_1 = 0$, $S_2 = \alpha$, $S_3 = \alpha^2$ and $S_4 = \alpha^6$. Hence (4.9) is reduced to

$$A\Lambda = \begin{bmatrix} 0 & \alpha \\ \alpha & \alpha^2 \end{bmatrix} \begin{bmatrix} \Lambda_2 \\ \Lambda_1 \end{bmatrix} = \begin{bmatrix} \alpha^2 \\ \alpha^6 \end{bmatrix}$$

Apparently, A is nonsingular, and there are two errors. From simple substitutions, $\Lambda_1 = \alpha$ and $\Lambda_2 = \alpha^3$. Hence the error locator polynomial is $\Lambda(x) = \alpha^3 x^2 + \alpha x + 1$. The roots of $\Lambda(x)$ are 1 and α^4. Therefore, the errors are located at the 0th and 3rd symbols.

Next, the error magnitudes are calculated. Substitute the error locators and syndromes into (4.10), then

$$\begin{bmatrix} 1 & \alpha^3 \\ 1 & \alpha^6 \end{bmatrix} \begin{bmatrix} e_{i_1} \\ e_{i_2} \end{bmatrix} = \begin{bmatrix} 0 \\ \alpha \end{bmatrix}.$$

It can be computed that $e_{i_1} = 1$ and $e_{i_2} = \alpha^4$. Adding the error magnitudes to the received polynomial, the codeword polynomial is recovered as

$$\begin{aligned} c(x) &= (1+1) + \alpha^4 x + \alpha^2 x^2 + (\alpha^2 + \alpha^4)x^3 + \alpha^4 x^4 + \alpha x^5 \\ &= \alpha^4 x + \alpha^2 x^2 + \alpha x^3 + \alpha^4 x^4 + \alpha x^5 \\ &= \alpha x g(x) \end{aligned}$$

When t is not very small, computing the determinant and solving for linear equations are quite hardware-consuming. In these cases, the error locator and magnitudes are computed more efficiently using the BMA discussed next.

4.3.2   Berlekamp-Massey algorithm

It was shown in (4.8) that a syndrome S_j can be expressed recursively by an equation involving the coefficients of the error locator polynomial and earlier syndromes $S_{j-1}, S_{j-2}, \ldots, S_{j-v}$. The BMA can be interpreted as finding a LFSR of minimum length whose first $2t$ output samples are the syndromes $S_1, S_2, \ldots, S_{2t}$ [41]. The coefficients of the taps in such a LFSR are the coefficients of the error locator polynomial $\Lambda(x)$. Let $\Lambda^{(r)}(x) = 1 + \Lambda_1^{(r)} x + \Lambda_2^{(r)} x^2 + \ldots$ be a temporary error locator polynomial whose coefficients define the taps of a LFSR. The BMA starts with finding $\Lambda^{(1)}(x)$ so that the first output of the LFSR is S_1. Then the second output of this LFSR is computed and compared to S_2. If they are the same, nothing needs to be done on the LFSR. Otherwise, the discrepancy is used to modify $\Lambda^{(1)}(x)$ to derive $\Lambda^{(2)}(x)$, so that the first two outputs of the LFSR corresponding to $\Lambda^{(2)}(x)$ are S_1, S_2. Such a process is repeated until a LFSR that generates all $2t$ syndromes is obtained.

Taking the syndromes S_j ($1 \leq j \leq 2t$) as inputs, the BMA can be carried out following the steps listed in Algorithm 2 to compute the error locator polynomial. In this algorithm, L is used to keep track of the length of the LFSR, and r denotes the iteration number. At the end of iteration r, the first r outputs from the updated LFSR equal the first r syndromes. Moreover, $\delta^{(r)}$ is the discrepancy between the LFSR output and syndrome, and $B(x)$ is a correction polynomial used to update $\Lambda^{(r)}(x)$.

Algorithm 2 Berlekamp-Massey Algorithm
input: S_j $(1 \leq j \leq 2t)$
initialization: $r = 0$, $\Lambda^{(0)}(x) = 1$, $L = 0$, $B(x) = x$
begin:
1. $r \Leftarrow r + 1$;
 Compute the discrepancy between S_r and the rth LFSR output

$$\delta^{(r)} = S_r + \sum_{i=1}^{L} \Lambda_i^{(r-1)} S_{r-i}$$

2. If $\delta^{(r)} = 0$, go to step 6
3. $\Lambda^{(r)}(x) = \Lambda^{(r-1)}(x) + \delta^{(r)} B(x)$
4. If $2L \geq r$, go to step 6
5. $L \Leftarrow r - L$; $B(x) = \Lambda^{(r-1)}(x)/\delta^{(r)}$
6. $B(x) \Leftarrow xB(x)$
7. If $r < 2t$, go to step 1
output: $\Lambda(x) = \Lambda^{(2t)}(x)$

Similarly, root computation needs to be done on the error locator polynomial, $\Lambda(x)$, generated by the BMA to find the error locations. A valid error locator polynomial of degree v must have v distinct roots over $GF(2^q)$. In the case that the number of errors exceed t, the BMA still yields an error locator polynomial. However, the number of roots of this polynomial does not equal its degree. The root number needs to be checked to ensure that the roots correspond to actual error locations and the decoding is successful.

To compute the error magnitudes, the syndromes are considered to be the coefficients of an infinite-degree syndrome polynomial

$$S(x) = \sum_{j=1}^{\infty} S_j x^{j-1}.$$

From (4.5), $S_j = \sum_{l=1}^{v} e_{i_l} X_l^j$. Hence

$$
\begin{aligned}
S(x) &= \sum_{j=1}^{\infty}(\sum_{l=1}^{v} e_{i_l} X_l^j) x^{j-1} \\
&= \sum_{l=1}^{v} e_{i_l} \sum_{j=1}^{\infty} (X_l x)^j x^{-1} \\
&= \sum_{l=1}^{v} e_{i_l} \left(\frac{X_l}{1 - X_l x}\right).
\end{aligned}
$$

Define $\Omega(x) = \Lambda(x)S(x)$. $\Omega(x)$ can be considered as an error evaluator polynomial. Also

$$
\begin{aligned}
\Omega(x) &= \prod_{j=1}^{v}(1 - X_j x)\left(\sum_{l=1}^{v} e_{i_l}\left(\frac{X_l}{1 - X_l x}\right)\right) \\
&= \sum_{l=1}^{v}\left(e_{i_l} X_l \prod_{j=1, j\neq l}^{v}(1 - X_j x)\right).
\end{aligned}
\tag{4.11}
$$

Since only the first $2t$ coefficients of $S(x)$ are known, $\Omega(x) = \Lambda(x)S(x)$ is reduced to

$$
\Lambda(x)S(x) = \Omega(x) \quad \mathrm{mod}\ x^{2t}.
\tag{4.12}
$$

This is called the key equation for RS decoding.

To solve for e_{i_l} from (4.11), the formal derivative is used. For a polynomial $a(x) = a_0 + a_1 x + a_2 x^2 + a_3 x^3 + \ldots$, its formal derivative is $a'(x) = a_1 + 2a_2 x + 3a_3 x^2 + \ldots$. Note that for finite fields of characteristic two, adding up even copies of an element leads to zero, and the sum of odd copies of an element is the element itself. Hence $a'(x) = a_1 + a_3 x^2 + a_5 x^4 + \ldots$ over $GF(2^q)$. It can be derived that

$$
\Lambda'(x) = \left(\prod_{l=1}^{v}(1 - X_l x)\right)' = \sum_{l=1}^{v}\left(X_l \prod_{j=1, j\neq l}^{v}(1 - X_j x)\right).
$$

Accordingly,

$$
\Lambda'(X_k^{-1}) = \sum_{l=1}^{v}\left(X_l \prod_{j=1, j\neq l}^{v}(1 - X_j X_k^{-1})\right) = X_k \prod_{j=1, j\neq k}^{v}(1 - X_j X_k^{-1}).
$$

Now substitute x by X_k^{-1} in (4.11),

$$\Omega(X_k^{-1}) = \sum_{l=1}^{v} \left(e_{i_l} X_l \prod_{j=1, j \neq l}^{v} (1 - X_j X_k^{-1}) \right)$$

$$= e_{i_k} X_k \prod_{j=1, j \neq k}^{v} (1 - X_j X_k^{-1})$$

$$= e_{i_k} \Lambda'(X_k^{-1}).$$

As a result, the error magnitudes are computed from the error locator and evaluator polynomials as

$$e_{i_k} = \frac{\Omega(X_k^{-1})}{\Lambda'(X_k^{-1})}. \tag{4.13}$$

This is actually Forney's formula for computing the error magnitudes [42].

Example 30 *Consider the decoding of a 2-error-correcting (7,3) RS code using the BMA. Assume that this code has the same generator polynomial, and $GF(2^3)$ is constructed using the same primitive polynomial as in Example 29. Also the received polynomial is $1 + \alpha^4 x + \alpha^2 x^2 + \alpha^2 x^3 + \alpha^4 x^4 + \alpha x^5$, and the corresponding syndromes are $S_1 = 0$, $S_2 = \alpha$, $S_3 = \alpha^2$ and $S_4 = \alpha^6$. The values of the variables and polynomials at the end of each iteration of the BMA are listed in the following table.*

TABLE 4.2
Example variable and polynomial values
in the BMA

r	S_r	$\Lambda^{(r)}(x)$	$\delta^{(r)}$	L	$B(x)$
0	-	1	-	0	x
1	0	1	0	0	x^2
2	α	$1 + \alpha x^2$	α	2	$\alpha^6 x$
3	α^2	$1 + \alpha x + \alpha x^2$	α^2	2	$\alpha^6 x^2$
4	α^6	$1 + \alpha x + \alpha^3 x^2$	α	-	-

The same error locator polynomial $\Lambda(x) = 1 + \alpha x + \alpha^3 x^2$ as in Example 29 is derived at the end of the BMA. This polynomial has two distinct roots, 1 and α^4, and their inverses are real error locators. Using $\Lambda(x)$, the error evaluator polynomial is computed as

$$\Omega(x) = \Lambda(x)S(x) \mod x^4$$
$$= (1 + \alpha x + \alpha^3 x^2)(\alpha x + \alpha^2 x^2 + \alpha^6 x^3) \mod x^4$$
$$= \alpha x \mod x^4$$

According to (4.13),

$$e_{i_k} = \frac{\Omega(X_k^{-1})}{\Lambda'(X_k^{-1})} = \frac{\alpha X_k^{-1}}{\alpha} = X_k^{-1}.$$

Substituting $X_1^{-1} = 1$ and $X_2^{-1} = \alpha^4$ into the above equation, $e_{i_1} = 1$ and $e_{i_2} = \alpha^4$. These error magnitudes are also the same as those computed using the Peterson-Gorenstein-Zierler Algorithm in Example 29.

4.3.3   Reformulated inversionless Berlekamp-Massey algorithm and architectures

In Algorithm 2, divisions by the discrepancy coefficient, $\delta^{(r)}$, may be needed to update $B(x)$ in iteration r. From the discussion in Chapter 1, a finite field inverter is expensive to implement and has long data path. Since the BMA is an iterative algorithm, the inverter is located in a feedback loop, and its long data path leads to low clock frequency. To eliminate the divisions, a scaled version of the error locator polynomial can be computed instead. Nevertheless, from the syndromes and the coefficients of $\Lambda^{(r-1)}(x)$, the computation of $\delta^{(r)}$ needs t multipliers and a binary adder tree of depth $\lceil \log_2(t+1) \rceil$. On the other hand, the critical path of the hardware units updating the error locator polynomial consists of one multiplier and one adder. Therefore, the $\delta^{(r)}$ computation is the speed bottleneck and limits the maximum achievable clock frequency of the BMA implementation.

A reformulated inversionless Berlekamp-Massey (riBM) decoding algorithm was proposed in [43] to eliminate the bottleneck of $\delta^{(r)}$ computation. Instead of computing a single discrepancy coefficient in each iteration, a discrepancy polynomial, $\Delta^{(r)}(x) = \Lambda^{(r)}(x)S(x) = \Delta_0^{(r)} + \Delta_1^{(r)}x + \Delta_2^{(r)}x^2 + \ldots$, is defined. This polynomial is initialized using the syndromes, and is updated iteratively with the aid of a polynomial $\Theta(x) = B(x)S(x)$ in a way similar to that of updating the error locator polynomial $\Lambda(x)$ using $B(x)$. The $\delta^{(r)}$ in Algorithm 2 actually equals $\Delta_r^{(r)}$. In the case that the coefficients of $\Delta^{(r)}(x)$ are updated in parallel in different processing elements (PEs), $\delta^{(r)}$ needs to be read from a different PE in each iteration. To eliminate this routing complexity, a modified discrepancy polynomial is defined as $\hat{\Delta}^{(r)}(x) = \Delta_{0+r}^{(r)} + \Delta_{1+r}^{(r)}x + \Delta_{2+r}^{(r)}x^2 + \ldots$, and $\Theta(x)$ is modified accordingly to $\hat{\Theta}(x)$. In this case, the discrepancy needed to update the error locator polynomial in each iteration is always $\hat{\Delta}_0^{(r)}$ and hence can be read from the same PE. As a result of these modifications, the riBM algorithm can be implemented by very regular structures, and its critical path is reduced to one multiplier and one adder.

The riBM algorithm is listed in Algorithm 3 [43]. Besides the modified discrepancy coefficient computation, the updating of the error locator polynomial has also been reorganized from Algorithm 2 to make the hardware structure more regular. Details about the derivations of the riBM algorithm can be found in [43].

Algorithm 3 riBM Algorithm
input: S_j $(1 \leq j \leq 2t)$
initialization: $\Lambda^{(0)}(x) = B^{(0)}(x) = 1$
$\qquad\qquad\quad k^{(0)} = 0,\ \gamma^{(0)} = 1$
$\qquad\qquad\quad \hat{\Delta}^{(0)}(x) = \hat{\Theta}^{(0)}(x) = S_1 + S_2 x + \cdots + S_{2t} x^{2t-1}$

begin:
$\qquad$ *for* $r = 0$ *to* $2t - 1$
$\qquad\qquad \Lambda^{(r+1)}(x) = \gamma^{(r)} \Lambda^{(r)}(x) + \hat{\Delta}_0^{(r)} x B^{(r)}(x)$
$\qquad\qquad \hat{\Delta}^{(r+1)}(x) = \gamma^{(r)} \hat{\Delta}^{(r)}(x)/x + \hat{\Delta}_0^{(r)} \hat{\Theta}^{(r)}(x)$
$\qquad\qquad$ *if* $\hat{\Delta}_0^{(r)} \neq 0$ *and* $k^{(r)} \geq 0$
$\qquad\qquad\qquad B^{(r+1)}(x) = \Lambda^{(r)}(x)$
$\qquad\qquad\qquad \hat{\Theta}^{(r+1)}(x) = \hat{\Delta}^{(r)}(x)/x$
$\qquad\qquad\qquad \gamma^{(r+1)} = \hat{\Delta}_0^{(r)}$
$\qquad\qquad\qquad k^{(r+1)} = -k^{(r)} - 1$
$\qquad\qquad$ *else*
$\qquad\qquad\qquad B^{(r+1)}(x) = x B^{(r)}(x)$
$\qquad\qquad\qquad \hat{\Theta}^{(r+1)}(x) = \hat{\Theta}^{(r)}(x)$
$\qquad\qquad\qquad \gamma^{(r+1)} = \gamma^{(r)}$
$\qquad\qquad\qquad k^{(r+1)} = k^{(r)} + 1$
output: $\Lambda(x) = \Lambda^{(2t)}(x)$;
$\qquad\qquad \hat{\Omega}(x) = \hat{\Delta}^{(2t)}(x) \bmod x^t$

For easy understanding and simplifying the mapping to hardware, Algorithm 3 is described in terms of polynomial updating. However, the multiplications and divisions of a polynomial by x in this algorithm are not done in the traditional way. The degrees of $\Lambda(x)$ and $B(x)$ are kept at t, and those of $\hat{\Delta}(x)$ and $\hat{\Theta}(x)$ are $2t - 1$ in all iterations. If the actual degree is lower, zeros are padded as the higher-degree terms. To multiply $B^{(r)}(x)$ by x, the term of x^{t+1} in the product is deleted, and the constant coefficient becomes zero. To divide $\hat{\Delta}^{(r)}(x)$ by x, the coefficient of x^{2t-1} in the quotient is set to zero, and the term of x^{-1} is deleted. Assume that $B^{(r)}(x) = B_0^{(r)} + B_1^{(r)} x + \cdots + B_t^{(r)} x^t$ and $\hat{\Delta}^{(r)}(x) = \hat{\Delta}_0^{(r)} + \hat{\Delta}_1^{(r)} x + \cdots + \hat{\Delta}_{2t-1}^{(r)} x^{2t-1}$. Then $x B^{(r)}(x) = B_0^{(r)} x + B_1^{(r)} x^2 + \cdots + B_{t-1}^{(r)} x^t$ and $\hat{\Delta}^{(r)}(x)/x = \hat{\Delta}_1^{(r)} + \hat{\Delta}_2^{(r)} x + \cdots + \hat{\Delta}_{2t-1}^{(r)} x^{2t-2}$ in Algorithm 3.

Another advantage of the riBM algorithm is that the lowest t coefficients of $\hat{\Delta}^{(2t)}$ can be used as an error evaluator polynomial $\hat{\Omega}(x)$. Unlike using (4.12), $\hat{\Delta}^{(r)}(x)$ is updated simultaneously as $\Lambda^{(r)}(x)$, and hence does not require additional time to compute. It has been derived in [43] that the error magnitudes are computed as

$$e_{i_l} = \frac{X_l^{-2t}\hat{\Omega}(X_l^{-1})}{\Lambda'(X_l^{-1})},\tag{4.14}$$

where $X_l = \alpha^{i_l}$ is an error locator.

Example 31 *Consider the same 2-error-correcting $(7,3)$ RS code constructed over $GF(2^3)$ in Examples 29 and 30, and $p(x) = x^3+x+1$ is used to construct $GF(2^3)$. Also assume that the received word polynomial is $1 + \alpha^4 x + \alpha^2 x^2 + \alpha^2 x^3 + \alpha^4 x^4 + \alpha x^5$. Accordingly, the syndromes are $S_1 = 0$, $S_2 = \alpha$, $S_3 = \alpha^2$ and $S_4 = \alpha^6$. The values of the key variables for each iteration of Algorithm 3 are listed in the following table.*

TABLE 4.3
Example variable values in the riBM algorithm.

r	$\Lambda^{(r)}(x)$	$B^{(r)}(x)$	$k^{(r)}$	$\gamma^{(r)}$	$\hat{\Delta}^{(r)}(x)$	$\hat{\Theta}^{(r)}(x)$
0	1	1	0	1	$\alpha x + \alpha^2 x^2 + \alpha^6 x^3$	$\alpha x + \alpha^2 x^2 + \alpha^6 x^3$
1	1	x	1	1	$\alpha + \alpha^2 x + \alpha^6 x^2$	$\alpha x + \alpha^2 x^2 + \alpha^6 x^3$
2	$1 + \alpha x^2$	1	-2	α	$\alpha^2 + x + \alpha^3 x^2 + x^3$	$\alpha^2 + \alpha^6 x$
3	$\alpha + \alpha^2 x + \alpha^2 x^2$	x	-1	α	$\alpha^2 + \alpha^2 x + \alpha x^2$	$\alpha^2 + \alpha^6 x$
4	$\alpha^2 + \alpha^3 x + \alpha^5 x^2$	-	-	-	$\alpha^6 + \alpha^4 x$	-

From the above table, $\Lambda(x) = \Lambda^{(2t)} = \alpha^2 + \alpha^3 x + \alpha^5 x^2$. It equals the error locator polynomial computed in Example 29 and 30 scaled by α^2, and has the same roots 1 and α^4. $\hat{\Omega}(x) = \hat{\Delta}^{(2t)}(x) \mod x^t = \alpha^6 + \alpha^4 x$. Substituting the two roots as X_l^{-1} into (4.14), it can be computed that the two error magnitudes are 1 and α^4, respectively, and they are the same as those calculated in Examples 29 and 30.

The architecture for implementing the riBM algorithm has two major parts, the error locator update (ELU) block for updating $\Lambda(x)$ and $B(x)$ and reformulated discrepancy computation (rDC) block for updating $\hat{\Delta}(x)$ and $\hat{\Theta}(x)$. Each of them consists of identical PEs. The PEs in the ELU block are different from those in the rDC block, and they are denoted by PE0 and PE1, respectively. To finish one iteration of the riBM algorithm in one clock cycle, the ELU block has $t+1$ copies of PE0, and the rDC block has $2t$ PE1s. A PE of either type updates one pair of coefficients in each clock cycle. The details of PE0 and PE1 are shown in Fig. 4.2.

Fig. 4.3 depicts the overall architecture for the riBM algorithm. γ and k are updated in the control block. The same γ, $\hat{\Delta}_0$, and multiplexor select signal are used in every PE in each clock cycle. The initial values of the registers in each PE are also labeled in this figure. At the end of iteration $2t$, the coefficients of the error locator polynomial $\Lambda(x)$ can be read from the registers in the $t+1$ PE0s, and those of $\hat{\Omega}(x)$ are located at the first t PE1s. Note that the PE0s

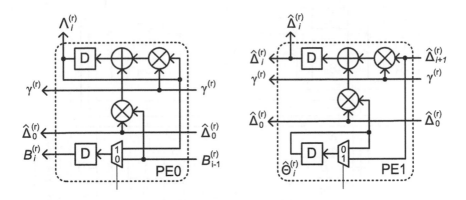

FIGURE 4.2
Architectures for the processing elements (PEs) (modified from [43])

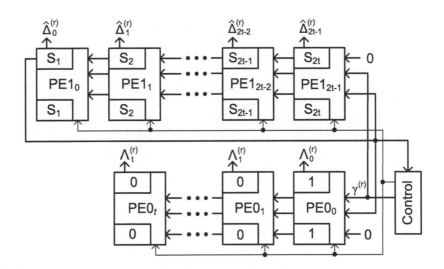

FIGURE 4.3
The riBM architecture (modified from [43])

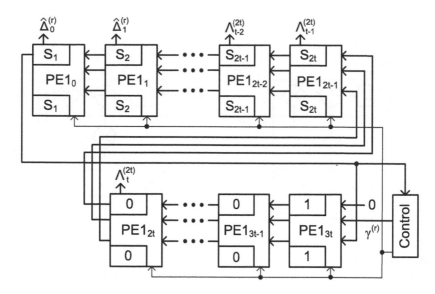

FIGURE 4.4
The RiBM architecture (modified from [43])

are ordered from right to left, and the PE1s are ordered from left to right if the PE0 and PE1 architectures in Fig. 4.2 are employed.

Due to the similarity between the PE0 and PE1 architectures, it is possible to use the same PEs to compute the error locator and evaluator polynomials. In Fig. 4.3, the coefficients updated by a PE in each clock cycle are associated with the same power of x. For example, the values updated in PE0$_i$ are always the coefficients of x^i in $\Lambda^{(r)}(x)$ and $B^{(r)}(x)$. If the $\Lambda^{(r)}(x)$ coefficient derived in a PE0 is shifted to the next PE0 on its left, then the coefficients of $\hat{B}^{(r)}(x)$ do not need to propagate. As a result, PE1s can be used instead of PE0s to update $\Lambda^{(r)}(x)$ and $B^{(r)}(x)$. However, it is inefficient to add another t copies of PE1s just to accommodate the shifting of $\Lambda^{(r)}(x)$. Fortunately, since $\deg(\Delta^{(r)}(x)) = \deg(S(x)\Lambda^{(r)}(x)) = \deg(S(x)) + \deg(\Lambda^{(r)}(x))$, $\deg(\hat{\Delta}^{(r)}(x)) = 2t - 1 - r + \deg(\Lambda^{(r)}(x))$. When the received word is decodable, $\deg(\Lambda^{(r)}(x)) \leq t$, and hence $\deg(\hat{\Delta}(x)) < 3t - r$. In addition, the degree of $\Lambda(x)$ is initially zero and increases by at most one in each iteration. Therefore, the coefficients of $\Lambda(x)$ and $B(x)$ can be shifted into the PEs initially used for updating $\hat{\Delta}(x)$ and $\hat{\Theta}(x)$ without overwriting useful coefficients as shown in Fig. 4.4. This architecture is referred to as the RiBM architecture [43]. It also consists of $3t + 1$ PEs. However, all of them are PE1s. The registers for updating the four polynomials are initialized in the same way as those in the riBM architecture, and the first t PE1s have the coefficients of $\hat{\Omega}(x) = \hat{\Delta}^{(2t)} \mod x^t$ at the end of the $2t$th clock cycle. Nevertheless, since the coefficients of $\Lambda(x)$ have been

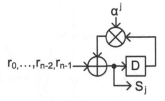

FIGURE 4.5
Serial architecture for syndrome computation

shifted to the left in each clock cycle, $\Lambda^{(2t)}(x)$ is found in PE1_t through PE1_{2t} at the end as labeled in Fig. 4.4.

Since the RiBM architecture consists of identical PEs, the folding technique described in Chapter 2 can be applied to achieve throughput-area tradeoff. Recall that an N-folded architecture reduces the number of hardware units by N times at the cost of requiring N times clock cycles to finish the same computation. In an extreme case, all the $3t + 1$ PE1s in the RiBM architecture are folded onto one single PE to achieve an ultra-ultra folded iBM (UiBM) architecture [44]. The UiBM architecture takes $2t(3t + 1)$ clock cycles to find the error locator and evaluator.

4.3.4 Syndrome, error location and magnitude computation architectures

The implementation of the key equation solver (KES) for computing the error locator and evaluator polynomials in the BMA has been detailed. Next, the architectures for the syndrome, error location and magnitude computations are presented. The same syndrome computation architecture can be used in other RS decoding algorithms. Moreover, the BMA, Euclidean and Berlekamp-Welch algorithms are only different in the KES step. The computations of the error locations and magnitudes in these algorithms share the same architecture.

As mentioned previously, the syndromes are defined as $S_j = r(\alpha^j)$, and the coefficients of $r(x) = r_{n-1}x^{n-1} + r_{n-2}x^{n-2} + \cdots + r_1 x + r_0$ are the received symbols. Applying Horner's rule, $r(x)$ is rewritten as

$$r(x) = (\ldots((r_{n-1}x + r_{n-2})x + r_{n-3})x + \ldots)x + r_0. \qquad (4.15)$$

Accordingly, the syndrome computation can be implemented by the serial architecture illustrated in Fig. 4.5. At the beginning, the register is initialized as zero. The received symbols are fed into this architecture serially, starting with r_{n-1}. In each clock cycle, the value store in the register is multiplied with α^j and the product is added up with the input symbol. After n clock cycles,

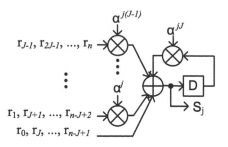

FIGURE 4.6
J-parallel architecture for syndrome computation

S_j is available. $2t$ copies of the serial architecture in Fig. 4.5 can be adopted to compute the $2t$ syndromes in parallel.

To speed up the syndrome computation, the J-parallel architecture in Fig. 4.6 can be employed. It is derived by first grouping J coefficients of $r(x)$ together and then applying Horner's rule. To simplify the notations, it is assumed that n is divisible by J in Fig. 4.6. Otherwise, $J - (n \mod J)$ zeros are padded to the most significant coefficients of $r(x)$. This architecture computes one syndrome in $\lceil n/J \rceil$ clock cycles.

The syndromes are evaluation values, so are the values computed by the Chien search. Hence the Chien search architecture shown in Fig. 3.1 can also compute the syndromes. In this Chien search architecture, all the polynomial coefficients are available at the inputs simultaneously. If the parallel processing factor J in Fig. 4.6 is set to n, then it no longer needs the feedback loop and becomes one column of multipliers in the Chien search architecture, which computes a syndrome in one clock cycle. One restriction of the Chien search architecture in Fig. 3.1 is that the evaluation is done over consecutive elements in the order of $1, \alpha, \alpha^2, \ldots$. Even if the first evaluation value needed is over α^b, the computations over $1, \alpha, \ldots, \alpha^{b-1}$ still need to be done. This causes long latency when b is large, which may happen when the RS code is not narrow-sense. To address this issue, a multiplexor can be added to select α^{ib} as the constant multiplicand in the ith feedback loop in Fig. 3.1 in the first clock cycle.

The roots of the error locator polynomial $\Lambda(x)$ need to be computed to find the error locations. If $\Lambda(x)$ has a root α^{-i}, then the ith received symbol is erroneous. When the degree of $\Lambda(x)$ is two or three, the roots can be directly computed as discussed in Chapter 3. It should be noted that the roots derived using those methods are in basis representations. They need to be converted to power representations since the exponents tell the error locations. The look-up table compression technique in Chapter 3 for degree-3 polynomial root computation can also be used to compress the conversion table between basis representation and power representation.

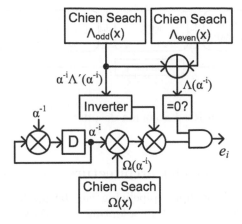

FIGURE 4.7
Error locator and magnitude computation architecture

If the degree of $\Lambda(x)$ is higher than three, Chien search is used to find the roots. When the original BMA is adopted, the error magnitudes are computed by using (4.13), which needs the evaluation values of $\Lambda'(x)$. Over finite fields of characteristic two, $\Lambda'(x) = \Lambda_1 + \Lambda_3 x^2 + \ldots$. Let $\Lambda_{even}(x) = \Lambda_0 + \Lambda_2 x^2 + \ldots$ and $\Lambda_{odd}(x) = \Lambda_1 x + \Lambda_3 x^3 + \ldots$. Then $\Lambda_{odd}(x) = x\Lambda'(x)$. Therefore, if the even and odd terms of $\Lambda(x)$ are separated into two groups in the Chien search, the evaluation values over the odd terms can be reused in computing the error magnitudes. Fig. 4.7 shows the block diagram for computing the error locations and magnitudes. If the finite field elements are tested in the order of $1, \alpha^{-1}, \alpha^{-2}, \ldots, \alpha^{-(n-1)}$, which can be done by changing each constant multiplicand to its inverse in the Chien search architecture of Fig. 3.1, then the error magnitudes generated are in the order of $e_0, e_1, e_2, \ldots, e_{n-1}$. The evaluation values of the even and odd parts of $\Lambda(x)$ are added up to compute $\Lambda(\alpha^{-i})$ $(i = 0, 1, \ldots, n-1)$. $\Omega(\alpha^{-i})$ is calculated simultaneously. If $\Lambda(\alpha^{-i}) = 0$, then the real error magnitude is passed to the output. Otherwise, $e_i = 0$. When J-parallel Chien search is adopted, a vector of J error magnitudes is output at a time. In this case, J finite field inverters are needed, and the constant multiplicand for the multiplier in the feedback loop of Fig. 4.7 is changed to α^{-J} in order to generate $1, \alpha^{-J}, \alpha^{-2J}, \ldots$ in successive clock cycles. Moreover, another $J-1$ constant multipliers should be added to generate $\alpha^{-Jl+1}, \alpha^{-Jl+2}, \ldots, \alpha^{-Jl+J-1}$ in each clock cycle.

When the riBM or RiBM architecture is adopted, the error magnitude computation formula is changed to (4.14). Compared to (4.13), this modified formula has an extra term $(\alpha^{-i})^{2t}$ when the polynomials are evaluated over α^{-i}. This term can be incorporated into the error magnitudes by changing the constant multiplicand in the feedback loop of Fig. 4.7 accordingly.

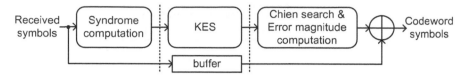

FIGURE 4.8
Pipelined hard-decision Reed-Solomon decoder

4.3.5   Pipelined decoder architecture

To achieve higher throughput, pipelining can be applied to divide a hard-decision RS decoder, such as that adopting the BMA, into three stages as shown in Fig. 4.8. In a pipelined architecture, the overall throughput is decided by the pipelining stage that has the longest computation time. Even if the other stages are running at a faster speed, the overall throughput is not improved. Hence, to increase the hardware utilization efficiency, the computations in each pipelining stage should be completed in about the same amount of time. The numbers of clock cycles to spend on the syndrome computation stage and the Chien search and error magnitude computation stage can be easily adjusted by changing their parallel processing factors. However, the KES step that finds the error locator and evaluator polynomials from the syndromes is an iterative process. Although a throughput-area tradeoff can also be achieved in this step by applying folding, it requires at least $2t$ clock cycles for a t-error-correcting RS code using the BMA. If higher throughput is required, multiple copies of the KES architectures can be employed to handle the computations for different words received in parallel.

For high-rate long RS codes, the complexity of hard-decision RS decoders is dominated by the syndrome computation, Chien search and error magnitude computation. The complexities of these steps increase linearly with n, while the complexity of the KES step is largely decided by t. In the case that the KES step requires substantially fewer clock cycles than the other steps even if the UiBM architecture [44] is adopted, the KES architecture can be shared in a time-multiplexed manner among multiple decoders to reduce the overall silicon area requirement.

4.4   Error-and-erasure Reed-Solomon decoders

The hard-decision RS decoders discussed so far are error-only decoders. They find both the locations and magnitudes of the errors. For an (n, k) RS code, the minimum distance is $n - k + 1$, and the number of errors that can be corrected is at most $t = \lfloor (n-k)/2 \rfloor$. When a received symbol is unreliable, making a hard decision on the symbol would probably result in an error. Instead of making hard decisions, the unreliable symbols can be set as erasures. In other words, the symbols are marked as possibly erroneous. Since the locations of these possible errors are known, the total number of correctable erasures and errors is increased; v errors and u erasures can be corrected as long as $2v + u < (n-k+1)$. Erasure-and-error decoding is the simplest type of soft-decision decoding. Extensive discussions on soft-decision decoders will be provided in the next chapter. However, erasure-and-error decoding is briefly introduced in this chapter since it can be implemented by modifying the inputs to the BMA that has been discussed in previous sections.

Assume that there are v errors and u erasures. The v errors and u erasures are located at positions $i_1, i_2, \ldots, i_v$ and $i'_1, i'_2, \ldots, i'_u$, respectively. $i_1, i_2, \ldots, i_v$ are unknown and $i'_1, i'_2, \ldots, i'_u$ are decided from channel information. Error locators are still $X_1 = \alpha^{i_1}, X_2 = \alpha^{i_2}, \ldots, X_v = \alpha^{i_v}$ as defined in the error-only decoders discussed before. Similarly, erasure locators are defined as $Y_1 = \alpha^{i'_1}, Y_2 = \alpha^{i'_2}, \ldots, Y_u = \alpha^{i'_u}$, and an erasure locator polynomial is

$$\Gamma(x) = \prod_{l=1}^{u}(1 - Y_l x).$$

The first step of error-and-erasure decoding is still syndrome computation. The received symbols, $r_{i'_1}, r_{i'_2}, \ldots, r_{i'_u}$, at those erasure positions can be set to any value in the syndrome computation. However, to make the computations simpler, they are usually set to zero. The syndromes are also computed as $S_j = r(\alpha^j)$ for $j = 1, 2, \ldots, 2t$. Assume that the magnitudes of the errors and erasures are $e_{i_1}, e_{i_2}, \ldots, e_{i_v}$ and $e_{i'_1}, e_{i'_2}, \ldots e_{i'_u}$, respectively. Since the evaluation values of a codeword over α^j ($j = 1, 2, \ldots, 2t$) are zero, the syndromes can be written as

$$S_j = r(\alpha^j) = \sum_{l=1}^{v} e_{i_l} X_l^j + \sum_{l=1}^{u} e_{i'_l} Y_l^j.$$

Define the key equation for the error-and-erasure decoding as

$$\Lambda(x)\Gamma(x)S(x) = \Omega(x) \mod x^{2t}, \qquad (4.16)$$

where $S(x) = S_1 + S_2 x + \cdots + S_{2t} x^{2t-1}$. In order to utilize the BMA to solve the above equation, it needs to be converted to the format of (4.12). Since $\Gamma(x)$ is known, a modified syndrome polynomial is defined by

$$S'(x) = \Gamma(x)S(x) \mod x^{2t}.$$

Then (4.16) becomes

$$\Lambda(x)S'(x) = \Omega(x) \mod x^{2t}.$$

After $\Lambda(x)$ is computed by the BMA using $S'(x)$ as the input syndrome polynomial, $\Omega(x)$ is calculated using the above equation. Then an error-and-erasure locator polynomial $\Psi(x)$ is derived as

$$\Psi(x) = \Lambda(x)\Gamma(x).$$

Following an analysis similar to that for the error-only case, it can be derived that the error and erasure magnitudes can be computed as

$$e_{i_l} = \frac{\Omega(X_l^{-1})}{\Psi'(X_l^{-1})}$$

and

$$e_{i'_l} = \frac{\Omega(Y_l^{-1})}{\Psi'(Y_l^{-1})}$$

respectively.

Example 32 *Consider the (7,3) narrow-sense 2-error-correcting RS code over $GF(2^3)$ adopted in Examples 29 and 30, and $p(x) = x^3 + x + 1$ is used to construct $GF(2^3)$. Assume that $r(x) = \alpha^3 + \alpha x + x^2 + \square x^4$, where $\square$ denotes an erasure. Then the erasure locator is $Y_1 = \alpha^4$, and the erasure locator polynomial is*

$$\Gamma(x) = (1 + \alpha^4 x).$$

Setting the erasure to zero, it can be computed that the syndromes are $S_1 = \alpha^3, S_2 = \alpha^4, S_3 = 0$ and $S_6 = \alpha^4$. Accordingly,

$$S'(x) = S(x)\Gamma(x) \mod x^4 = \alpha^4 x^3 + \alpha x^2 + \alpha^5 x + \alpha^3.$$

Applying the BMA in Algorithm 2 to $S'_1 = \alpha^3, S'_2 = \alpha^5, S'_3 = \alpha$ and $S'_4 = \alpha^4$, the results in the following table can be derived.

r	S'_r	$\Lambda^{(r)}(x)$	$\delta^{(r)}$	L	$B(x)$
0	-	1	-	0	x
1	α^3	$1 + \alpha^3 x$	α^3	1	$\alpha^4 x$
2	α^5	$1 + \alpha^2 x$	α	1	$\alpha^4 x^2$
3	α	$1 + \alpha^2 x + x^2$	α^3	2	$\alpha^4 x + \alpha^6 x^2$
4	α^4	$1 + \alpha^3 x$	α	-	-

Accordingly, $\Lambda(x) = 1 + \alpha^3 x$, and the third received symbol is in error. Since $\Lambda(x)S'(x) = \Omega(x) \mod x^4$,

$$\Omega(x) = \alpha x + \alpha^3.$$

Also

$$\Psi'(x) = (\Lambda(x)\Gamma(x))' = (x^2 + \alpha^6 x + 1)' = \alpha^6.$$

Therefore, the error magnitude is $\Omega((\alpha^3)^{-1})/\Psi'((\alpha^3)^{-1}) = \alpha^2/\alpha^6 = \alpha^3$, *and the erasure magnitude is* $\Omega((\alpha^4)^{-1})/\Psi'((\alpha^4)^{-1}) = \alpha^6/\alpha^6 = 1$. *As a result, the correct codeword polynomial is* $c(x) = \alpha^3 + \alpha x + x^2 + \alpha^3 x^3 + x^4$, *which equals the generator polynomial of this code.*

As mentioned previously, for an (n, k) RS code, v errors and u erasures can be corrected as long as $2v + u \le (n-k)$. Hence $n-k$ erasures are correctable if there is no error. This means that a codeword can be found given the symbols at any k code positions. Therefore, erasure decoding can be used to achieve systematic encoding, where the systematic code positions can be any k code positions. In the case that there are $n-k$ erasures, $\Gamma(x)$ is directly computed from the erasure locations. $\Lambda(x) = 1$ and the BMA does not need to be carried out. Hence, $\Omega(x) = S(x)\Gamma(x) \mod x^{2t}$ and $\Psi(x) = \Gamma(x)$. Then the erasure magnitudes are computed using $\Omega(x)$, $\Psi(x)$, and the erasure locations.

5

Algebraic soft-decision Reed-Solomon decoder architectures

CONTENTS

In hard-decision Reed-Solomon (RS) decoders, decisions are made first on which finite field element each received symbol should take, based on channel observations, before the decoding is carried out. For an (n, k) RS code, traditional hard-decision decoding algorithms, such as the Berlekamp-Massey algorithm (BMA) [4] and the Euclidean algorithm [37], can correct up to $(n-k)/2$

errors. Nevertheless, the probability information from the channel is utilized in soft-decision decoders to correct more errors. In the past fifty years, information theorists and mathematicians have spent tremendous efforts searching for efficient and high-gain soft-decision decoding algorithms of RS codes. However, many available algorithms have overwhelming hardware complexity. The most hardware-friendly soft-decision RS decoding algorithms are the generalized minimum-distance (GMD) algorithm [45] and the Chase algorithm [46]. The GMD decoder assigns erasures to the least reliable code positions and tries multiple test vectors with erasures. In the Chase decoder, the test vectors are formed by flipping the least reliable bits. Although the achievable coding gains of these two algorithms are limited, they remain popular because their complexities are relatively lower. Substantial advancements have been made on algebraic soft-decision (ASD) decoding of RS codes [47, 48, 49, 50, 51, 52] in the past decade. Through incorporating the reliability information from the channel into the algebraic interpolation process developed by Guruswami and Sudan [53, 54], these algorithms can achieve significant coding gain over hard-decision decoding with a complexity that is polynomial with respect to the codeword length.

In this chapter, the ASD algorithms are first introduced. Then the implementation architectures for each step are detailed.

5.1   Algebraic soft-decision decoding algorithms

ASD algorithms are better explained by interpreting the codeword symbols as evaluation values of the message polynomial. However, systematic encoding can still be used when ASD decoding is adopted as will be discussed later in this chapter. Without loss of generality, RS codes over finite field $GF(2^q)$ are considered. For an (n, k) RS code, the k message symbols, $f_0, f_1, \cdots, f_{k-1}$, can be viewed as the coefficients of a message polynomial

$$f(x) = f_0 + f_1 x + \cdots + f_{k-1} x^{k-1}.$$

The encoding can be carried out by evaluating the message polynomial at n distinct non-zero elements of $GF(2^q)$. For primitive RS codes, the evaluation points include all the non-zero elements of $GF(2^q)$. Choosing the fixed-ordered set $\{\alpha_0, \alpha_1, \cdots, \alpha_{n-1}\}$ as the n distinct evaluation points, the codeword corresponding to $[f_0, f_1, \cdots, f_{k-1}]$ is $[f(\alpha_0), f(\alpha_1), \cdots, f(\alpha_{n-1})]$.

Some definitions necessary for understanding the interpolation-based decoding algorithms are given below.

Definition 25 *A bivariate polynomial $Q(x, y)$ passes a point (α, β) with multiplicity m if the shifted polynomial $Q(x + \alpha, y + \beta)$ contains a monomial $x^a y^b$ with degree $a + b = m$, and does not contain any monomial with degree less than m.*

Definition 26 *For non-negative integers w_x and w_y, the (w_x, w_y)-weighted degree of a monomial $x^r y^s$ is $rw_x + sw_y$. The (w_x, w_y)-weighted degree of a bivariate polynomial $Q(x, y) = \sum_{i=0}^{\infty} \sum_{j=0}^{\infty} q_{r,s} x^r y^s$, is the maximum of $rw_x + sw_y$ such that $q_{r,s} \neq 0$.*

Assume that the hard decisions of the received symbols are $y_0, y_1, \cdots, y_{n-1}$. A set of n ordered pairs $(x_0, y_0), (x_1, y_1), \cdots, (x_{n-1}, y_{n-1})$ are formed by associating each received symbol with its evaluation point. The hard-input hard-output Sudan's list decoding algorithm [53] tries to reconstruct the message polynomial by interpolating through this set of points. It only achieves better error-correcting capability than the hard-decision decoders when the code rate k/n is less than $1/3$. Later Guruswami and Sudan [54] extended this algorithm by forcing each interpolation point to have a higher multiplicity. However, the same multiplicity is used for each interpolation point, and it is still a hard-decision decoding algorithm. Although performance improvements have been observed for codes with higher rates, the achievable coding gain is very limited. In contrast, ASD algorithms incorporate soft information into the interpolation process. The interpolation points and their multiplicities are determined by the reliability information received from the channel.

The codeword symbols may be corrupted by channel noise. Hence, in ASD algorithms, the interpolation points for code position i may include $(\alpha_i, \beta_{i,j})$, where $\beta_{i,j}$ can be any finite field element. ASD algorithms try to recover the message polynomial through assigning larger multiplicities to more reliable points according to the channel information. This is actually the first step of ASD algorithms. Although different ASD algorithms have different multiplicity assignment schemes, they share the same interpolation and factorization steps.

• **Interpolation step**: Given the set of interpolation points $(\alpha_i, \beta_{i,j})$ with corresponding multiplicities $m_{i,j}$, compute the nontrivial bivariate polynomial $Q(x, y)$ of minimal $(1, k-1)$-weighted degree that passes each interpolation point with its associated multiplicity.

• **Factorization step**: Determine all factors of the bivariate polynomial $Q(x, y)$ in the form of $y + f(x)$ with $deg(f(x)) < k$. Each $f(x)$ corresponds to one message polynomial in the list.

The most likely message polynomial in the list can be chosen as the decoder output. The multiplicity assignment step not only decides the error-correcting performance of the decoder, but also affects the complexity of the interpolation and factorization steps. Quite a few multiplicity assignment schemes have been proposed. For practical applications, the multiplicity assignment scheme needs to be easy to implement. In the Kötter-Vardy (KV) ASD algorithm [47], the multiplicities are computed by constant multiplications followed by the floor function. Assuming the probability that the ith received symbol equals $\beta_{i,j}$ is $p_{i,j}$, the multiplicity assigned to the point $(\alpha_i, \beta_{i,j})$ is $m_{i,j} = \lfloor \lambda p_{i,j} \rfloor$, where λ is a positive constant. Hence the maximum multiplicity in the KV algorithm is $\lfloor \lambda \rfloor$. In the bit-level generalized minimum distance (BGMD) decoder [52], the reliability of each bit is compared to a threshold. If it is lower than the threshold, then the bit is said to be erased. If none of the bits in the ith received symbol is erased, (α_i, β_i) is assigned multiplicity m_{max}. If there is only one erased bit, then (α_i, β_i) and (α_i, β_i') are both assigned $m_{max}/2$. Otherwise, no point is assigned for the ith code position. Here β_i and β_i' are the hard-decision and second most-likely field element, respectively, for the ith received symbol. In addition, multiple decoding iterations with different thresholds can be carried out to correct more errors in the BGMD algorithm. Instead of tuning the multiplicities, the low-complexity Chase (LCC) ASD algorithm [51] incorporates the soft information into test vectors. Each of the η least reliable code positions has two possible points (α_i, β_i) and (α_i, β_i'), and each of the rest code positions has one single point (α_i, β_i). The multiplicities of all the points are one. Test vectors are formed by picking one point from each code position. Hence, there are 2^η test vectors and the decoding is tried on each of them. The LCC algorithm is a re-interpretation of the Chase algorithm [46] by using the interpolation-based decoding. It has the same performance as the Chase algorithm.

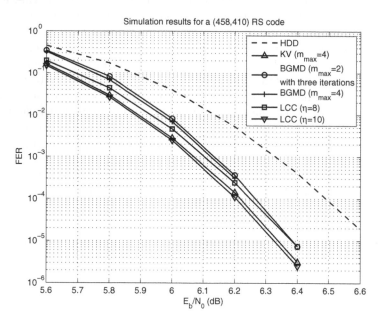

FIGURE 5.1
FERs of decoding algorithms over AWGN channel for a (458, 410) RS code over $GF(2^{10})$

The complexities of the interpolation and factorization steps increase with the maximum multiplicity and/or the number of test vectors. On the other hand, larger maximum multiplicity or more test vectors lead to better error-

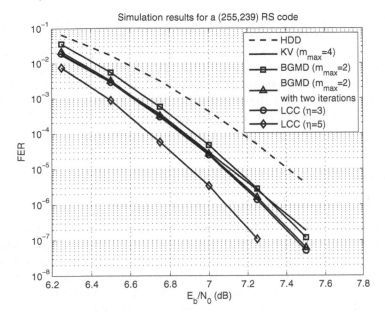

FIGURE 5.2
FERs of decoding algorithms over AWGN channel for a (255, 239) RS code over $GF(2^8)$ (re-plotted from the data in [55])

correcting performance. Hence, the maximum multiplicity or test vector number can be used as tuning nobs to achieve performance-complexity tradeoff. Fig. 5.1 shows the frame error rates (FERs) of different ASD algorithms for a (458, 410) RS code constructed over $GF(2^{10})$ under the additive white Gaussian noise (AWGN) channel with binary phase-shift keying (BPSK) modulation. For the purpose of comparison, the FER of hard-decision decoding (HDD) using the BMA is also included in this figure. Assume that the maximum multiplicity is m_{max}. The complexity of the interpolation is $O(m_{max}^5)$. Therefore, m_{max} should be small in order to keep the hardware complexity at a practical level. From Fig. 5.1, the KV and BGMD algorithms with small m_{max}, as well as the LCC algorithm with small η, can achieve significant coding gain over the HDD. The performance of the LCC algorithm improves with η. Also larger m_{max} or more iterations lead to better error-correcting capability in the BGMD algorithm. However, the improvement achieved by increasing m_{max} from 2 to 4 is marginal. It has also been discovered from simulations that carrying out more than three iterations in the BGMD decoding only results in negligible coding gain. To achieve similar error-correcting performance as the KV algorithm, the η needed in the LCC algorithm increases with $n - k$. For the (458, 410) RS code, the error-correcting capability of the

LCC decoding with $\eta = 10$ is about the same as that of the KV decoding with $m_{max} = 4$. Simulation results for a $(255, 239)$ RS code over $GF(2^8)$ are shown in Fig. 5.2. For this code, which has smaller $n-k$, the LCC decoder with $\eta = 3$ has about the same performance as the KV algorithm with $m_{max} = 4$. Although the achievable coding gain of the BGMD algorithm is less than that of the KV algorithm for the $(458, 410)$ code, the BGMD algorithm performs slightly better than the KV algorithm for the shorter $(255, 239)$ RS code.

5.2   Re-encoded algebraic soft-decision decoder

The interpolation is a very hardware-demanding step. To reduce the complexity of the interpolation, re-encoding and coordinate transformation techniques have been proposed in [56, 57]. The re-encoding is to compute a codeword ϕ that equals the hard decisions of the received symbols r in the k most reliable code positions. Denote these positions by the set R, and the rest of the positions by $\bar{R}$. ϕ can be found by erasure decoding. Assume that $r = c + e$, where c is the correct codeword, and e is the error vector. Then $\bar{r} = r + \phi = (c + \phi) + e$ is another codeword $\bar{c} = c + \phi$ corrupted by the same error vector. Therefore, decoding can be done on $\bar{r}$ instead to find the error vector, and accordingly recover c. If systematic encoding is used, then the message symbols can be directly read from the codeword c.

Assume that the interpolation points are originally $(\alpha_i, r_{i,j})$ for each code position, where $r_{i,0} = r_i$ is the hard-decision symbol. When re-encoding is adopted, the interpolation points become $(\alpha_i, r_{i,j} + \phi_i)$. To reduce the hardware complexity, only the interpolation point with the largest multiplicity, *i.e.* $(\alpha_i, r_{i,0} + \phi_i)$, is kept for each $i \in R$. Such a modification causes only negligible performance loss. Since $r_{i,0} = r_i = \phi_i$ for $i \in R$, the interpolation points are reduced to $(\alpha_i, 0)$ for $i \in R$, and $(\alpha_i, r_{i,j} + \phi_i)$ for $i \in \bar{R}$. The interpolation over the points in the format of $(\alpha_i, 0)$ can be pre-computed as will be shown later in this chapter. In addition, by applying a coordinate transformation, a factor $v(x) = \prod_{i \in R}(x + \alpha_i)$ is taken out of the pre-computed results and accordingly from the entire interpolation process. To accommodate the coordinate transformation, the interpolation points in code positions $i \in \bar{R}$ are transformed to $(\alpha_i, \beta_{i,j})$, where $\beta_{i,j} = (r_{i,j} + \phi_i) / \prod_{l \in R}(\alpha_l + \alpha_i)$.

Employing the re-encoding and coordinate transformation, the expensive bivariate interpolation only needs to be applied over the points in the $n - k$ least reliable code positions. The points in the code positions in R have the highest multiplicity. As will be shown later in this chapter, the interpolation is an iterative process. The number of iterations needed to interpolate over a point with multiplicity m is $m(m + 1)/2$. Hence, the number of interpolation iterations is reduced by at least $n/(n - k)$ times as a result of re-encoding. Moreover, by factoring out $v(x) = \prod_{i \in R}(x + \alpha_i)$, the degrees of the inter-

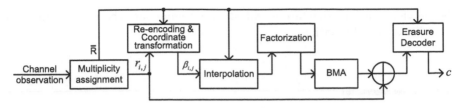

FIGURE 5.3
Re-encoded and coordinate transformed ASD decoder (modified from [58])

polation polynomials are reduced by $\deg(v(x)) = k$. The degree and hence the number of coefficients in a polynomial translates to the number of clock cycles needed in a serial design or the number of computation units required in a parallel architecture. Hence, the coordinate transformation also helps to substantially reduce the interpolation complexity. For high-rate codes, the re-encoding and coordinate transformation are necessary to reduce the complexity of ASD decoders to a practical level.

When re-encoding and coordinate transformation are adopted, the ASD decoding is carried out according to Fig. 5.3. Originally, if the received word is decodable, the interpolation output $Q(x, y)$ would have factors $y + f(x)$, where each $f(x)$ is a possible message polynomial. Then the factorization step computes $f(x)$ from $Q(x, y)$. When the coordinate transformation is employed, $v(x)$ has been taken out of the interpolation process. It can be multiplied back to the interpolation output in order to recover the $y + f(x)$ factors. Nevertheless, $\deg(v(x)) = k$, and multiplying $v(x)$ back causes large hardware overhead. Instead, the factorization can be done directly on the re-encoded and coordinate transformed interpolation output. In this case, what the factorization step computes are $y + \gamma(x)$ factors, where $\gamma(x)$ can be used as the syndromes in the BMA to correct the errors in code positions belonging to R. 2τ syndromes are needed to correct τ errors. Since it is less likely to have errors in the k most reliable code positions in R, 2τ can be set to a number much less than k. As a result, the complexity of the factorization is also substantially reduced. After the errors in R are corrected, another erasure decoding is needed to recover the entire codeword. If systematic encoding has been adopted, then the message symbols are read directly from the recovered codeword. ASD algorithms are also list decoding algorithms. They generate multiple possible codewords, one corresponding to each $y + \gamma(x)$ factor. The correct codeword can be chosen based on reliability.

The implementation of the BMA has been discussed in the previous chapter. Next, the architectures for the other blocks in the re-encoded ASD decoder are presented. The interpolation and factorization algorithms are also given before their implementation architectures are introduced. Details will be provided on how they are modified to accommodate the re-encoding and coordinate transformation.

5.3 Re-encoding algorithms and architectures

As a result of the re-encoding and coordinate transformation, the interpolation points in $\bar{R}$ are transformed to $(\alpha_i, \beta_{i,j})$, where

$$\beta_{i,j} = \frac{r_{i,j} + \phi_i}{\prod_{l \in R}(\alpha_l + \alpha_i)}. \tag{5.1}$$

ϕ is a codeword such that $\phi_i = r_{i,0}$ for $i \in R$. It can be computed by erasure decoding, which is also used to recover the entire codeword at the end after the errors in R are corrected.

By modifying the BMA, an iterative process was proposed in [59] to compute the erasure locator and evaluator polynomials. Then the erasure magnitudes are computed using the Forney's formula. Such a re-encoder has long latency due to the iterative process of the key equation solving. Moreover, mapping (5.1) directly to hardware would require multiplying $|R| = k$ terms in the denominator. This results in a large area overhead. Alternatively, as proposed in [58, 60], the erasure locator and evaluator polynomials can be computed by direct multiplications and the computations in (5.1) can be re-formulated. Next, this simplified re-encoding and coordinate transformation method and the corresponding VLSI architectures are presented.

5.3.1 Simplified re-encoding and coordinate transformation schemes

For narrow-sense RS codes, α^l is the lth root of the generator polynomial. The syndromes, S_l $(1 \le l \le n - k)$ are computed as

$$S_l = \sum_{i=0}^{n-1} r_{i,0}\alpha^{li}.$$

Define a syndrome polynomial $s(x) = \sum_{l=0}^{n-k-1} S_{n-k-l}x^l$. The $n - k$ erasure positions are those in $\bar{R}$. An erasure locator polynomial, whose roots are the erasure locations, is computed as

$$\tau(x) = \prod_{l \in \bar{R}}(x + \alpha_l).$$

The code positions in $\bar{R}$ have already been found in the multiplicity assignment step. Therefore, $\tau(x)$ can be calculated at the same time as the syndromes.

After $s(x)$ and $\tau(x)$ are derived, the erasure evaluator polynomial, $\omega(x)$, is computed directly through a polynomial multiplication followed by a modulo reduction [61] as

$$\omega(x) = s(x)\tau(x) \mod x^{n-k}. \tag{5.2}$$

Then the erasure magnitudes, ϕ_i, for $i \in \bar{R}$, are computed by using a formula similar to that of the Forney's algorithm [61],

$$\phi_i = \frac{\alpha_i^{-(n-k+1)}\omega(\alpha_i)}{\tau'(\alpha_i)} + r_{i,0}. \tag{5.3}$$

Although $\tau'(\alpha_i)$ can be derived by the evaluation values of the odd terms of $\tau(x)$, the computation of $\tau(x)$ itself takes iterative polynomial multiplications. Instead, since $\tau(x)$ is a product of $(x + \alpha_l)$ for $l \in \bar{R}$,

$$\tau'(\alpha_i) = \prod_{l \in \bar{R}, l \neq i} (\alpha_i + \alpha_l). \tag{5.4}$$

Substituting (5.4) into (5.3), ϕ_i is computed according to

$$\phi_i = \frac{\omega(\alpha_i)}{\alpha_i^{n-k}(\alpha_i \prod_{l \in \bar{R}, l \neq i}(\alpha_i + \alpha_l))} + r_{i,0}. \tag{5.5}$$

The transformed coordinate, $\beta_{i,j}$, is computed according to (5.1) after ϕ_i is calculated. However, the denominator of (5.1), $\prod_{l \in R}(\alpha_l + \alpha_i)$, contains k terms, and a finite field inversion is required. For high-rate codes constructed over large finite fields, these computations are expensive to implement. Alternatively, the product of all distinct nonzero finite field elements equals the identity. In other words, $\alpha_i \prod_{l \in R \cup \bar{R}, l \neq i}(\alpha_i + \alpha_l) = 1$. Therefore, $1/\prod_{l \in R}(\alpha_i + \alpha_l) = \alpha_i \prod_{l \in \bar{R}, l \neq i}(\alpha_i + \alpha_l)$ for $i \in \bar{R}$ and the computation of $\beta_{i,j}$ can be simplified as

$$\beta_{i,j} = (r_{i,j} + \phi_i)\alpha_i \prod_{l \in \bar{R}, l \neq i} (\alpha_i + \alpha_l). \tag{5.6}$$

There are only $n - k$ terms in the product $\alpha_i \prod_{l \in \bar{R}, l \neq i}(\alpha_i + \alpha_l)$. Compared to the computations in (5.1), not only the number of multiplications has been reduced by a factor of $k/(n - k)$, the inversion is also eliminated.

The final goal of the re-encoding and coordinate transformation is to compute the transformed coordinate $\beta_{i,j}$ for $i \in \bar{R}$. ϕ_i are intermediate results and do not need to be generated explicitly. Substituting (5.5) into (5.6), the transformed coordinates are computed as

$$\beta_{i,j} = \frac{\omega(\alpha_i)}{\alpha_i^{n-k}} + (r_{i,j} + r_{i,0})\alpha_i \prod_{l \in \bar{R}, l \neq i} (\alpha_i + \alpha_l). \tag{5.7}$$

The above formula can be further simplified. According to (5.2), $\omega(x)$ is the

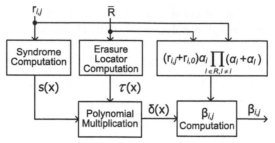

FIGURE 5.4
Re-encoder architecture (modified from [60])

remainder polynomial of dividing $s(x)\tau(x)$ by x^{n-k}. Assume that $\delta(x)$ is the quotient polynomial of this division. Then $s(x)\tau(x)$ can be written as

$$s(x)\tau(x) = \delta(x)x^{n-k} + \omega(x). \tag{5.8}$$

The degrees of $s(x)$ and $\tau(x)$ do not exceed $n-k-1$ and $n-k$, respectively. Hence $\deg(\delta(x)) \leq n-k-1$. For $i \in \bar{R}$, $\tau(\alpha_i) = 0$. As a result, the evaluation values of (5.8) over α_i with $i \in \bar{R}$ are

$$s(\alpha_i)\tau(\alpha_i) = \delta(\alpha_i)\alpha_i^{n-k} + \omega(\alpha_i) = 0.$$

From the above equation, it can be derived that for $i \in \bar{R}$

$$\delta(\alpha_i) = \frac{\omega(\alpha_i)}{\alpha_i^{n-k}}. \tag{5.9}$$

As a result, (5.7) is reduced to

$$\beta_{i,j} = \delta(\alpha_i) + (r_{i,j} + r_{i,0})\alpha_i \prod_{l \in \bar{R}, l \neq i} (\alpha_i + \alpha_l). \tag{5.10}$$

Compared to (5.7), the term $1/\alpha_i^{n-k}$ is eliminated in (5.10). $\delta(x)$ has the same maximum degree as $\omega(x)$, and can be derived from the same polynomial multiplication process, as will be shown next. Nevertheless, $\delta(x)$ consists of the terms with higher degrees in $s(x)\tau(x)$. In a polynomial multiplication process that generates the most significant coefficient first, which is required in order to apply Horner's rule in the syndrome computation and polynomial evaluation, the coefficients of $\delta(x)$ are available before those of $\omega(x)$. As a result, the latency can be also reduced by using $\delta(x)$ instead of $\omega(x)$.

5.3.2   Simplified re-encoder architectures

Fig. 5.4 shows the block diagram of the re-encoder architecture according to (5.10). Applying Horner's rule, the syndrome computation can be implemented using the serial architecture in Fig. 4.5. It takes n clock cycles to

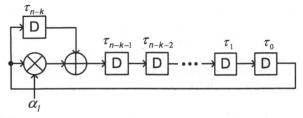

FIGURE 5.5
Architecture for $\tau(x)$ computation (modified from [60])

compute a syndrome. Multiple syndromes can be computed simultaneously using multiple copies of this architecture.

A fully folded architecture for computing the erasure locator polynomial $\tau(x) = \prod_{l \in \bar{R}}(x + \alpha_l)$ is shown in Fig. 5.5. The registers store the polynomial coefficients. Since $\deg(\tau(x)) \leq n - k$, $n - k + 1$ registers are needed. At the beginning, the registers labeled by τ_1 and τ_0 are initialized as 1 and α_l, respectively. They are the coefficients of the first $(x + \alpha_l)$ term. All the other registers are initialized to zero. Then each additional $(x + \alpha_l)$ term is multiplied by connecting the corresponding α_l to the multiplier. The multiplication of each term takes $n - k$ clock cycles. Therefore, the computation of $\tau(x)$ takes $(n - k - 1)(n - k)$ clock cycles. After all the terms are multiplied, the coefficients of $\tau(x)$ are located at the registers, and the constant coefficient is in the register labeled τ_0. To achieve faster speed, the unfolding technique discussed in Chapter 2 can be applied to Fig. 5.5. The computation takes $\lceil (n - k)/N_\tau \rceil (n - k - 1)$ clock cycles in an N_τ-unfolded architecture.

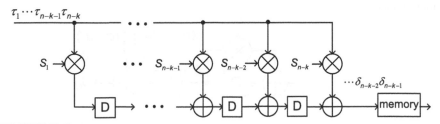

FIGURE 5.6
Architecture for $\delta(x)$ computation (modified from [60])

$\delta(x)$ is computed from $s(x)$ and $\tau(x)$ through polynomial multiplication, whose implementation architecture is shown in Fig. 5.6. From (5.8), the co-efficients of $\delta(x)$ are those of $s(x)\tau(x)$ with degree at least $n - k$. Since $\deg(s(x)) \leq n - k - 1$, τ_0 is not needed in the computation of $\delta(x)$. In Fig. 5.6, the coefficients of $\tau(x)$ are shifted in serially starting with τ_{n-k}, and each coefficient of $s(x)$ is input to a multiplier. The coefficients of $\delta(x)$ are generated one after another at the output of the right-most adder in the order of

$\delta_{n-k-1}, \delta_{n-k-2}, \ldots, \delta_0$, and the last coefficient is available after $n - k$ clock cycles. To reduce the latency, once a coefficient of $\delta(x)$ is available, it is sent to the $\beta_{i,j}$ calculation engine immediately. Since $\delta(x)$ is also needed later for other computations, it is stored into a $(n-k) \times q$-bit memory. The folding technique presented in Chapter 2 can be applied to reduce the area requirement of the $\delta(x)$ computation architecture at the cost of longer latency.

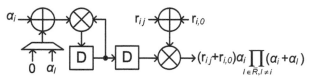

FIGURE 5.7
Architecture for $(r_{i,j} + r_{i,0})\alpha_i \prod_{l \in \bar{R}, l \neq i}(\alpha_i + \alpha_l)$ computation (modified from [60])

The $(r_{i,j} + r_{i,0})\alpha_i \prod_{l \in \bar{R}, l \neq i}(\alpha_i + \alpha_l)$ computation can be done by the architecture in Fig. 5.7. In this figure, $\alpha_i \prod_{l \in \bar{R}, l \neq i}(\alpha_i + \alpha_l)$ is computed by the adder and multiplier loop on the left in $n - k$ clock cycles. This product is loaded into the register in the middle right away to be multiplied by $(r_{i,j} + r_{i,0})$, while the adder and multiplier loop starts to compute the product for the next code position. Let J be the maximum number of non-trivial interpolation points for each code position in $\bar{R}$. Usually, J can be forced to a very small number, such as two, without causing any noticeable performance loss. In the worst case, it takes $(n - k)^2 + J - 1$ clock cycles to compute $(r_{i,j} + r_{i,0})\alpha_i \prod_{l \in \bar{R}, l \neq i}(\alpha_i + \alpha_l)$ for all $i \in \bar{R}$. If higher speed is needed, multiple copies of this engine can be employed to compute the values for different code positions in parallel. If N_1 copies are used, the latency of the computation is $\lceil (n - k)/N_1 \rceil (n - k) + J - 1$ clock cycles.

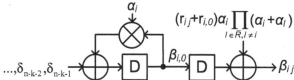

FIGURE 5.8
Architecture for $\beta_{i,j}$ computation (modified from [60])

Horner's rule is applied to derive the architecture shown in Fig. 5.8 for computing $\beta_{i,j}$. In this architecture, $\beta_{i,0} = \delta(\alpha_i)$ is computed by the feedback loop in $n - k$ clock cycles. After that, $\beta_{i,j}$ for $1 \leq j < J$ are computed in the following $J - 1$ clock cycles. Similarly, higher speed is achieved if multiple copies of such units are used to compute $\beta_{i,j}$ for different code positions in parallel. Assume that there are N_β copies. The computation of all $\beta_{i,j}$ can be done in $\lceil (n - k)/N_\beta \rceil (n - k + J - 1)$ clock cycles.

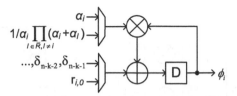

FIGURE 5.9
Architecture for ϕ_i computation

With minor modifications, the aforementioned architectures can also implement the erasure decoder needed at the end of the ASD decoding process. In the case of erasure decoding, ϕ_i instead of $\beta_{i,j}$ needs to be computed as the outputs. By substituting (5.9) into (5.5), the computation of ϕ_i is simplified as

$$\phi_i = \frac{\delta(\alpha_i)}{\alpha_i \prod_{l \in \bar{R}, l \neq i}(\alpha_i + \alpha_l)} + r_{i,0}. \tag{5.11}$$

$1/(\alpha_i \prod_{l \in \bar{R}, l \neq i}(\alpha_i + \alpha_l))$ can be computed by an adder and multiplier feedback loop as shown in Fig. 5.7, followed by a finite field inverter. Then ϕ_i is generated by using the architecture illustrated in Fig. 5.9. In this architecture, $\delta(\alpha_i)$ is first computed in $n - k$ clock cycles by passing the upper inputs of the multiplexors. After that, the lower inputs are passed to calculate ϕ_i.

One way to increase the re-encoder speed is to compute the $\beta_{i,j}$ for all code positions in $\bar{R}$ in parallel. This would require $n - k$ copies of the architecture in Fig. 5.8. In this case, the units for $\beta_{i,j}$ computation occupy a significant part of the overall re-encoder area. Next, two schemes are introduced to reduce the area requirement of $\beta_{i,j}$ computation. To simplify the notations, assume that $\bar{R} = \{0, 1, \cdots, n - k - 1\}$. Accordingly, the erasure positions and their corresponding transformed coordinates are denoted by $\{\alpha_0, \alpha_1, \cdots, \alpha_{n-k-1}\}$ and $\{\beta_{0,j}, \beta_{1,j}, \cdots, \beta_{n-k-1,j}\}$, respectively.

Finite field multipliers are expensive to implement. One way to reduce the area requirement of computing $\beta_{i,j}$ is to reuse the multipliers in other blocks of the re-encoder. The $\beta_{i,j}$ computation mainly requires multiplier-adder loops. From Fig. 5.6, the polynomial multiplication architecture for building $\delta(x)$ also consists of multiplier-adder pairs. Therefore, a single architecture as shown in Fig. 5.10 can be used for both purposes in a time-multiplexed way. In this figure, $\delta(x)$ is computed when the coefficients of $\tau(x)$ and $s(x)$ are passed to the multipliers, and the multiplexors connected to the adders choose the lower inputs. The derived coefficients of $\delta(x)$ are written to the memory. Then the multiplexors choose the alternative inputs, and the coefficients of $\delta(x)$ are read out serially from the memory and fed back to the architecture. Accordingly, $\delta(\alpha_i) = \beta_{i,0}$ for all $i \in \bar{R}$ are computed at the same time. In addition, $\beta_{i,j}$ $(1 \le j < J)$ are derived by adding $\delta(\alpha_i)$ to $(r_{i,j} + r_{i,0})\alpha_i \prod_{l \in \bar{R}, l \neq i}(\alpha_i + \alpha_l)$ that have been computed by the architecture in Fig. 5.7. Using the fully-parallel shared architecture in Fig. 5.10, $n - k$ multiplier-adder pairs have been saved

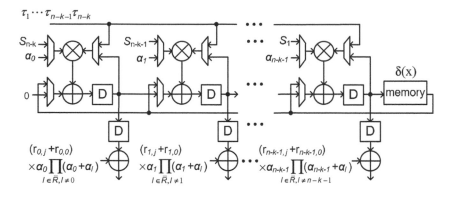

FIGURE 5.10
Shared architecture for computing both $\delta(x)$ and $\beta_{i,j}$ (modified from [60])

by adding $3(n-k) \times q$ extra multiplexors if the RS code is over $GF(2^q)$. Since a $GF(2^q)$ finite field multiplier requires many more logic gates than a q-bit multiplexor, the shared architecture leads to substantial area reduction. Adding the extra multipliers does not increase the critical path. On the other hand, in the shared architecture, the computation of $\beta_{i,j}$ can not start until that of $\delta(x)$ is completed. Hence, the re-encoder latency will be increased by $n-k-1$ clock cycles. The shared architecture in Fig. 5.10 is also scalable, and area-speed tradeoff can be achieved by adjusting the number of multiplier-adder pairs employed.

The $\beta_{i,j}$ computed from the re-encoder are used to form interpolation points, $(\alpha_i, \beta_{i,j})$, which are sent to the interpolation step. The interpolation problem can be solved by the Kötter's algorithm [62, 63], in which the points are processed one after another. In this case, the computation of a point only needs to be completed before it is required in the interpolation. Considering this, not all $\beta_{i,j}$ need to be computed before the interpolation starts. A smaller number of $\beta_{i,j}$ are calculated at a time to reduce the required hardware units. The interpolation starts right after the first group of $\beta_{i,j}$ are available. The second group of $\beta_{i,j}$ are computed simultaneously as the interpolation over the first group of points is carried out. The interpolation latency will not be affected as long as $\beta_{i,j}$ are derived before the corresponding points are needed in the interpolation. Moreover, the top three blocks of the re-encoder shown in Fig. 5.4 use separate hardware units and take a much longer time than the $\beta_{i,j}$ computation. Even if the hardware units for $\beta_{i,j}$ computation are occupied by multiple rounds, they will be available by the time that the $\delta(x)$ for the next received word is calculated. Therefore, in a pipelined ASD decoder, the re-encoding for the next received word can start at the same time as the interpolation for the current word, and only the latency for deriving the first group of $\beta_{i,j}$ contributes to the re-encoder latency.

Information on the scheduling of the computations involved in the interpolation is necessary to decide when $\beta_{i,j}$ are needed in the interpolation. Many architectures have been developed to implement the interpolation [55, 64, 65, 66, 67, 68], and more details will be given later in this chapter. These architectures share similar computation scheduling, and a generalized analysis can be carried out to decide when $\beta_{i,j}$ are required and accordingly the minimum number of units needed for the $\beta_{i,j}$ computation. The interpolation is an iterative process, and each iteration consists of discrepancy coefficient computation followed by polynomial updating. To reduce the area requirement, the available interpolation architectures process the coefficients of each polynomial serially. Hence, as will be shown later in this chapter, the discrepancy coefficient computation takes $dx_p + 1$ clock cycles plus some extra clock cycles caused by pipelining. Here dx_p is the maximum x-degree of the polynomials involved in the p-th iteration of the interpolation. If $(\alpha_i, \beta_{i,j})$ is interpolated in iteration p, $\beta_{i,j}$ is actually needed in the $(dx_p + 2)$th clock cycle of this iteration. In addition, the polynomial updating of iteration p is overlapped with the discrepancy coefficient computation of iteration $p + 1$. Therefore, the number of clock cycles needed for interpolation iteration p is $(dx_p + 1 + \xi)$, where ξ includes the latency of the pipelining for both discrepancy coefficient computation and polynomial updating. The interpolation polynomials are initialized with a simple format, and the first interpolation iteration only takes ξ clock cycles. The number of interpolation iterations spent on a point is decided by the multiplicity of the point. If every point has multiplicity one, then each interpolation iteration requires a different $\beta_{i,j}$. In this case, the number of clock cycles, T_i, passed before $\beta_{i,j}$ is needed in the interpolation is

$$T_i = \xi + \sum_{j=1}^{i-1}(dx_j + 1 + \xi) + dx_i + 1.$$

Using N_β copies of the architecture in Fig. 5.8, the computation of each group of $\beta_{i,j}$ takes $n - k + J - 1$ clock cycles. After the first group is calculated, the interpolation over the corresponding points is carried out while the second group of $\beta_{i,j}$ is computed. To avoid disrupting the interpolation, the computation for the second group has to be completed before any of them is needed in the interpolation. Accordingly, N_β is set to

$$N_\beta = \arg\min_i\{T_i | T_i \geq n - k + J - 1\}. \tag{5.12}$$

The polynomials involved in the interpolation get longer as more points are interpolated. Hence, the interpolation over the points corresponding to the second and later group of $\beta_{i,j}$ has longer latency. Therefore, it is guaranteed that $\beta_{i,j}$ are always generated before they are needed in the interpolation if the N_β derived from (5.12) is adopted.

5.4   Interpolation algorithms and architectures

For an (n, k) RS code, the goal of the interpolation is to find a polynomial $Q(x, y)$ with minimum $(1, k - 1)$ weighted degree that passes each interpolation point with its associated multiplicity. To pass a point $(\alpha_i, \beta_{i,j})$ with multiplicity $m_{i,j}$, the coefficient for each monomial $x^a y^b$ with $a + b < m_{i,j}$ in the shifted polynomial $Q(x + \alpha_i, y + \beta_{i,j})$ needs to be zero. Therefore, a point with multiplicity $m_{i,j}$ adds $m_{i,j}(m_{i,j} + 1)/2$ constraints to the interpolation. The polynomial satisfying the interpolation constraints can be found by solving a set of $C = \sum_{i=0}^{n-1} \sum_j m_{i,j}(m_{i,j}+1)$ linear equations. For RS codes used in many practical systems, n is not small. Solving a large set of linear equations has very high hardware complexity. Alternatively, various algorithms have been proposed to more efficiently address the interpolation problem. Among available schemes, the Kötter's [62] and Lee-O'Sullivan algorithms [69] are most suitable for hardware implementations. These two algorithms and their implementation architectures are discussed in detail in this chapter.

5.4.1   Kötter's interpolation algorithm

Kötter's algorithm iteratively forces a set of polynomials to pass each interpolation point with the corresponding multiplicity. At the end, the polynomial with the minimum $(1, k - 1)$ weighted degree is chosen to be the interpolation output. In this algorithm, the number of polynomials, t, equals the maximum y-degree of the polynomials. To guarantee that there exists a solution for a set of C linear equations, there should be at least $C + 1$ variables. A monomial $x^r y^s$ is said to have lower weighted lexicographical order than $x^{r'} y^{s'}$ if the weighted degree of $x^r y^s$ is lower than that of $x^{r'} y^{s'}$, or the weighted degrees are the same but $s < s'$. Arranging all monomials in increasing $(1, k - 1)$ weighted lexicographical order, then the interpolation polynomial should include the first $C + 1$ monomials in order to have minimum $(1, k - 1)$ weighted degree. From this, the maximum y-degree of the monomials and hence the number of polynomials involved in Kötter's interpolation, t, can be decided as [56]

$$t = \left\lfloor \frac{(k - 1) + \sqrt{(k - 1)^2 + 8C(k - 1)}}{2(k - 1)} \right\rfloor.$$

For high-rate codes, t usually equals the maximum multiplicity of the interpolation points.

Algorithm 4 Kötter's Interpolation Algorithm
input: $(\alpha_i, \beta_{i,j})$ *with multiplicity* $m_{i,j}$
initialization: $Q^{(0)}(x,y) = 1, Q^{(1)}(x,y) = y, \ldots, Q^{(t)}(x,y) = y^t$
$$w_0 = 0, w_1 = k - 1, \ldots, w_t = t(k - 1)$$
begin:

$\quad\quad$ *for each point* $(\alpha_i, \beta_{i,j})$ *with multiplicity* $m_{i,j}$
$\quad\quad\quad$ *for* $a = 0$ *to* $m_{i,j} - 1$ *and* $b = 0$ *to* $m_{i,j} - a - 1$
$\quad\quad\quad\quad$ *compute* $d_{a,b}^{(l)}(\alpha_i, \beta_{i,j})$, $0 \le l \le t$
$\quad\quad\quad\quad$ *if* $\{l | d_{a,b}^{(l)}(\alpha_i, \beta_{i,j}) \ne 0, 0 \le l \le t\} \ne \emptyset$
$\quad\quad\quad\quad\quad$ $l^* = \arg\min_l\{w_l | d_{a,b}^{(l)}(\alpha_i, \beta_{i,j}) \ne 0, 0 \le l \le t\}$
$\quad\quad\quad\quad\quad$ *for* $l = 0$ *to* t, $l \ne l^*$
$\quad\quad\quad\quad\quad\quad$ $Q^{(l)}(x,y) \Leftarrow d_{a,b}^{(l^*)}(\alpha_i, \beta_{i,j})Q^{(l)}(x,y)$
$\quad\quad\quad\quad\quad\quad\quad\quad$ $+d_{a,b}^{(l)}(\alpha_i, \beta_{i,j})Q^{(l^*)}(x,y)$
$\quad\quad\quad\quad\quad$ $Q^{(l^*)}(x,y) \Leftarrow Q^{(l^*)}(x,y)(x + \alpha_i)$
$\quad\quad\quad\quad\quad$ $w_{l^*} \Leftarrow w_{l^*} + 1$
output: $Q^{(l^*)}(x,y)(l^* = \arg\min_l\{w_l | 0 \le l \le t\})$

Kötter's interpolation algorithm [62] can be carried out as in Algorithm 4. In this algorithm, $w_0, w_1, \ldots, w_t$ are used to track the weighted degrees of the polynomials, and $d_{a,b}^{(l)}(\alpha_i, \beta_{i,j})$ is called the discrepancy coefficient. It is the coefficient of $x^a y^b$ in $Q^{(l)}(x + \alpha_i, y + \beta_{i,j})$. Assume that $q_{r,s}^{(l)}$ is the coefficient of $x^r y^s$ in $Q^{(l)}(x,y)$.

$$d_{a,b}^{(l)}(\alpha_i, \beta_{i,j}) = \sum_{r \ge a} \sum_{s \ge b} \binom{r}{a}\binom{s}{b} q_{r,s}^{(l)}(\alpha_i)^{r-a}(\beta_{i,j})^{s-b}. \tag{5.13}$$

Kötter's algorithm forces each of the $t + 1$ polynomials to satisfy one additional interpolation constraint in each iteration. As shown in Algorithm 4, the discrepancy coefficient, $d_{a,b}^{(l)}(\alpha_i, \beta_{i,j})$, for each polynomial is computed first. If all discrepancy coefficients are zero, then all polynomials already satisfy the interpolation constraint and the rest of the computations are skipped. Otherwise, the polynomials are updated as linear combinations or multiplied by $(x + \alpha_i)$. As a result of the updating, the discrepancy coefficient of each polynomial becomes zero, and hence the interpolation constraint of $x^a y^b$ is satisfied. In addition, the polynomial updating does not affect the interpolation constraints that have been satisfied in previous iterations, as long as the constraint of $x^a y^b$ is taken care of before $x^{a+1} y^b$. Also, the weighted degree of the polynomials is only increased by one, which is the minimum possible increase. Therefore, after the iteration is repeated for each interpolation constraint, a set of t polynomials satisfying all the interpolation constraints is derived and the one with the minimum weighted degree is chosen as the interpolation output.

Next, consider how the re-encoding and coordinate transformation affect Kötter's interpolation. In the existing ASD algorithms, most likely there is only one interpolation point with the maximum multiplicity $m_{max} = t$ in each of the k reliable code positions in R. If there are other points in these positions, they can be forced to have zero multiplicity without bringing any noticeable error-correcting performance loss. Because of the re-encoding, the interpolation points in the code positions in R are in the format of $(\alpha_i, 0)$. The polynomials at the beginning of the Kötter's interpolation are $Q^{(0)}(x, y) = 1, Q^{(1)}(x, y) = y, Q^{(2)}(x, y) = y^2 \ldots$. In the first iteration, only the discrepancy coefficient of $Q^{(0)}(x, y) = 1$ over $(\alpha_i, 0)$ is nonzero. Therefore, $l^* = 0$ in the first interpolation iteration and $Q^{(0)}(x, y)$ becomes $(x + \alpha_i)$ at the end of this iteration. Since the discrepancy coefficient of the other polynomials are zero, $Q^{(l)}(x, y)$ for $l \neq l^*$ should be updated as $d_{0,0}^{(l^*)}(\alpha_i, 0)Q^{(l)}(x, y)$ according to Algorithm 4. However, as will be explained later in this chapter, multiplying a polynomial with a nonzero scaler does not change the $y + f(x)$ factors in the interpolation output and hence the decoding result. Therefore, those $Q^{(l)}(x, y)$ with zero discrepancy coefficients do not need to be updated. Accordingly, the polynomials at the end of the first iteration are $Q^{(0)}(x, y) = (x + \alpha_i), Q^{(1)}(x, y) = y, Q^{(2)}(x, y) = y^2 \ldots$. In the second iteration, $a = 0$ and $b = 1$. Only the discrepancy coefficient of $Q^{(1)}(x, y) = y$ is nonzero. Similarly, at the end of the second iteration, the polynomials become $Q^{(0)}(x, y) = (x + \alpha_i), Q^{(1)}(x, y) = y(x + \alpha_i), Q^{(2)}(x, y) = y^2 \ldots$. Following Kötter's algorithm, it can be derived that after the interpolation is repeated for $a = 0, 1, \ldots t - 1$ and $b = 0, 1, \ldots, t - a - 1$ for $(\alpha_i, 0)$ with multiplicity t, the polynomials are $Q^{(0)}(x, y) = (x + \alpha_i)^t, Q^{(1)}(x, y) = y(x + \alpha_i)^{t-1}, Q^{(2)}(x, y) = y^2(x + \alpha_i)^{t-2} \ldots$. After the interpolation is done for every point in code positions in R, the polynomials become $Q^{(0)}(x, y) = (\prod_{i \in R}(x + \alpha_i))^t, Q^{(1)}(x, y) = y(\prod_{i \in R}(x + \alpha_i))^{t-1}, Q^{(2)}(x, y) = y^2(\prod_{i \in R}(x + \alpha_i))^{t-2} \ldots$. Let $v(x) = \prod_{i \in R}(x + \alpha_i)$, and apply coordinate transformation $z = y/v(x)$. Then the interpolation polynomials are changed to $Q^{(0)}(x, z) = (v(x))^t, Q^{(1)}(x, y) = z(v(x))^t, Q^{(2)}(x, y) = z^2(v(x))^t \ldots$. Taking the common factor $(v(x))^t$ out, the interpolation over the remaining points, which are the points in code positions of $\bar{R}$, can continue on $Q^{(0)}(x, z) = 1, Q^{(1)}(x, z) = z, Q^{(2)}(x, z) = z^2 \ldots$. These polynomials have the same format as the initial polynomials. However, since the coordinate transformation $z = y/v(x)$ is adopted, $(1, k - 1 - \deg(v(x))) = (1, -1)$ weighted degree should be used. To simplify the notations, the rest of the book still uses $Q^{(0)}(x, y) = 1, Q^{(1)}(x, y) = y, Q^{(2)}(x, y) = y^2 \ldots$ to denote the initial polynomials for the interpolation over the points in $\bar{R}$ when re-encoding and coordinate transformation are employed.

TABLE 5.1
Terms need to be added to compute $u_{a,s}^{(l)}(\alpha_i)$

$u_{a,s}^{(l)}(\alpha_i)$	$q_{0,s}^{(l)}$	$q_{1,s}^{(l)}\alpha_i$	$q_{2,s}^{(l)}\alpha_i^2$	$q_{3,s}^{(l)}\alpha_i^3$	$q_{4,s}^{(l)}\alpha_i^4$	$q_{5,s}^{(l)}\alpha_i^5$	$q_{6,s}^{(l)}\alpha_i^6$	$q_{7,s}^{(l)}\alpha_i^7$	$\cdots$
$u_{0,s}^{(l)}(\alpha_i)$	$q_{0,s}^{(l)}$	$q_{1,s}^{(l)}\alpha_i$	$q_{2,s}^{(l)}\alpha_i^2$	$q_{3,s}^{(l)}\alpha_i^3$	$q_{4,s}^{(l)}\alpha_i^4$	$q_{5,s}^{(l)}\alpha_i^5$	$q_{6,s}^{(l)}\alpha_i^6$	$q_{7,s}^{(l)}\alpha_i^7$	$\cdots$
$u_{1,s}^{(l)}(\alpha_i)$		$q_{1,s}^{(l)}$		$q_{3,s}^{(l)}\alpha_i^2$		$q_{5,s}^{(l)}\alpha_i^4$		$q_{7,s}^{(l)}\alpha_i^6$	$\cdots$
$u_{2,s}^{(l)}(\alpha_i)$			$q_{2,s}^{(l)}$	$q_{3,s}^{(l)}\alpha_i$			$q_{6,s}^{(l)}\alpha_i^4$	$q_{7,s}^{(l)}\alpha_i^5$	$\cdots$
$u_{3,s}^{(l)}(\alpha_i)$				$q_{3,s}^{(l)}$				$q_{7,s}^{(l)}\alpha_i^4$	$\cdots$
$\vdots$	$\vdots$	$\vdots$	$\vdots$	$\vdots$	$\vdots$	$\vdots$	$\vdots$	$\vdots$	$\ddots$

5.4.2 Architectures for Kötter's interpolation

The architecture for implementing Kötter's interpolation has two major blocks: the discrepancy coefficient computation block and the polynomial updating block.

5.4.2.1 Discrepancy coefficient computation architecture

The discrepancy coefficient can be rewritten as

$$
\begin{aligned}
d_{a,b}^{(l)}(\alpha_i, \beta_{i,j}) &= \sum_{s \geq b} \binom{s}{b}(\beta_{i,j})^{s-b} \sum_{r \geq a} \binom{r}{a} q_{r,s}^{(l)}(\alpha_i)^{r-a} \\
&= \sum_{s \geq b} \binom{s}{b}(\beta_{i,j})^{s-b} u_{a,s}^{(l)}(\alpha_i)
\end{aligned}
\tag{5.14}
$$

Hence the computation of a discrepancy coefficient is broken down into two layers. Partial discrepancy coefficients $u_{a,s}^{(l)}(\alpha_i) = \sum_{r \geq a} \binom{r}{a} q_{r,s}^{(l)}(\alpha_i)^{r-a}$ are calculated in the first layer. Then similar computations are repeated over $u_{a,s}^{(l)}(\alpha_i)$ and $\beta_{i,j}$ in the second layer. For computations over finite field $GF(2^q)$, the binomial coefficients are reduced modulo two. Hence, they are either '1' or '0'. The terms needed to be added up to calculate $u_{a,s}^{(l)}(\alpha_i)$ are listed in Table 5.1.

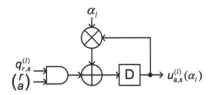

FIGURE 5.11
Architecture for computing partial discrepancy coefficients

Applying Horner's rule, the partial discrepancy coefficient, $u_{a,s}^{(l)}(\alpha_i)$, can be computed using the architecture illustrated in Fig. 5.11. Assume that $Q^{(l)}(x, y) = q_0^{(l)}(x) + q_1^{(l)}(x)y + q_2^{(l)}(x)y^2 + \ldots$, and $q_s^{(l)}(x) = q_{0,s}^{(l)} + q_{1,s}^{(l)}x + q_{2,s}^{(l)}x^2 + \ldots$. The coefficients of $q_s^{(l)}(x)$ are input to the architecture in Fig. 5.11 serially starting with the most significant one. After the coefficient $q_{a,s}^{(l)}$ is input, $u_{a,s}^{(l)}(\alpha_i)$ will be available in the register at the next clock edge. As shown in Table 5.1, $\binom{r}{a}$ have regular patterns, and hence can be derived by simple logic. The same architecture as in Fig. 5.11 can be used to carry out the second layer of the discrepancy coefficient computation according to the second equation in (5.14), after $u_{a,s}^{(l)}(\alpha_i)$ are available. On the other hand, although the x-degrees of the polynomials increase with the interpolation iterations and can become quite high for long code, the maximum y-degree equals t in the interpolation process and is usually small for practical applications. Hence, the second layer of the discrepancy coefficient computation can also be done by calculating and adding up the $\binom{s}{b}(\beta_{i,j})^{s-b}u_{a,s}^{(l)}(\alpha_i)$ terms using a multiplier-adder tree.

5.4.2.2   Polynomial updating architecture

Fig. 5.12 shows an architecture for implementing the polynomial updating in Kötter's interpolation. This architecture updates the univariate polynomials $q_s^{(l)}(x)$ for $l = 0, 1, \ldots, t$ in parallel, and copies of this architecture are needed to simultaneously update the univariate polynomials with different s, *i.e.* the univariate polynomials corresponding to different y-degrees. The coefficients in a polynomial are updated serially. The minimum polynomial $q_s^{(l^*)}(x)$ and the corresponding discrepancy coefficient are switched to the top. The non-minimum polynomials are updated as linear combinations, and the minimum polynomial is multiplied by $(x + \alpha_i)$. Registers at the outputs of the linear combinations are added to align the coefficients. If the polynomial coefficients are input with the most significant one first, the multiplication by $(x + \alpha_i)$ can be implemented by unit in the dashed block of Fig. 5.12. If the least significant coefficient is input first, the register needs to be moved to the top branch before the adder in this block. After the updating, the polynomial coefficients are written back into fixed memory blocks to avoid using another set of switches. In other words, the first updated polynomial is written back into the memory labeled by $q_s^{(0)}(x)$, and the second updated polynomial is stored in the memory block for $q_s^{(1)}(x)$, etc. To enable this, the respective weighted degrees of the polynomials are tracked in a control unit, which is not shown in Fig. 5.12 for the purpose of conciseness.

5.4.2.3   Computation scheduling for Kötter's interpolation

Fig. 5.13 shows the scheduling of the computations in Kötter's interpolation. Assume that the maximum x-degree of the polynomials is dx_p in iteration p. The computations of the partial discrepancy coefficients $u_{a,s}^{(l)}(\alpha_i)$ for $0 \leq l \leq t$

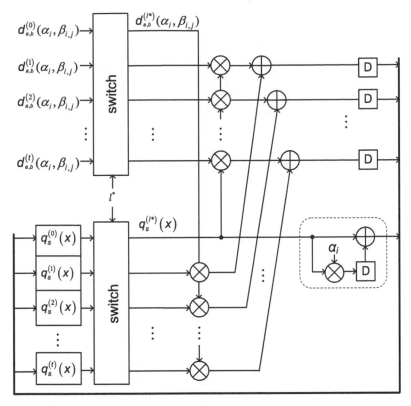

FIGURE 5.12
Polynomial updating architecture for interpolation

and $0 \leq s \leq t$ are carried out using $(t+1)^2$ copies of the architecture in Fig. 5.11. They are completed in $dx_p + 1$ clock cycles. The maximum multiplicity, and hence the maximum y-degree of the polynomials, are usually low in ASD decoders with practical hardware complexity. In this case, it is more efficient to multiply $u_{a,s}^{(l)}(\alpha_i)$ with proper powers of $\beta_{i,j}$ and add up the products according to (5.14) using a multiplier-adder tree to complete the second layer of the discrepancy coefficients. Accordingly, the second layer would take much fewer clock cycles. The control logic decides l^* from the discrepancy coefficients and weighted degrees. Then $t+1$ copies of the architecture in Fig. 5.12 are employed to update the $t+1$ set of polynomials $q_0^{(l)}(x), q_1^{(l)}(x), \ldots, q_t^{(l)}(x)$ $(0 \leq l \leq t)$ in parallel. Once a polynomial coefficient is updated, it is consumed in the discrepancy coefficient computation of the next iteration right away. The polynomial updating of iteration p and discrepancy coefficient computation for iteration $p+1$ are overlapped. Therefore, the number of clock cycles needed

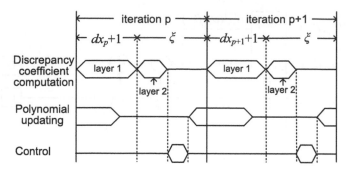

FIGURE 5.13
Scheduling of the computations in Kötter's interpolation

for interpolation iteration p is $dx_p + 1 + \xi$, where ξ equals the sum of pipelining latencies caused by the computation units and control logic.

5.4.2.4 Discrepancy coefficient properties and interpolation complexity reduction

The discrepancy coefficient computation is an expensive component in the interpolation. Nevertheless, the discrepancy coefficients have quite a few properties, and they can be exploited to reduce the complexity of the interpolation [70].

Following the polynomial updating in the interpolation iteration with constraint (a, b) in Algorithm 4, it can be derived that

$$\begin{cases} Q^{(l)}(x + \alpha_i, y + \beta_{i,j}) \Leftarrow d_{a,b}^{(l^*)}(\alpha_i, \beta_{i,j})Q^{(l)}(x + \alpha_i, y + \beta_{i,j}) \\ \qquad\qquad + d_{a,b}^{(l)}(\alpha_i, \beta_{i,j})Q^{(l^*)}(x + \alpha_i, y + \beta_{i,j}) \\ Q^{(l^*)}(x + \alpha_i, y + \beta_{i,j}) \Leftarrow Q^{(l^*)}(x + \alpha_i, y + \beta_{i,j})x \end{cases}$$

As a result, the coefficients of $x^r y^s$ in $Q^{(l)}(x + \alpha_i, y + \beta_{i,j})$ $(0 \le l \le t)$ are updated as

$$\begin{cases} d_{r,s}^{(l)}(\alpha_i, \beta_{i,j}) \Leftarrow d_{a,b}^{(l^*)}(\alpha_i, \beta_{i,j})d_{r,s}^{(l)}(\alpha_i, \beta_{i,j}) + d_{a,b}^{(l)}(\alpha_i, \beta_{i,j})d_{r,s}^{(l^*)}(\alpha_i, \beta_{i,j}) \\ d_{r,s}^{(l^*)}(\alpha_i, \beta_{i,j}) \Leftarrow d_{r-1,s}^{(l^*)}(\alpha_i, \beta_{i,j}) \end{cases}$$

$$(5.15)$$

If the interpolation has been carried out over the constraint (r, s) in a previous iteration, then all $d_{r,s}^{(l)}(\alpha_i, \beta_{i,j})$ for $0 \le l \le t$ are zero at the beginning of iteration (a, b). Hence, $d_{r,s}^{(l)}(\alpha_i, \beta_{i,j})$ for $l \ne l^*$ are still zero after the polynomial updating according to (5.15). In addition, if the interpolation iteration for the constraint $(r-1, s)$ is always done before that for (r, s), such as in Algorithm 4, then $d_{r,s}^{(l^*)}(\alpha_i, \beta_{i,j})$ is also zero after the polynomial updating. In this case, the polynomial updating does not break the constraints that have been covered in previous iterations. Let $r = a$ and $s = b$. $d_{a,b}^{(l)}(\alpha_i, \beta_{i,j})$ for $l \neq l^*$ become zero after the polynomial updating according to (5.15). $d_{a,b}^{(l^*)}(\alpha_i, \beta_{i,j})$ would also become zero if $d_{a-1,b}^{(l^*)}(\alpha_i, \beta_{i,j})$ is zero. Hence, as long as the iteration for $(a - 1, b)$ has been carried out before, $d_{a,b}^{(l)}(\alpha_i, \beta_{i,j})$ for $0 \leq l \leq t$ are all forced to zero in the iteration for (a, b). Accordingly, the interpolation constraints can be satisfied in an order different from that in Algorithm 4. For example, the interpolation constraints can be also satisfied in the order of $(a, b) = (0, 0), (1, 0), (2, 0), \ldots, (m - 1, 0), (0, 1), (1, 1), (2, 1), \ldots$

From (5.15), the discrepancy coefficients for each interpolation iteration can be also derived by updating initial values. This is analogous to the updating of the discrepancy polynomial in the reformulated inversionless Berlekamp-Massey (riBM) algorithm discussed in Chapter 4. At the beginning of the interpolation for a point $(\alpha_i, \beta_{i,j})$ with multiplicity $m_{i,j}$, all initial values for discrepancy coefficients $d_{a,b}^{(l)}(\alpha_i, \beta_{i,j})$ with $0 \leq a + b < m_{i,j}$ and $0 \leq l \leq t$ are computed. Then the initial values are updated according to (5.15) along with the polynomials in each iteration to derive the real discrepancy coefficients. This is referred to as the point-serial interpolation [71]. To keep track of the discrepancy coefficient updating, discrepancy polynomials $D^{(l)}(x, y)$ are used. The coefficient of $x^r y^s$ in $D^{(l)}(x, y)$ is the coefficient of the same monomial in $Q^{(l)}(x + \alpha_i, y + \beta_{i,j})$. Unlike $Q^{(l)}(x, y)$, which is carried over the interpolation of all points, $D^{(l)}(x, y)$ is re-initialized at the beginning of the interpolation for each point. Hence $D^{(l)}(x, y)$ is much shorter than $Q^{(l)}(x, y)$. The number of monomials in $D^{(l)}(x, y)$ for an interpolation point with multiplicity m is $m(m + 1)/2$, and only those coefficients that will be used in later interpolation iterations need to be updated. However, the computations of the initial discrepancy coefficients cause extra overhead.

In the following, $Q_{(a,b)}^{(l)}(x, y)$ denote the interpolation polynomials at the beginning of the interpolation iteration for constraint (a, b), and $d_{(a,b)r,s}^{(l)}(\alpha_i, \beta_{i,j})$ is the coefficient of $x^r y^s$ in $Q_{(a,b)}^{(l)}(x + \alpha_i, y + \beta_{i,j})$. Employing these notations, the discrepancy coefficients $d_{a,b}^{(l)}(\alpha_i, \beta_{i,j})$ used in Algorithm 4 can be re-written as $d_{(a,b)a,b}^{(l)}(\alpha_i, \beta_{i,j})$, and those on the left-hand side of the equations in (5.15) are actually $d_{(a,b)r,s}^{(l)}(\alpha_i, \beta_{i,j})$. In the remainder, by discrepancy coefficients, it means $d_{a,b}^{(l)}(\alpha_i, \beta_{i,j})$ or $d_{(a,b)r,s}^{(l)}(\alpha_i, \beta_{i,j})$ with $r = a$ and $s = b$. Since α_i and $\beta_{i,j}$ remain unchanged in all interpolation iterations over this point, they are dropped from the notations when no ambiguity occurs.

Lemma 5 *Assume that the interpolation over a point (α, β) with multiplicity m is carried out for the constraints in the order of $a = 0$ to $m - 1$, $b = 0$ to $m - 1 - a$, and $l^* = w$ in the iteration for $(a = a_0, b = b_0)$. Then $d_{a,b}^{(w)} = 0$ for $(a = a_0, b_0 < b \leq m - 1 - a_0)$ and $(a = a_0 + 1, 0 \leq b < b_0)$ [70].*

Proof In the iteration for the constraint (a, b), the polynomial updating forces the coefficients of $x^a y^b$ in $Q^{(l)}(a + \alpha, y + \beta)$ $(0 \leq l \leq t)$ to zero, while the coefficients that have been forced to zero in previous iterations remain zero. If the interpolation is carried out in the order of $a = 0$ to $m - 1$, $b = 0$ to $m - 1 - a$, then before the iteration for (a_0, b_0), the constraints of $(0 \leq a < a_0, 0 \leq b \leq m - 1 - a_0)$ and $(a = a_0, 0 \leq b < b_0)$ have already been covered. Therefore, at the beginning of the interpolation iteration for the constraint (a_0, b_0), $d^{(w)}_{(a_0,b_0)r,s} = 0$ for $(0 \leq r < a_0, 0 \leq s \leq m - 1 - a_0)$ and $(r = a_0, 0 \leq s < b_0)$. The constraint to be covered in the next iteration is $(a_0, b_0 + 1)$ if $b_0 < m - 1 - a_0$ or $(a_0 + 1, 0)$ if $b_0 = m - 1 - a_0$. To simplify the notations, assume that the next iteration is for the constraint $(a_0, b_0 + 1)$. The following proof applies similarly to the case of $(a_0 + 1, 0)$. Since $l^* = w$ in the iteration for (a_0, b_0), $Q^{(w)}_{(a_0,b_0+1)}(x + \alpha, y + \beta) = Q^{(w)}_{(a_0,b_0)}(x + \alpha, y + \beta) \times x$, and hence $d^{(w)}_{(a_0,b_0+1)r,s} = d^{(w)}_{(a_0,b_0)r-1,s}$. Therefore $d^{(w)}_{(a_0,b_0+1)r,s} = 0$ for $(0 \leq r \leq a_0, 0 \leq s \leq m - 1 - a_0)$ and $(r = a_0 + 1, 0 \leq s < b_0)$. As a result, $d^{(w)}_{a_0,b_0+1} = d^{(w)}_{(a_0,b_0+1)a_0,b_0+1} = 0$, and $Q^{(w)}_{(a_0,b_0+1)}(x,y)$ is not selected as the minimal polynomial in the iteration for $(a_0, b_0 + 1)$. If the iteration after $(a_0, b_0 + 1)$ is $(a_0, b_0 + 2)$, then

$$Q^{(w)}_{(a_0,b_0+2)}(x,y) = d^{(l^*)}_{(a_0,b_0+1)} Q^{(w)}_{(a_0,b_0+1)}(x,y) + d^{(w)}_{(a_0,b_0+1)} Q^{(l^*)}_{(a_0,b_0+1)}(x,y).$$

Since $d^{(w)}_{(a_0,b_0+1)} = 0$, the second term on the right-hand side of the above equation is zero. Accordingly

$$Q^{(w)}_{(a_0,b_0+2)}(x + \alpha, y + \beta) = d^{(l^*)}_{(a_0,b_0+1)}(\alpha, \beta) Q^{(w)}_{(a_0,b_0+1)}(x + \alpha, y + \beta),$$

and

$$d^{(w)}_{(a_0,b_0+2)r,s} = d^{(l^*)}_{(a_0,b_0+1)} d^{(w)}_{(a_0,b_0+1)r,s}. \tag{5.16}$$

As mentioned previously, $d^{(w)}_{(a_0,b_0+1)r,s} = 0$ for $(0 \leq r \leq a_0, 0 \leq s \leq m-1-a_0)$ and $(r = a_0 + 1, 0 \leq s < b_0)$ at the beginning of the iteration $(a_0, b_0 + 1)$. Therefore, $d^{(w)}_{(a_0,b_0+2)r,s} = 0$ for $(0 \leq r \leq a_0, 0 \leq s \leq m - 1 - a_0)$ and $(r = a_0 + 1, 0 \leq s < b_0)$. From this, $d^{(w)}_{a_0,b_0+2} = d^{(w)}_{(a_0,b_0+2)a_0,b_0+2} = 0$.

The discrepancy coefficients of $Q^{(w)}(x, y)$ in the iterations for $(a = a_0, b_0 + 2 < b \leq m - 1 - a_0)$ and $(a = a_0 + 1, 0 \leq b < b_0)$ are computed iteratively by an equation similar to (5.16). Since the selection of $Q^{(w)}(x, y)$ as the minimal polynomial in the iteration for (a_0, b_0) leads to $d^{(w)}_{(a_0,b_0+1)r,s} = 0$ for $(0 \leq r \leq a_0, 0 \leq s \leq m-1-a_0)$ and $(r = a_0+1, 0 \leq s < b_0)$, the discrepancy coefficient in each iteration of $(a = a_0, b_0+2 < b \leq m-1-a_0)$ and $(a = a_0+1, 0 \leq b < b_0)$ equals to a nonzero value multiplied by zero, which is still zero. $\square$

From Lemma 5, if a candidate polynomial is selected as the minimal polynomial in the interpolation iteration for constraint (a_0, b_0), then its discrepancy coefficients in the following $m-1-a_0$ iterations are zero. Fig. 5.14 shows an example case of the discrepancy coefficients during the interpolation iterations over a point with $m = 3$. For high-rate codes, $m + 1 = 4$ polynomials

FIGURE 5.14
Example of discrepancy coefficients for $m = 3$

are involved in the interpolation. In Fig. 5.14, a cross and a solid dot represent a nonzero coefficient and a coefficient that can be either zero or nonzero, respectively. If there is a square around a cross, it means that the corresponding polynomial is selected as the minimal polynomial in that iteration. Let $l^* = 0$ in the iteration for $(a = 0, b = 0)$. Then the discrepancy coefficients for $Q^{(0)}(x, y)$ in the following $m - 1 - a = 3 - 1 - 0 = 2$ iterations are zero. These are the iterations for the constraints $(a = 0, b = 1)$ and $(a = 0, b = 2)$. Assume that $l^* = 3$ in the iteration for $(a = 1, b = 1)$. Then the discrepancy coefficients for $Q^{(3)}(x, y)$ in the next $m - 1 - a = 1$ iteration, which is the iteration for $(a = 2, b = 0)$, is zero. The minimum polynomial for each iteration can be different from those shown in Fig. 5.14. Nevertheless, for a point with multiplicity m, the total number of zero discrepancy coefficients resulting from multiplying $(x + \alpha_i)$ to the minimal polynomials is $\sum_{i=2}^{m} i(i - 1)$.

Lemma 6 *Assume that the interpolation over a point (α, β) with multiplicity m is carried out for the constraints in the order of $a = 0$ to $m - 1$, $b = 0$ to $m - 1 - a$, and $l^* = w$ in the iteration for $(a = a_0, b = b_0)$. Then in the case of $b_0 \neq m - 1 - a_0$, $d_{a_0+1,b_0}^{(w)} = d_{a_0,b_0}^{(w)} \times \prod_{\substack{a=a_0,b_0 < b \leq m-1-a_0 \\ a=a_0+1, 0 \leq b < b_0}} d_{a,b}^{(l^*)}$, where $d_{a,b}^{(l^*)}$ is the discrepancy coefficient of the minimal polynomial during the iteration for (a, b). If $b_0 = m - 1 - a_0$, then $d_{a_0+2,0}^{(w)} = d_{(a_0,b_0)a_0+1,0}^{(w)} \times \prod_{\substack{a=a_0,b_0 < b \leq m-1-a_0 \\ a=a_0+1, 0 \leq b < b_0}} d_{a,b}^{(l^*)}$ [70].*

Proof From Lemma 5, $d_{a,b}^{(w)} = 0$ for $(a = a_0, b_0 < b \leq m - 1 - a_0)$ and $(a = a_0 + 1, 0 \leq b < b_0)$. Hence, in each of the iterations for $(a = a_0, b_0 < b \leq m - 1 - a_0)$ and $(a = a_0 + 1, 0 \leq b < b_0)$, $l^* \neq w$ and the updating of $Q^{(w)}(x, y)$ is carried out as

$$Q^{(w)}(x, y) \Leftarrow d_{a,b}^{(l^*)} Q^{(w)}(x, y).$$

Moreover, since $Q_{(a_0,b_0)}^{(w)}(x, y)$ was selected as the minimal polynomial in the iteration of (a_0, b_0), it was updated by multiplying $(x + \alpha)$ in that iteration.

Consequently, at the end of the iteration for (a_0+1, b_0-1), $Q^{(w)}(x, y)$ equals

$$Q^{(w)}_{(a_0, b_0)}(x, y)(x+\alpha) \times \prod_{\substack{a=a_0, b_0 < b \leq m-1-a \\ a=a_0+1, 0 \leq b < b_0}} d^{(l^*)}_{a,b}. \tag{5.17}$$

The first iteration afterwards is for (a_0+1, b_0) if $b_0 \neq m-1-a_0$, or $(a_0+2, 0)$ if $b_0 = m-1-a_0$. The coefficient of $x^r y^s$ in (5.17) shifted by the point (α, β) is $d^{(w)}_{(a_0, b_0)r-1, s} \times \prod_{\substack{a=a_0, b_0 < b \leq m-1-a_0 \\ a=a_0+1, 0 \leq b < b_0}} d^{(l^*)}_{a,b}$, and the results follow. $\square$

In any interpolation iteration, the polynomial that is selected as the minimal polynomial, $Q^{(l^*)}(x, y)$, should have nonzero discrepancy coefficient. Hence, it can be derived from Lemma 6 that if $b_0 \neq m-1-a_0$, the minimal polynomial selected in the iteration of (a_0, b_0) has nonzero discrepancy coefficient in the $(m-a_0)^{th}$ iteration after. On the other hand, if $b_0 = m-1-a_0$, then whether the discrepancy coefficient in the $(m-a_0)^{th}$ iteration after is zero or nonzero is dependant on the value of $d^{(w)}_{(a_0, b_0)a_0+1, 0}$. This has been reflected in Fig. 5.14. In the iteration of $(a_0 = 0, b_0 = 0)$, $l^* = 0$. Hence, $d^{(0)}_{1,0} = d^{(0)}_{(0,0)} \times \prod_{a=0, b=1, 2} d^{(l^*)}_{a,b} \neq 0$ and it is denoted by a cross. $l^* = 1$ in the iteration of $(a_0 = 0, b_0 = 2)$ and $b_0 = m-1-a_0$ in this case. $d^{(1)}_{2,0} = d^{(1)}_{(0,2)1,0} \times \prod_{a=1, b=0, 1} d^{(l^*)}_{a,b}$, and $d^{(1)}_{(0,2)1,0}$ can be either zero or nonzero. Hence, $d^{(1)}_{(2,0)}$ can be either zero or nonzero, and is denoted by a solid dot in Fig. 5.14.

Lemma 7 *Scaling the interpolation polynomials by nonzero factors does not affect the decoding output, as long as the corresponding discrepancy coefficients are scaled by the same factors in the polynomial updating [70].*

Proof Consider the interpolation iteration for the constraint (a_0, b_0). Assume that $Q^{(l^*)}(x, y)$ and $d^{(l^*)}_{a_0, b_0}$ are both scaled by $s_{l^*} \neq 0$. Moreover, $Q^{(l)}(x, y)(l \neq l^*)$ and $d^{(l)}_{a_0, b_0}$ are scaled by $s_l \neq 0$. According to Algorithm 4, these two polynomials are updated as

$$\begin{cases} Q^{(l)}(x, y) \Leftarrow (s_{l^*} d^{(l^*)}_{a_0, b_0})(s_l Q^{(l)}(x, y)) + (s_l d^{(l)}_{a_0, b_0})(s_{l^*} Q^{(l^*)}(x, y)) \\ Q^{(l^*)}(x, y) \Leftarrow s_{l^*} Q^{(l^*)}(x, y)(x+\alpha_i) \end{cases}$$

Hence, scaling the polynomials and their discrepancy coefficients is equivalent to multiplying nonzero constants to the updated polynomials. Scaling an updated polynomial by a nonzero value does not force zero discrepancy coefficients to nonzero or vise versa. Moreover, the weighted degree of the updated polynomial is not changed. Hence, scaling the updated polynomial does not affect the selection of the minimal polynomial in later iterations either. Although the final interpolation output polynomial may be different by a nonzero scaler, its $y + f(x)$ factors and accordingly the decoding results remain the same. $\square$

The above three properties of the discrepancy coefficients can be exploited to reduce the complexity of the interpolation. According to Lemma 5, during the interpolation over a point with multiplicity m, if a polynomial is selected as the minimal polynomial in the iteration of constraint (a_0, b_0), then the corresponding discrepancy coefficients in the following $m - 1 - a_0$ iterations are zero. Therefore, the larger the multiplicity of a point, the more zero discrepancy coefficients in the corresponding interpolation iterations. Generally, ASD algorithms try to assign those more reliable interpolation points with higher multiplicities. When the signal-to-noise ratio (SNR) is higher, there are more points with higher multiplicities. It has been shown in [70] that, when the KV algorithm with $m_{max} = 4$ is adopted to decode a $(15, 7)$ and a $(255, 239)$ RS code, more than 50% of the discrepancy coefficients are zero when the FER is lower than 10^{-2}. This ratio further increases as FER decreases.

If a polynomial has zero discrepancy coefficient in an interpolation iteration, then it is updated as multiplying itself by a nonzero scaler according to Algorithm 4. From Lemma 7, the scaler multiplication can be skipped. Moreover, Lemma 5 says that for a polynomial $Q^{(w)}(x, y)$ that is selected as the minimal polynomial in the iteration of constraint (a_0, b_0), the discrepancy coefficients in the following $m - 1 - a_0$ iterations are zero. As a result, no computation needs to be done on $Q^{(w)}(x, y)$ in those iterations. In an architecture that carries out the computations over all polynomials in parallel, this means that the number of units for the discrepancy coefficient computation and polynomial updating can be reduced. Nevertheless, this saving may be offset by the complex control logic and switching network required to route the polynomials to hardware engines. Serial architectures benefit more from the zero discrepancy coefficients. In a serial architecture that processes one polynomial at a time, skipping the computations over the polynomials with zero discrepancy coefficients significantly reduces the latency and energy consumption of the interpolation process.

5.4.2.5 Slow-downed interpolation architecture

The partial discrepancy coefficient computation architecture in Fig. 5.11 has a feedback loop that can not be pipelined without changing the function. The critical path of this architecture consists of one finite field multiplier and one adder. The coefficients of each polynomial are updated serially using the architecture in Fig. 5.12. Different from the discrepancy coefficient computation using Horner's rule, the polynomial updating is not iterative. Pipelining the polynomial updating architecture in Fig. 5.12 by ξ stages only adds ξ extra clock cycles in each iteration. This latency overhead is small since the x-degree of the polynomials can become quite large, especially when $n - k$ and/or the interpolation multiplicities are not small. Therefore, the minimum achievable clock period and the throughput of the overall interpolation architecture are limited by the feedback loops in the discrepancy coefficient computation architecture.

To address the speed bottleneck caused by these feedback loops, it was proposed in [64] to use power representations of finite field elements. In power representation, finite field multiplications are implemented as additions of the exponents. Moreover, carry-save adders are adopted so that only one 1-bit full adder is in the data path of each feedback loop. This architecture can achieve high clock frequency. However, additions of finite field elements are also required in the interpolation. To carry out finite field additions, the power representation needs to be converted to a basis representation using look-up tables (LUTs). Although it was proposed in [64] to divide the bits representing the exponent into two groups and make use of two smaller LUTs, the conversions still lead to a large silicon area overhead.

Alternatively, the clock frequency of the interpolation architecture can be increased by applying the slow-down and re-timing techniques discussed in Chapter 2. In an N-slow architecture, each register is replaced by N registers. If retiming is applied to evenly distribute the N registers in the feedback loops so that the data path is cut into N segments of equal length, then the maximum achievable clock frequency of the architecture is increased by almost N times. To maintain the same function and maximize the hardware utilization efficiency, N independent input sequences need to be interleaved. Apparently, different interpolation polynomials can be interleaved since their discrepancy coefficient computations are independent. The feedback loop shown in Fig. 5.11 computes the partial discrepancy coefficient, $u_{a,s}^{(l)}(\alpha_i)$. The computations for partial coefficients with different s are also independent. Therefore, $q_s^{(l)}(x)$ $(s = 0, 1, 2, \ldots)$ with different s from the same interpolation polynomial $Q^{(l)}(x, y) = q_0^{(l)}(x) + q_1^{(l)}(x)y + q_2^{(l)}(x)y^2 + \ldots$ can be also interleaved as the input sequences to a slow-downed version of the architecture in Fig. 5.11.

Next, an example is given for applying 2-slowdown to the interpolation architecture with $m_{max} = 3$. For high-rate codes, there are $t + 1 = m_{max} + 1 = 4$ polynomials involved in the interpolation. Fig. 5.15 shows the original architecture for the discrepancy coefficient computation when $t + 1 = 4$. Since the maximum y-degree is only three, the partial discrepancy coefficients, $u_{a,s}^{(l)}(\alpha_i)$, are multiplied by proper powers of $\beta_{i,j}$ according to the second equation in (5.14), and the products are added up using an adder tree. The powers of $\beta_{i,j}$ are computed by a multiplier not included in Fig. 5.15. Since pipelining can be always applied to cut the data paths of a non-recursive architecture, the minimum achievable clock period of the architecture in Fig. 5.15 is lower bounded by the multiplier-adder path in the feedback loops. Assume that the maximum x-degree of the polynomials $q_s^{(l)}(x)$ $(s = 0, 1, 2, 3)$ is dx. Then the partial discrepancy coefficients are available after $dx + 1$ clock cycles, and they are multiplied with powers of $\beta_{i,j}$ in the next clock cycle.

Fig. 5.16 shows the discrepancy coefficient computation architecture after applying 2-slowdown. Each register in the feedback loops of the original architecture in Fig. 5.15 is replaced by two registers. Finite field multipliers

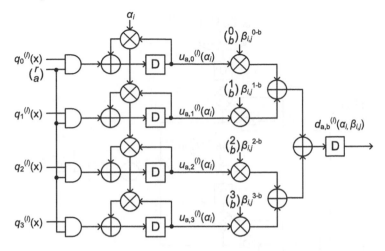

FIGURE 5.15
Discrepancy coefficient computation architecture with four polynomials (modified from [65])

are implemented by combinational logic, and registers can be moved into the data paths inside the multipliers. To minimize the achievable clock period, retiming is applied to move the two registers in each feedback loop so that the multiplier-adder path is cut into two equal segments. For the purpose of clarity, this retiming is not shown in Fig. 5.16. Two independent input sequences need to be interleaved in the 2-slow architecture in order to avoid wasting computations on zero inputs. In Fig. 5.16, the coefficients from two $q_s^{(l)}(x)$ with different s are interleaved. Accordingly, two partial discrepancy coefficients are computed simultaneously by sharing one copy of the multiplier-adder feedback loop, and they are available after $2dx+1$ and $2dx+2$ clock cycles, respectively. Although these computations take twice the number of clock cycles, the clock period is reduced to almost half as a result of the retiming. Hence, two partial discrepancy coefficients are computed by one feedback loop in about the same amount of time as that required for computing one partial discrepancy coefficient in the original architecture. In the 2-slow architecture, the multipliers in the feedback loops are also reused to multiply $u_{a,s}^{(l)}(\alpha_i)$ with $\binom{s}{b}\beta_{ij}^{s-b}$. These multiplications are only needed once in each interpolation iteration. With the penalty of two multiplexors and one extra clock cycle for each interpolation iteration, two multipliers have been saved by sharing the multipliers in the feedback loops. In addition, to accommodate the sharing, the signal c is set to $\binom{r}{a}$ in the first $2dx+2$ clock cycles, and is changed to zero in the following clock cycle. Since $u_{a,0}^{(l)}(\alpha_i)$ and $u_{a,1}^{(l)}(\alpha_i)$, and also $u_{a,2}^{(l)}(\alpha_i)$ and $u_{a,3}^{(l)}(\alpha_i)$, are

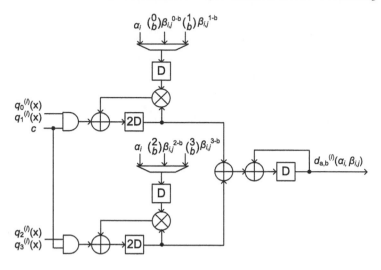

FIGURE 5.16
2-slow discrepancy coefficient computation architecture with four interleaved polynomials (modified from [65])

generated in two adjacent clock cycles, the last adder feedback loop is needed to add up $u_{a,s}^{(l)}(\alpha_i)\binom{s}{b}\beta_{ij}^{s-b}$ to derive $d_{a,b}^{(l)}(\alpha_i,\beta_j)$.

The polynomial updating architecture needs to be modified to suit the slow-downed and interleaved discrepancy coefficient computation. As can be seen from Kötter's interpolation algorithm, there is no data dependency among the coefficients with different y-degree during the polynomial updating. Hence, the coefficients of $q_s^{(l)}(x)$ with different s are interleaved. Fig. 5.17 shows an architecture that updates $q_0^{(l)}(x)$ and $q_1^{(l)}(x)$ in an interleaved manner. Another copy of this architecture is needed to update $q_2^{(l)}(x)$ and $q_3^{(l)}(x)$. Similar to the original polynomial updating architecture in Fig. 5.12, the discrepancy coefficients and interpolation polynomials are first switched according to l^*, so that $d_{a,b}^{(l^*)}(\alpha_i,\beta_{i,j})$ and $q_s^{(l^*)}(x)$ become the top outputs of the switches. Then $q_s^{(l)}(x)$ for $l \neq l^*$ are updated as linear combinations, and $q_s^{(l^*)}(x)$ is multiplied by $(x + \alpha_i)$. Since the coefficients of two polynomials are interleaved, two registers are needed in the dashed block to implement the multiplication by $(x + \alpha_i)$. Also two registers are added to the output of each linear combination to synchronize all updated polynomials. Similarly, the updated polynomial coefficients are written back to fixed memory blocks to avoid using another set of switches, and the respective weighted degrees of the polynomials are tracked in a control unit. Retiming is also applied to move the registers so that the critical path of this polynomial updating architecture becomes half of a multiplier and an adder. Compared to the original polynomial

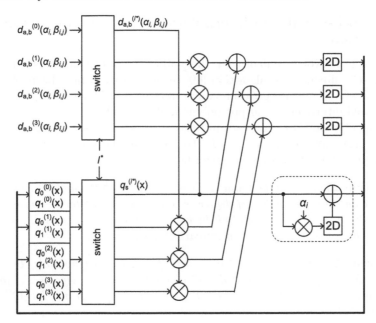

FIGURE 5.17
2-slow polynomial updating architecture for interpolation (modified from [65])

updating architecture in Fig. 5.12, the 2-slow architecture only requires some extra registers. However, it completes the updating of twice the coefficients in about the same amount of time. The outputs from the interleaved 2-slow polynomial updating architecture are sent to the interleaved 2-slow discrepancy coefficient computation architecture in Fig. 5.16 right after, so that these two steps of the interpolation can still be overlapped.

To avoid wasting computations in slow-downed interpolation architectures, the coefficients that can be interleaved are not limited to those of different y-degrees in the same bivariate polynomial. Coefficients with the same y-degree from different polynomials can be also interleaved. In addition, to enable a higher level of slow down, the coefficients that belong to the same bivariate polynomial and have the same y-degree can be decomposed into shorter sequences and interleaved at the cost of slight modifications in the interpolation architecture. Higher clock frequency is enabled by adopting a larger slow-down factor N. Also, high hardware utilization efficiency is achieved if there are enough polynomial coefficient sequences to be interleaved. However, as N increases, the setup and hold time of registers become non-negligible and the number of registers also increases. Accordingly, the achievable clock frequency is lower than N times of that in the original architecture, and the hardware efficiency gain becomes less significant.

5.4.3   Lee-O'Sullivan interpolation algorithm and simplifications

Different from Kötter's algorithm, the goal of the interpolation is achieved in the Lee-O'Sullivan (LO) algorithm [69] through converting a basis of polynomials satisfying all interpolation constraints to a Gröbner basis, from which the desired interpolation output polynomial is found.

The polynomials with a certain maximum y-degree satisfying all interpolation constraints form an ideal $\mathcal{I}$. To facilitate the discussions on the LO algorithm, a few definitions are given below.

Definition 27 *Among the monomials of $Q(x, y)$ with the highest weighted degree, the one with the highest y-degree is called the leading term, and is denoted by $LT(Q(x,y))$. The coefficient of the leading term is denoted by $LC(Q(x,y))$. Similar definitions also apply to univariate polynomials.*

Definition 28 *A basis formed by a set of nonzero polynomials, $Q^{(0)}(x,y)$, $Q^{(1)}(x,y), \ldots, Q^{(t)}(x,y)$ is called a Gröbner basis of the ideal $\mathcal{I}_t$, if for each polynomial $F(x,y)$ in $\mathcal{I}_t$, there exists $l \in \{0, 1, 2, ..., t\}$ such that $LT(Q^{(l)}(x,y))$ divides $LT(F(x,y))$.*

A monomial $x^r y^s$ is said to divide $x^{r'} y^{s'}$ only when $s = s'$ and $r \leq r'$ [72]. Assume that the maximum y-degree of the polynomials in $\mathcal{I}_t$ is t. Then the y-degree of the leading terms of the $t+1$ polynomials in a Gröbner basis of $\mathcal{I}_t$ are distinct, and they are $0, 1, \ldots, t$. For each polynomial in an ideal, there always exists a polynomial in the Gröbner basis that has the same or lower weighted degree. Hence, the polynomial with the minimum weighted degree in the ideal, which is the polynomial that needs to be computed by the interpolation, is the polynomial in the Gröbner basis with the minimum weighted degree. Note that the polynomials of minimum weighted degree can be different by a nonzero scaler, and the scaler does not affect the overall decoding output. The Gröbner bases of modules of polynomials are defined similarly and have similar properties.

Like Kötter's algorithm, a set of $t+1$ bivariate polynomials are involved in the LO algorithm for high-rate codes when the maximum multiplicity of the interpolation points is t. The LO algorithm consists of two steps: initial basis construction and basis conversion. In the first step, a basis of polynomials, whose maximum y-degree is t, and satisfying all interpolation constraints, is constructed. Then the second step converts this initial basis to a Gröbner basis.

Let α_i be the ith element for evaluation map encoding. The following polynomials are used in the LO algorithm to simplify the notations.

$$\tilde{h}_j(x) = \prod_{i=0, i \neq j}^{n-1} (x + \alpha_i)$$

$$h_j(x) = \tilde{h}_j(\alpha_j)^{-1} \tilde{h}_j(x). \tag{5.18}$$

Assume that the multiplicity of the interpolation point $(\alpha_i, \beta_{i,j})$ is $m_{i,j}$ $(0 \le i < n, 0 \le j < 2^q)$. Also use L to denote the set of all i such that $m_i \ge 1$. The initial basis $\{G^{(0)}(x,y), G^{(1)}(x,y), \cdots, G^{(t)}(x,y)\}$ is constructed using the following procedure in the LO algorithm [69].

Algorithm 5 Initial Basis Construction in the Lee-O'Sullivan Algorithm

input: $(\alpha_i, \beta_{i,j})$ *with multiplicity* $m_{i,j}$
initialization: $m_i = \max_{0 \le j < 2^q}\{m_{i,j}\}$; $p_i = \arg\max_j\{m_{i,j}|0 \le j < 2^q\}$
begin:
$\quad G^{(0)}(x,y) = \prod_{i \in L}(x + \alpha_i)^{m_i}$
$\quad$ *for l=1 to t*
$\quad\quad h^{(l-1)}(x) = \sum_{i \in L} \beta_{i,p_i} h_i(x)$
$\quad\quad$ *for* $i \in L$
$\quad\quad\quad m_{i,p_i} = m_{i,p_i} - 1$
$\quad\quad$ *update* m_i, p_i *for* $i \in L$, *then update* L
$\quad\quad G^{(l)}(x,y) = \prod_{0 \le j < l}(y + h^{(j)}(x)) \prod_{i \in L}(x + \alpha_i)^{m_i}$

At the beginning of the initial basis construction, L consists of all code positions that have non-trivial interpolation points. Since $G^{(0)}(x + \alpha_i, y + \beta_{i,j}) = \prod_{i \in L} x^{m_i}$, it is a polynomial passing each of the point $(\alpha_i, \beta_{i,j})$ with multiplicity m_i. Apparently, it satisfies all interpolation constraints. It actually satisfies more constraints than needed for those points in the code position of α_i, but with multiplicity less than m_i. $h^{(l-1)}(x)$ is the Lagrange polynomial passing each of the point $(\alpha_i, \beta_{i,p_i})$ with $i \in L$. As l increases from 1 to t in Algorithm 5, m_i is decreased by one each time until it becomes zero. Hence, the $\prod_{i \in L}(x + \alpha_i)^{m_i}$ part of $G^{(l)}(x,y)$ in the last line of Algorithm 5 passes each point with lower multiplicity as l increases. This multiplicity decrease is compensated for by the Lagrange polynomials. Accordingly, each of the $G^{(l)}(x,y)$ for $l = 0, 1, \ldots, t$ computed by Algorithm 5 satisfies all the interpolation constraints. They have different y-degrees, and hence form a basis of polynomials with maximum y-degree t satisfying all the interpolation constraints.

Assume that $G^{(l)}(x,y)$ is written as $g_0^{(l)}(x) + g_1^{(l)}(x)y + g_2^{(l)}(x)y^2 + \ldots$. The pseudo codes in Algorithm 6 [69] convert the initial polynomials constructed by Algorithm 5 to a Gröbner basis. From Algorithm 5, the y-degree of the leading term in each constructed initial polynomial is zero when $(1, k-1)$ weighted degree is adopted. The basis conversion process in Algorithm 6 works on one polynomial at a time starting from $G^{(1)}(x,y)$. For each polynomial $G^{(r)}(x,y)$, the leading term is iteratively canceled out by steps 4 and 5, until a term whose y-degree is r becomes the leading term. Using the initial basis constructed by Algorithm 5, the variable s in Algorithm 6 is always less than r before s becomes equal to r. Multiplying a polynomial with a positive power of x does not decrease the multiplicities of the points it passes. Also linearly combining two polynomials satisfying the same constraints does not change the constraints either. Hence, after steps 4 and 5 in Algorithm 6 are carried out, each polynomial still passes the same interpolation points with the same multiplicities. In addition, these steps do not change the y-degrees of the leading terms in polynomials except $G^{(r)}(x, y)$. As a result, after the conversion is completed for $r = 1, 2, 3, \ldots t$, the $t + 1$ polynomials have distinct y-degrees in the leading terms, and form a Gröbner basis.

Algorithm 6 Basis Conversion in the Lee-O'Sullivan Algorithm
input: $G^{(l)}(x, y)$
initialization: $r = 0$

1: $r = r + 1;$
  proceed if $r \leq t$; Otherwise, go to step 5
2: find $s = deg_y(LT(G^{(r)}(x, y)))$
  if $s == r$, go to step 1
3: $d = deg(g_s^{(r)}(x)) - deg(g_s^{(s)}(x))$
  $c = LC(g_s^{(r)}(x))/LC(g_s^{(s)}(x))$
  if $d \geq 0$
4:   $G^{(r)}(x, y) = G^{(r)}(x, y) + cx^d G^{(s)}(x, y)$
  else
5:   $tmp(x, y) = G^{(s)}(x, y);$
   $G^{(s)}(x, y) = G^{(r)}(x, y);$
   $G^{(r)}(x, y) = x^{-d}G^{(r)}(x, y) + c \cdot tmp(x, y)$
  goto step 2
Output: $G^{(l)}(x, y)$ *with the lowest weighted degree*

It can be observed from Algorithm 5 and 6 that the computations involved in the LO algorithm are mainly coefficient multiplications, polynomial multiplications and polynomial additions. The LO algorithm does not require complicated discrepancy coefficient computations as in the Kötter's algorithm, and hence may potentially lead to simpler hardware implementation when t is small. However, simplification schemes need to be developed for the LO algorithm to reduce its complexity similar to what the re-encoding and coordinate transformation have done for Kötter's algorithm.

It has been shown in Section 5.4.1 that, when the re-encoding and coordinate transformation techniques are employed, Kötter's algorithm still starts with the same initial polynomials. However, it only needs to be carried out over the points in the $n - k$ least reliable code positions in $\bar{R}$. Kötter's algorithm adds points to interpolation curves iteratively. However, the LO algorithm starts with a set of polynomials passing all points. Hence, the LO algorithm can not take advantage of the re-encoding in the same way. Nevertheless, by exploiting the zero $\beta_{i,j}$ coordinates of the interpolation points after the re-encoding, simplification schemes have been developed in [73] to reduce the complexity of the LO algorithm by a similar scale. Moreover, the factorization step also only needs to compute 2τ coefficients from the LO interpolation output after applying those simplifications schemes.

As a result of re-encoding, $\beta_{i,p_i} = 0$ for $i \in R$. Therefore, the computation of $h^{(l-1)}(x)$ in Algorithm 5 is simplified as

$$h^{(l-1)}(x) = \sum_{i \in L, i \notin R} \beta_{i,p_i} h_i(x). \tag{5.19}$$

Split the second term of $G^{(l)}(x, y)$ in Algorithm 5 into two parts: one for the code positions in R and one for those in $\bar{R}$.

$$G^{(l)}(x, y) = \prod_{0 \leq j < l} (y + h^{(j)}(x)) \prod_{i \in L, i \in R} (x + \alpha_i)^{m_i} \prod_{i \in L, i \notin R} (x + \alpha_i)^{m_i}.$$

As it was mentioned before, the k positions in R are the most reliable. Forcing each of the positions in R to have only one interpolation point with the maximum multiplicity, m_{max}, does not lead to noticeable error-correcting performance loss. For $i \in R$, m_i is equal to m_{max} in $G^{(0)}(x, y)$, and decreases by one in each iteration of Algorithm 5 until it becomes zero. Hence, R is always a subset of L in the iterations of $l = 1, \ldots, m_{max} - 1$, and $G^{(l)}(x, y)$ from these iterations can be rewritten as

$$G^{(l)}(x, y) = \prod_{i \in R} (x + \alpha_i)^{m_{max}} \prod_{0 \leq j < l} \frac{y + h^{(j)}(x)}{\prod_{i \in R}(x + \alpha_i)} \prod_{i \in L, i \notin R} (x + \alpha_i)^{m_i}. \tag{5.20}$$

The maximum multiplicity, m_{max}, equals t when the code rate is high, and may be less than t when the code rate is low. In Algorithm 5, L becomes empty at the end of iteration $l = m_{max}$ and the term $\prod_{i \in L, i \notin R} (x + \alpha_i)^{m_i}$ in (5.20) can be eliminated accordingly. Starting from iteration $l = m_{max} + 1$, $h^{(l-1)}(x)$ is zero, and hence $G^{(l)}(x, y)$ is computed as $yG^{(l-1)}(x, y)$. As a result, for $m_{max} \leq l \leq t$, $G^{(l)}(x, y)$ can be rewritten as

$$G^{(l)}(x, y) = \prod_{i \in R} (x + \alpha_i)^{m_{max} + (l - m_{max})} \left(\frac{y}{\prod_{i \in R}(x + \alpha_i)} \right)^{l - m_{max}}$$

$$\times \prod_{0 \leq j < m_{max}} \frac{y + h^{(j)}(x)}{\prod_{i \in R}(x + \alpha_i)}. \tag{5.21}$$

Taking out a common polynomial in x from each bivariate polynomial in a basis for the LO interpolation does not affect the relative weighted degrees nor the $y + f(x)$ factors of the polynomials. It can be seen from (5.20) and (5.21) that all initial basis polynomials have a common factor $\prod_{i \in R}(x + \alpha_i)^{m_{max}}$. It can be eliminated from each $Q^{(l)}(x, y)$ ($0 \le l \le t$) without affecting the decoding results.

From (5.18), if $j \notin R$, then $h_j(x)$ has a factor $\prod_{i \in R}(x + \alpha_i)$. Hence the $h^{(l-1)}(x)$ computed by (5.19) has this factor too. As a result, $\prod_{i \in R}(x + \alpha_i)$ can be also taken out from $h^{(j)}(x)$ in (5.20) and (5.21). Let

$$\hat{h}^{(j)}(x) = h^{(j)}(x) / \prod_{i \in R}(x + \alpha_i),$$

and use the coordinate transformation $y = z \prod_{i \in R}(x + \alpha_i)$. The initial basis polynomials for the LO interpolation are transformed to

$$G^{(l)}(x, z) = \prod_{0 \le j < l}(z + \hat{h}^{(j)}(x)) \prod_{i \in L, i \notin R}(x + \alpha_i)^{m_i}$$

for $0 \le l < m_{max}$, and

$$G^{(l)}(x, z) = z^{l - m_{max}} \prod_{i \in R}(x + \alpha_i)^{l - m_{max}} \prod_{0 \le j < m_{max}}(z + \hat{h}^{(j)}(x)).$$

for $m_{max} \le l \le t$. Accordingly, the initial basis construction for the LO algorithm can be simplified as in Algorithm 7. $G^{(l)}(x, y)$ instead of $G^{(l)}(x, z)$ are used in this algorithm to simplify the notations.

Algorithm 7 Simplified Initial Basis Construction in the Lee-O'Sullivan Algorithm

input: $(\alpha_i, \beta_{i,j})$ *with multiplicity* $m_{i,j}$
initialization: $m_i = \max_{0 \le j < 2^q}\{m_{i,j}\}$; $p_i = \arg\max_j\{m_{i,j} | 0 \le j < 2^q\}$
begin:
    $G^{(0)}(x, y) = \prod_{i \in L, i \notin R}(x + \alpha_i)^{m_i}$
    for l=1 to t
       $\hat{h}^{(l-1)}(x) = \sum_{i \in L} \beta_{i,p_i} h_i(x) / \prod_{i \in R}(x + \alpha_i)$
       for $i \in L$
          $m_{i,p_i} = m_{i,p_i} - 1$
       update m_i, p_i *for* $i \in L$, *then update L*
       if $l \le m_{max}$
          $G^{(l)}(x, y) = \prod_{0 \le j < l}(y + \hat{h}^{(j)}(x)) \prod_{i \in L, i \notin R}(x + \alpha_i)^{m_i}$
       else
          $G^{(l)}(x, y) = y^{l - m_{max}} \prod_{i \in R}(x + \alpha_i)^{l - m_{max}}$
             $\times \prod_{0 \le j < m_{max}}(y + \hat{h}^{(j)}(x))$

Compared to the initial polynomials constructed by Algorithm 5, the initial polynomials computed by Algorithm 7 have a much lower x-degree. A lower degree means less memory for storing the polynomials. Also the number of iterations needed to convert each polynomial in Algorithm 6 is largely decided by its initial x-degree. Using the modified initial polynomials constructed by Algorithm 7 not only leads to significant memory reduction but also substantial speedup in the LO algorithm. Similarly, due to the transformation $y = z \prod_{i \in R}(x + \alpha_i)$, $(1, -1)$ instead of $(1, k-1)$ weighted degree should be adopted to decide the leading terms in the basis conversion process when Algorithm 7 is used to construct the initial basis. The factorization can be still carried out directly on the output of the LO interpolation starting from the modified initial basis. Since the coordinate transformation employed here for the LO algorithm is the same as that for the Kötter's interpolation algorithm, the 2τ coefficients computed from the factorization can be also used as the syndromes to correct τ errors in the code positions in R.

5.4.4 Architectures for Lee-O'Sullivan interpolation

Although the computations involved in the LO algorithm looks simpler than those in the Kötter's algorithm, they are more irregular and the multiplications of polynomials are hardware-consuming when their degree is high. Moreover, the number of iterations needed in the LO algorithm increases significantly with t. Hence the implementation architectures to be presented next for the LO algorithm focus on the case of $m_{max} = 2$ for high-rate codes. There architectures have been introduced in [74]. When the multiplicities of the interpolation points are higher and there are more polynomials involved, the LO algorithm becomes less efficient. One example of ASD algorithm that can achieve good error-correcting performance with $m_{max} = 2$ is the BGMD algorithm [52].

In the BGMD decoding with $m_{max} = 2$ for high-rate codes, the maximum y-degree of the interpolation polynomials is $t = 2$, and there are $t + 1 = 3$ polynomials involved. From the BGMD multiplicity assignment, there are three types of code positions: i) if none of the bit is erased in the received symbol of code position i, then the interpolation point for code position i is (α_i, β_i) with multiplicity $m_{max} = 2$. Here β_i is the hard-decision symbol; ii) if only one bit is erased, then there are two interpolation points, (α_i, β_i) and (α_i, β_i'), for code position i. Each of them has multiplicity $m_{max}/2 = 1$ and β_i' is the second most likely symbol for code position i; iii) if more than one bit is erased, then there is no interpolation point for code position i. Let A, B, and C be the sets of the indexes for these three types of code positions in $\bar{R}$. Apparently $A \cup B \cup C = \bar{R}$.

5.4.4.1   Initial basis construction architecture

According to Algorithm 7, the simplified initial basis polynomials for the LO algorithm are constructed as

$$G^{(0)}(x,y) = \prod_{i \in A}(x + \alpha_i)^2 \prod_{i \in B}(x + \alpha_i)$$

$$G^{(1)}(x,y) = (y + \hat{h}^{(0)}(x)) \prod_{i \in A \cup B}(x + \alpha_i) \qquad (5.22)$$

$$G^{(2)}(x,y) = (y + \hat{h}^{(0)}(x))(y + \hat{h}^{(1)}(x)),$$

Also

$$\hat{h}_j(x) = \frac{\prod_{i=0, i \notin R, i \neq j}^{n-1}(x + \alpha_i)}{\prod_{i=0, i \neq j}^{n-1}(\alpha_j + \alpha_i)}, \qquad (5.23)$$

and

$$\hat{h}^{(0)}(x) = \sum_{i \in A} \beta_i \hat{h}_i(x) + \sum_{i \in B} \beta_i \hat{h}_i(x)$$

$$\hat{h}^{(1)}(x) = \sum_{i \in A} \beta_i \hat{h}_i(x) + \sum_{i \in B} \beta'_i \hat{h}_i(x). \qquad (5.24)$$

For an arbitrary set J, $\prod_{i \in J}(x + \alpha_i)$ can be computed by an architecture similar to that in Fig. 5.5. The number of registers in this architecture should be $|J| + 1$. At the beginning, the two right-most registers are initialized as the first α_i and 1 respectively, and all the other registers are initialized as zero. The multiplication of a $(x + \alpha_i)$ term takes $|J|$ clock cycles, during which α_i is fed to the multiplier. After $|J|(|J|-1)$ clock cycles, the coefficients of $\prod_{i \in J}(x+\alpha_i)$ are located at the $|J| + 1$ registers, and the least significant coefficient is in the right most register. If higher speed is required, the unfolding technique introduced in Chapter 2 can be applied.

The denominator of $\hat{h}_j(x)$ in (5.23) has $n - 1$ finite field elements to multiply. Fortunately, the product of all nonzero finite field elements is the identity. Hence $1/\prod_{i=0, i \neq j}^{n-1}(\alpha_j + \alpha_i)$ equals α_j, and the computation of $\hat{h}_j(x)$ is simplified accordingly. The numerator of $\hat{h}_j(x)$ can be also derived by an architecture similar to that in Fig. 5.5. Nevertheless, $n - k$ different $\hat{h}_j(x)$ need to be calculated in the worst case to derive $\hat{h}^{(0)}(x)$ and $\hat{h}^{(1)}(x)$ in (5.24). This requires either very long latency if a serial architecture is used or large area if a parallel architecture is employed. Alternatively, the numerator can be rewritten as

$$\prod_{i=0, i \notin R, i \neq j}^{n-1}(x + \alpha_i) = \frac{\prod_{i=0, i \notin R}^{n-1}(x + \alpha_i)}{(x + \alpha_j)}.$$

Therefore, a single product, $\prod_{i=0, i \notin R}^{n-1}(x + \alpha_i)$, which equals to $\prod_{i \in \bar{R}}(x + \alpha_i)$ is derived first. Then it is divided by each different $(x + \alpha_j)$.

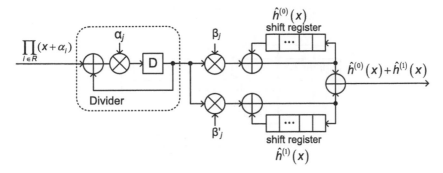

FIGURE 5.18

Architecture for computing $\hat{h}^{(0)}(x)$ and $\hat{h}^{(1)}(x)$ (modified from [74])

An architecture for computing $\hat{h}^{(0)}(x)$ and $\hat{h}^{(1)}(x)$ is illustrated in Fig. 5.18. In this figure, the division by $(x + \alpha_j)$ is done by the part in the dashed block on the left. The coefficients of $\prod_{i \in \bar{R}}(x + \alpha_i)$ are input serially starting with the most significant one. The output of the adder in this block is multiplied by α_j before it is passed to the register. Hence, the coefficients of $\alpha_j \prod_{i \in \bar{R}, i \neq j}(x + \alpha_i)$, which is $\hat{h}_j(x)$, is generated at the output of the register serially without using any extra unit for multiplying α_j. Since $\deg(\prod_{i \in \bar{R}}(x + \alpha_i)) = n - k$, it takes $n - k$ clock cycles to compute each $\hat{h}_j(x)$. For $j \in A$, $\hat{h}_j(x)$ only needs to be multiplied by β_j. However, $\hat{h}_j(x)$ for $j \in B$ is multiplied by β_j and β'_j to compute $\hat{h}^{(0)}(x)$ and $\hat{h}^{(1)}(x)$, respectively, according to (5.24). When a product is computed, it is added up with the sum of previously calculated products shifted through the registers. Since there are at most $n - k$ different $\hat{h}_j(x)$, the architecture in Fig. 5.18 takes at most $(n - k)^2$ clock cycles to compute $\hat{h}^{(0)}(x)$ and $\hat{h}^{(1)}(x)$. During the last $n - k$ clock cycles, $\hat{h}^{(0)}(x) + \hat{h}^{(1)}(x)$ needed for $G^{(2)}(x, y)$ construction is calculated by using the last adder in Fig. 5.18. If higher throughput is required, duplicated copies of the divider in Fig. 5.18 can be employed to compute multiple $\hat{h}_j(x)$ simultaneously. All of them are scaled by proper β_j and β'_j in parallel and the products are summed up by using adder trees. However, the shift registers record the outputs of the adder trees and do not need to be duplicated.

The product of $\hat{h}^{(0)}(x)$ and $\hat{h}^{(1)}(x)$ needs to be derived for $G^{(2)}(x, y)$ according to (5.22). Polynomial multiplications can be implemented by an architecture similar to that shown in Fig. 5.6. The coefficients of $\hat{h}^{(0)}(x)$ are connected to the multipliers, and those of $\hat{h}^{(1)}(x)$ are serially shifted in. Since $\deg(\hat{h}^{(0)}(x)) = n - k - 1$, $n - k$ taps are needed in the polynomial multiplication architecture. Because that $\deg(\hat{h}^{(1)}(x)) = n - k - 1$, one coefficient is available at the output of the architecture in each of the first $n - k$ clock cycles, starting with the most significant one. After $n - k$ clock cycles, the rest of the prod-

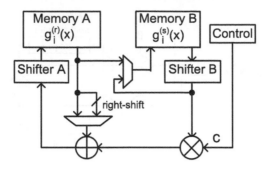

FIGURE 5.19
Top level architecture for basis conversion (modified from [74])

uct coefficients are located in the registers and shifted out. Similarly, folding and unfolding can be applied to the polynomial multiplication architecture to achieve a speed-complexity tradeoff.

5.4.4.2   Basis conversion architecture

The top-level architecture for the basis conversion is illustrated in Fig. 5.19. The coefficients of the interpolation polynomials are stored in two dual-port memories. In iteration r of Algorithm 6, $G^{(r)}(x,y)$ is stored in memory A and all $G^{(l)}(x,y)$ with $l < r$ are stored in memory B. Using the initial basis constructed by Algorithm 7, which is listed in (5.22), s is always less than r in the basis conversion process. Let $G^{(l)}(x,y) = g_0^{(l)}(x) + g_1^{(l)}(x)y + g_2^{(l)}(x)y^2 + \cdots$. The degree of $g_i^{(l)}(x)$ $(0 \le l, i \le 2)$ is at most $2(n-k)$. The coefficients of $g_i^{(l)}(x)$ that are updated in parallel are stored in the same memory line. Speed-area tradeoff is achieved by adjusting the number of coefficients updated simultaneously. In addition, the leading coefficient of $g_i^{(l)}(x)$ is stored as the first symbol in the first memory line for $g_i^{(l)}(x)$, and all coefficients of $g_i^{(l)}(x)$ are stored in consecutive memory cells. In this way, the leading coefficients can be accessed without search. Moreover, no shifting is required to make the coefficients of the same x-degree aligned when adding up $cx^d g_s^{(s)}(x)$ and $g_s^{(r)}(x)$ or $cg_s^{(s)}(x)$ and $x^{-d}g_s^{(r)}(x)$ during steps 4 and 5 of Algorithm 6. The leading coefficient and weighted degree of each $g_i^{(l)}(x)$ are also stored in the control unit. Besides generating control signals for the memory access, multiplexors and shifters in Fig. 5.19, the control unit computes $c = LC(g_s^{(r)}(x))/LC(g_s^{(s)}(x))$. To reduce the latency, the finite field inversion is implemented by a LUT. The adder and multiplier in Fig. 5.19 are actually vectors of adders and multipliers capable of updating the coefficients in one memory line at a time.

During iteration r of Algorithm 6, $g_s^{(r)}(x)$ is updated before $g_i^{(r)}(x)$ with $i \ne s$. When $r = 1$, the only possible value for s is 0. When $r = 2$, s can be

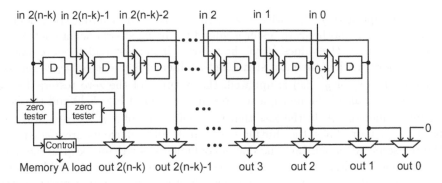

FIGURE 5.20
Architecture for Shifter A (modified from [74])

either 0 or 1. As mentioned in the previous paragraph, the leading coefficient of each $g_i^{(l)}(x)$ is stored as the first symbol in the corresponding memory line. Hence no further alignment is needed on $g_s^{(r)}(x)$ and $g_s^{(s)}(x)$ before step 4 or 5 of Algorithm 6 is carried out on them. Therefore, the coefficients of $g_s^{(s)}(x)$ stored in memory B are passed through Shifter B intact. After the updating in step 4 or 5, the leading term of $g_s^{(r)}(x)$ is canceled. From simulations, it was found that several other following terms may be canceled out as well. All these canceled terms are shifted out by Shifter A before the coefficients of the updated $g_s^{(r)}(x)$ are written back to the memory. When updating $g_i^{(r)}(x)$ with $i \neq s$, it is possible that the coefficients of $g_i^{(r)}(x)$ and $g_i^{(s)}(x)$ stored in the memory are not aligned for the computations in step 4 and 5. This issue is addressed by Shifter B. Although steps 4 and 5 are not designed to cancel out the leading term in the updated $g_i^{(r)}(x)$ ($i \neq s$), exceptions may happen. The accidental zero terms are also eliminated by Shifter A. In Algorithm 6, $G^{(s)}(x,y)$ is replaced by $G^{(r)}(x,y)$ when $d < 0$. This is taken care of by the multiplexor at the input of memory B.

The architecture of Shifter A is shown in Fig. 5.20. In the initial basis polynomials (5.22), the maximum degree of $g_i^{(l)}(x)$ ($0 \leq l, i \leq 2$) is $2(n-k)$. Following the computations in Algorithm 6, the maximum x-degree does not go beyond $2(n-k)$ during the basis conversion process. Hence, shifter A has $2(n-k)+1$ registers. The registers are connected through multiplexors so that all coefficients can be shifted by one position in each clock cycle. The updated coefficients of both $g_s^{(r)}(x)$ and $g_i^{(r)}(x)$ with $i \neq s$ are loaded into these registers to shift out the most significant zeros before they are written to memory A. The leading coefficient of the updated $g_s^{(r)}(x)$ is guaranteed to be zero. Hence the zero tester is connected to the second coefficient to save one clock cycle of shifting. The shifting is repeated until this coefficient becomes nonzero. At that time, the second and later coefficients are loaded into memory

A. In the updated $g_i^{(r)}(x)$ with $i \neq s$, most likely the leading coefficient is nonzero. Hence another zero tester is added to the most significant coefficient. Correspondingly, the values at the first through the last registers are loaded into memory A when the most significant coefficient is nonzero. Depending on whether $g_s^{(r)}(x)$ or $g_i^{(r)}(x)$ is updated, the outputs of the zero testers decide the select signals of the multiplexors on the bottom of Fig. 5.20 and the load signal for memory A. In the case that the coefficients of a $g_i^{(l)}(x)$ are stored in multiple memory lines, the shifting starts after all coefficients are loaded into the registers. After shifting, one line of coefficients is written back to memory A at a time by adding extra multiplexors to the bottom of Fig. 5.20. Moreover, after the coefficients are sent to memory A, the corresponding registers can be occupied by the coefficients of the next updated $g_i^{(l)}(x)$ for shifting.

To update $g_i^{(r)}(x)$ with $i \neq s$, the coefficients of $g_i^{(r)}(x)$ or $g_i^{(s)}(x)$ may need to be shifted to take into account their degree difference and the x^d or x^{-d} needs to be multiplied according to steps 4 and 5 of Algorithm 6. Simulations showed that most of the time, the coefficients of $g_i^{(r)}(x)$ require no shifting. Occasionally, they need to be shifted one position in the direction of the least significant coefficient. Hence, the right-shifted version of $g_i^{(r)}(x)$ is also provided as shown in Fig. 5.19. Shifter B takes care of the rest shifting needed for aligning the coefficients of $g_i^{(r)}(x)$ and $g_i^{(s)}(x)$. Simulations also showed that in most of the cases, $g_i^{(s)}(x)$ needs to be shifted by either one position to the left (in the direction of the most significant coefficient), one position to the right, or two positions to the right, in each iteration of the basis conversion. To reduce the shifting latency, shifter B is capable of making any of these shifts in one clock cycle. It can be implemented by a row of multiplexors. In the case that more shifting is needed, the outputs of shifter B are written back to memory B, and more positions are shifted in the next clock cycle by using shifter B again.

The number of iterations needed in the basis conversion process varies with the number of errors. An interesting observation is that for each r, the polynomial updating usually cancels out only the leading coefficient each time at the beginning. However, in the last iteration before the y-degree of the leading term in $G^{(r)}(x, y)$ becomes r, many coefficients are canceled out. The fewer the errors, the more coefficients are canceled out in the last iteration, and hence the fewer iterations needed for the overall basis conversion.

To reduce the hardware complexity, the number of multiplier and adder units adopted is less than the maximum x-degree of the involved polynomials. In this case, it takes multiple clock cycles to update each $g_i^{(l)}(x)$, and the number of clock cycles required is determined by the number of coefficients and hence the degree of each polynomial. To reduce the latency, $G'^{(1)}(x, y) = y + \hat{h}^{(0)}(x)$ and $G'^{(0)}(x, y) = \prod_{i \in A}(x + \alpha_i)$ can be used instead as the starting polynomials for the basis conversion. After $G'^{(1)}(x, y)$ is converted using $G'^{(0)}(x, y)$, the common factor $\prod_{i \in A \cup B}(x + \alpha_i)$ is multiplied back. This would lead to the same result as using $G^{(1)}(x,y)$ and $G^{(0)}(x,y)$ in (5.22) as the starting polynomials. However, by taking out the common factor $\prod_{i \in A \cup B}(x + \alpha_i)$, the maximum x-degree is reduced from $2(n-k)$ as in $G^{(1)}(x,y)$ and $G^{(0)}(x,y)$ to $(n-k)$ as in $G'^{(1)}(x,y)$ and $G'^{(0)}(x,y)$. Hence the number of clock cycles needed in each iteration for converting $G'^{(1)}(x,y)$ is reduced to around half compared to that for converting $G^{(1)}(x,y)$. The overall throughput of the LO interpolator can be improved by pipelining the basis construction and conversion steps. To increase the hardware utilization efficiency, unfolding or folding is applied to each component to adjust the latency so that these two steps take about the same number of clock cycles.

5.4.5 Kötter's and Lee-O'Sullivan interpolation comparisons

The initial polynomials $1, y, y^2, \ldots$ for the Kötter's algorithm form a Gröbner basis of polynomials. However, they do not pass the given interpolation constraints. The Kötter's algorithm iteratively updates these polynomials to satisfy one more interpolation constraint at a time. At the end of each iteration, these polynomials form a Gröbner basis of polynomials satisfying all constraints that have been covered. After the iteration is repeated for each interpolation constraint, a Gröbner basis of polynomials passing each interpolation point with its associated multiplicity is found. Interpolation points can be added and the multiplicities can be increased by carrying out more iterations of the Kötter's algorithm. Hence, it is considered as a 'forward' interpolation algorithm. The LO algorithm follows a different scheme. It starts with a polynomial basis that already satisfies all interpolation constraints. Then the initial basis is converted to a Gröbner basis through an iterative process. Hence, the points and their multiplicities can not be changed once the interpolation starts in the LO algorithm.

The computations involved in each iteration of the basis conversion step of the LO algorithm are simpler than those in Kötter's algorithm. However, the number of iterations needed in the LO basis conversion increases much faster with m_{max} and hence t. In addition, the basis construction of the LO algorithm involves polynomial multiplications and the computations of Lagrange polynomials, whose complexity becomes very high for large $n - k$. Therefore, the hardware efficiency of the LO algorithm is lower than that of the Kötter's algorithm except for high-rate codes with low interpolation multiplicity.

5.5 Factorization algorithm and architectures

Assume that the interpolation output polynomial is $Q(x,y)$, the goal of the factorization is to find all factors of $Q(x,y)$ in the form of $y + f(x)$ with $\deg(f(x)) < k$. When re-encoding is adopted, the decoding is carried out on

$\bar{r} = r + \phi = (c + e) + \phi = (c + \phi) + e = \bar{c} + e$. In this case, the interpolation output polynomial would have factors $y + \bar{f}(x)$, where $\bar{f}(x)$ is the message polynomial corresponding to $\bar{c}$. In other words, $\bar{f}(\alpha_i) = \bar{c}_i$ for $0 \le i < n$. Since $\bar{r}_i = 0$ for $i \in R$,

$$e_i = \bar{r}_i + \bar{c}_i = \bar{f}(\alpha_i)$$

in these code positions. If the *ith* code position has no error, then $e_i = 0$, and accordingly $\bar{f}(\alpha_i) = 0$. Therefore, $\bar{f}(x)$ has a factor $(x + \alpha_i)$ if the code position $i \in R$ has no error. Accordingly, $\bar{f}(x)$ is rewritten as

$$\bar{f}(x) = \Big(\prod_{i \in R, e_i = 0} (x + \alpha_i) \Big) \rho(x),$$

where $\rho(x)$ is a polynomial that does not have any root α_i for $i \in R$ and $e_i \ne 0$. $\prod_{i \in R}(x + \alpha_i)$ has been taken out by the coordinate transformation. If the factorization is carried out on the re-encoded and coordinate transformed interpolation output, it would yield factors $y + \gamma(x)$, where

$$\gamma(x) = \frac{\bar{f}(x)}{\prod_{i \in R}(x + \alpha_i)} = \frac{\rho(x)}{\prod_{i \in R, e_i \ne 0}(x + \alpha_i)}. \tag{5.25}$$

These analyses have been shown in [56, 57]. From (5.25), the coefficients of $\gamma(x)$ can be used as the syndromes in the BMA to find those $i \in R$ such that $e_i \ne 0$ and compute the corresponding e_i. To correct τ errors, 2τ syndromes are needed. Since it is less likely that the positions in R are erroneous, τ can be set to a very small number. Hence, the factorization step needs to compute much fewer coefficients if it is applied directly to the re-encoded and coordinate transformed interpolation output.

Regardless of whether the original or the re-encoded and transformed interpolation output polynomial is sent to the factorization, the same computations are carried out to find the factors. Among available factorization algorithms, the Roth-Ruckenstein Algorithm [75] leads to more efficient hardware implementations. Assume that the factorization input polynomial is $Q(x, y)$, and its coefficients are elements of $GF(2^q)$. The Roth-Ruckenstein Algorithm is described in Algorithm 8.

Assume that $\check{Q}(x, y)$ has a factor $y + \gamma(x)$, and hence it is written as $(y + \gamma(x))A(x, y)$, where $A(x, y)$ does not have any factor x^v with $v > 0$. Let $\gamma(x) = \gamma_0 + \gamma_1 x + \gamma_2 x^2 + \dots$. Then

$$\check{Q}(0, y) = (y + \gamma_0)A(0, y).$$

Since $A(0, y) \ne 0$, γ_0 can be found by computing the roots of $\check{Q}(0, y)$. It can be also derived that

$$\begin{aligned}
\hat{Q}(x, y) = \check{Q}(x, xy + \gamma_0) &= (xy + \gamma_0 + \gamma(x))A(x, xy + \gamma_0) \\
&= (xy + \gamma_1 x + \gamma_2 x^2 + \dots)A(x, xy + \gamma_0) \\
&= x(y + \gamma_1 + \gamma_2 x + \dots)A(x, xy + \gamma_0)
\end{aligned}$$

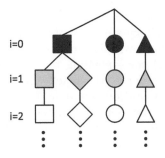

FIGURE 5.21
An example of the tree formed by the roots from the factorization.

Therefore, after x^v has been taken out, the $\check{Q}(0, y)$ in the next iteration level would have γ_1 as a root. Accordingly, all coefficients of $\gamma(x)$ can be found by iteratively repeating this process as shown in Algorithm 8. If the factorization is applied to the original interpolation output, the maximum iteration level in Algorithm 8 needs to be changed from 2τ to k.

Algorithm 8 Roth-Ruckenstein Factorization Algorithm
input: $Q(x, y)$
$\quad\quad 2\tau$: *number of coefficients to be computed for each factor*
initialization: *iteration level $i = 0$*
$\quad$ *Reconstruct*$(Q(x, y), i)$
$\quad$ {
$\quad\quad\quad$ *find the largest non-negative integer v, such that x^v divides $Q(x, y)$*
1:$\quad\quad$ $\check{Q}(x, y) = Q(x, y)/x^v$
2:$\quad\quad$ *find all the roots of $\check{Q}(0, y)$ in $GF(2^q)$*
$\quad\quad\quad\quad$ *for each root θ of $\check{Q}(0, y)$ do*
$\quad\quad\quad\quad\quad\quad$ $\gamma_i = \theta$
$\quad\quad\quad\quad\quad\quad$ *if $i = 2\tau - 1$*
$\quad\quad\quad\quad\quad\quad\quad\quad$ **output** $[\gamma_0, \gamma_1, \cdots, \gamma_{2\tau-1}]$
$\quad\quad\quad\quad\quad\quad$ *else*
3:$\quad\quad\quad\quad\quad\quad\quad\quad$ $\tilde{Q}(x, y) = \check{Q}(x, y + \theta)$
4:$\quad\quad\quad\quad\quad\quad\quad\quad$ $\hat{Q}(x, y) = \tilde{Q}(x, xy)$
$\quad\quad\quad\quad\quad\quad\quad\quad$ *call Reconstruct*$(\hat{Q}(x, y), i + 1)$

$\quad$ }

The roots computed from Algorithm 8 form a tree structure. Fig. 5.21 shows an example. Assume that in iteration level $i = 0$, $\check{Q}(0, y)$ has three roots, and they are denoted by the black square, circle and triangle in the first row of Fig. 5.21. When iteration level $i = 1$ is carried out for the first root denoted by the black square, the corresponding $\check{Q}(0, y)$ has two roots. Hence the black square has two children in iteration level $i = 1$. On the other hand, each of the $\check{Q}(0, y)$ corresponding to the other two roots in iteration level $i = 0$ has one root. Hence, the black circle and triangle only has one child each. After the computations are done for all iteration levels, the roots in each possible path from the root to a leaf of the tree form an output vector of the factorization. For example, if the tree in Fig. 5.21 has no more bifurcation, then there are four output vectors.

The length of the factorization input polynomial has been greatly reduced as a result of the re-encoding and coordinate transformation. In addition, only the polynomial coefficients that will be involved in the computations of later factorization iteration levels need to be updated. Hence, the number of coefficients to be calculated decreases as the iteration level increases. On the other hand, the root computation over finite field in step 2 of Algorithm 8 is usually done by exhaustive search. Its complexity does not change with the iteration level and is not affected by the total number of iteration levels. It is the most time-consuming step, especially when the RS code is constructed over high-order finite fields. To simplify the root computation, prediction-based schemes have been proposed in [7, 76] through exploiting the properties of the polynomials involved in the factorization. With more than 99% probability, the prediction is successful, in which case the roots are computed by simple multiplications. With lower hardware complexity, these designs can achieve much higher average throughput than the factorization architecture in [31] that directly computes the roots for polynomials with degree at most four. Additionally, the prediction-based architectures are not subject to the limitation that t should be at most four. Moreover, a partial-parallel factorization architecture has been developed in [77] to combine the computations from adjacent factorization iteration levels to achieve further speedup. Next, the prediction-based and partial-parallel factorization architectures are described.

5.5.1   Prediction-based factorization architecture

The root computation over finite fields needed for step 2 of the factorization algorithm is conventionally done by the exhaustive Chien search. If the polynomial is over $GF(2^q)$, a serial Chien search architecture takes $2^q - 1$ clock cycles to find all the nonzero roots in $GF(2^q)$. When q is not small, the serial Chien has very long latency. On the other hand, highly parallel Chien search is hardware-demanding. By exploiting the properties of the polynomials involved in the factorization algorithm, the root computation can be greatly simplified.

The equations involved in steps 1, 3, and 4 of Algorithm 8 are rewritten below for the purpose of clarity.

$$\check{Q}(x, y) = Q(x, y)/x^v,$$

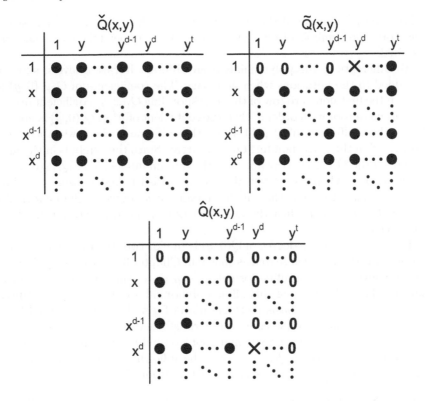

FIGURE 5.22
Transformations of polynomial coefficients in the factorization algorithm (modified from [7])

where v is the largest non-negative integer such that x^v divides $Q(x, y)$. Assume that θ is a root of $\check{Q}(x, y)$.

$$\tilde{Q}(x, y) = \check{Q}(x, y + \theta).$$

$$\hat{Q}(x, y) = \tilde{Q}(x, xy). \tag{5.26}$$

The coefficients of a bivariate polynomial can be represented by symbols in a two dimensional array as shown in Fig. 5.22. In this figure, a solid dot denotes a coefficient that can be either zero or nonzero, and a cross represents a nonzero coefficient. The symbols in the same row represent the coefficients of the monomials of the same x-degree, and those in the same column correspond to the coefficients with the same y-degree. Accordingly, the symbols in the first row of the $\check{Q}(x, y)$ array represent the coefficients of $\check{Q}(0, y)$. Assume that $\check{Q}(0, y)$ has a root θ with order d. Then $\check{Q}(0, y)$ has a factor $(y + \theta)^d$, and hence $\check{Q}(0, y + \theta)$ contains a factor y^d. As a result, the coefficients of the corresponding $\tilde{Q}(x, y)$ satisfy $\tilde{q}_{0,s} = 0$ ($0 \leq s < d$) and $\tilde{q}_{0,d} \neq 0$ as illustrated in Fig. 5.22. Moreover, the computations in (5.26), which is step 4 of Algorithm

8 shifts the jth column of the symbol array for $\tilde{Q}(x,y)$ down by j positions. Therefore, the symbols of $\hat{Q}(x,y)$ are zero in the upper triangle as shown in Fig. 5.22.

To differentiate the polynomials from different iteration levels, subscript i is added to the notations when necessary. The coefficients of $\check{Q}_{i+1}(0,y)$ are those in the first non-zero row of the corresponding $\hat{Q}_i(x,y)$ coefficient array. If $\check{Q}_i(0,y)$ has a root with order d, then the coefficient of y^d in $\tilde{Q}_i(0,y)$ is nonzero. Hence, the coefficients of $\check{Q}_{i+1}(0,y)$ are those in one of the rows labeled by x through x^d in the corresponding $\hat{Q}_i(x,y)$ array. Since the symbols in the upper triangle of the $\hat{Q}_i(x,y)$ array are zero, the maximum degree of $\check{Q}_{i+1}(0,y)$ is d, which happens when the x^d row of the $\hat{Q}_i(x,y)$ is the first nonzero row. From these, it is also apparent that, if the maximum y-degree of the factorization input polynomial is t, then the degree of $\check{Q}(0,y)$ in any iteration level does not exceed t.

In the case that the degree of $\check{Q}(0,y)$ is one or two, the roots can be easily computed without Chien search as shown in Chapter 3. A degree-2 polynomial can be converted to an affine format $y^2 + y = u$ using coordinate transformations. Then the roots of this affine polynomial are derived by multiplying the standard or normal basis representation of u with a pre-computed binary matrix, after which reverse coordinate transformations are carried out. The architecture for degree-2 polynomial root computation is illustrated in Fig. 3.2. When the degree of $\check{Q}(0,y)$ is one, the following lemma further helps to simplify the root computations in the factorization [7].

Lemma 8 *If $\check{Q}(0,y)$ has a simple root θ, then in all subsequent iteration levels, the degree of $\check{Q}(0,y)$ corresponding to θ is always one, and the coefficient of y in these $\check{Q}(0,y)$ does not change.*

Proof: Let θ be a simple root of $\check{Q}_i(0,y)$. Then $\check{Q}_i(0,y)$ can be written as:

$$\check{Q}_i(0,y) = a(y)(y+\theta), \tag{5.27}$$

where $a(y)$ is a non-zero polynomial that does not have θ as a root. Since

$$\tilde{Q}_i(0,y) = \check{Q}_i(0,y+\theta) = a(y+\theta)y,$$

the constant coefficient of $\tilde{Q}_i(0,y)$ is zero. The constant coefficient of $a(y+\theta)$ equals $a(\theta)$. Because θ is not a root of $a(y)$, the constant coefficient of $a(y+\theta)$ and hence the coefficient of y in $\tilde{Q}_i(0,y)$ is nonzero. Therefore, in the $\hat{Q}_i(x,y)$ coefficient array, the symbols in the first row are all zero, and the coefficient of xy, which equals the coefficient of y in $\tilde{Q}_i(0,y)$, is nonzero. As a result, the symbols in the second row of the $\hat{Q}_i(x,y)$ array are the coefficients of $\check{Q}_{i+1}(0,y)$. Let b be the coefficient of x in $\hat{Q}_i(x,y)$. Then $\check{Q}_{i+1}(0,y) = a(\theta)y + b = a(\theta)(y + a(\theta)^{-1}b)$. $\check{Q}_{i+1}(0,y)$ is in the same format as $\check{Q}_i(0,y)$ in (5.27), with $a(y)$ replaced by a constant $a(\theta)$. If these derivations are repeated, it can be shown that the corresponding $\check{Q}(0,y)$ is in the form of (5.27) with $a(y) = a(\theta)$ in all subsequent iteration levels. Hence, from iteration level $i+1$,

$\deg(\check{Q}(0,y))$ is always one, and the coefficient of y in $\check{Q}(0,y)$ remains the same.
□

From Lemma 8, if a simple root is found in iteration level i, then the root of the corresponding $\check{Q}(0,y)$ in all subsequent iteration levels is computed as $a(\theta)^{-1}b$. In iteration level $i+1$, it takes one finite field inversion and one field multiplication. Inversion is more expensive to implement than other finite field operations if basis representations are adopted for finite field elements. $a(\theta)^{-1}$ can be saved in registers and reused. In this case, the root computation in each of the $i+2$ and later iteration levels only needs one field multiplication.

Although the roots of a degree-3 polynomial can be also computed directly without Chien search as presented in Chapter 3, it involves large LUTs or has much longer latency. When a polynomial has degree 4 or higher, it may be converted to an affine format, from which the roots are computed by solving simultaneously linear equations. However, these calculations bring very high hardware overhead as evident from the design in [31]. To increase the average speed and reduce the area requirement of the factorization, prediction-based approaches were developed in [7, 76] to compute the roots when the degree of $\check{Q}(0,y)$ is higher than two.

As it was analyzed previously, the degree of $\check{Q}(0,y)$ is at most the order of the corresponding root, θ, in the previous iteration level. However, simulation results in [7] show that for θ with order $d > 2$, it happens with very high probability ($> 99\%$) that the corresponding $\check{Q}(0,y)$ has degree d and a single root with order d in each of the subsequent iteration levels when the FER is less than 10^{-2}. In addition, this probability increases when decreasing FER. From these observations, the design in [7] predicts that, if $\check{Q}(0,y)$ has a root with order d, then the corresponding $\check{Q}(0,y)$ in the next iteration level has degree d and a single root of degree d. Such a root can be computed directly using simple hardware. Although the chance that this prediction fails is extremely low, ignoring them leads to non-negligible performance loss. Hence, after the root of $\check{Q}(0,y)$ is computed according to the prediction, testing is carried out to see if it is indeed a root with order d. If the prediction fails, the roots of $\check{Q}(0,y)$ are re-computed.

If $\deg(\check{Q}(0,y)) = d$, and $\check{Q}(0,y)$ has a single root, θ, of order d, then $\check{Q}(0,y)$ can be written as:

$$\check{Q}(0,y) = a(y+\theta)^d = a(\theta^d + \binom{d}{1}\theta^{d-1}y + \cdots + \binom{d}{d-1}\theta y^{d-1} + y^d), \quad (5.28)$$

where a is a nonzero element of $GF(2^q)$. Assume that the coefficients of $\check{Q}(0,y)$ are $\check{q}_0, \check{q}_1, \ldots$, *i.e.*

$$\check{Q}(0,y) = \check{q}_0 + \check{q}_1 y + \cdots + \check{q}_{d-1}y^{d-1} + \check{q}_d y^d. \quad (5.29)$$

These coefficients have been computed from the previous iteration level. If d is odd, $\binom{d}{d-1} = d$ is odd. Hence, for odd d, $\binom{d}{d-1}\theta y^{d-1} = \theta y^{d-1}$ over fields of characteristic 2. Comparing (5.28) and (5.29), the root of $\check{Q}(0,y)$ is computed as

$$\theta = \check{q}_{d-1}\check{q}_d^{-1}. \tag{5.30}$$

θ can be also computed using other coefficients of $\check{Q}(0, y)$. However, if the predictions are correct, the most significant coefficient of $\check{Q}(0, y)$ in each subsequent iteration level remains the same. Therefore, $\check{q}_d^{-1}$ only needs to be computed once. It is stored and reused in later iteration levels. If d is even, $\binom{d}{d-1}\theta y^{d-1} = 0$ over fields of characteristic 2, and hence the root can not be computed by (5.30) in this case. Alternatively, the root θ satisfies the following equation when d is even.

$$\theta^w = \check{q}_{d-w}\check{q}_d^{-1}, \tag{5.31}$$

where w is the minimum positive integer such that $\binom{d}{d-w}$ is odd. Such a w can be derived from the binary representation of d.

Let d and s be two m-bit positive integers. Denote their binary representations by $d_{m-1}d_{m-2}\cdots d_0$ and $s_{m-1}s_{m-2}\cdots s_0$, respectively. From the Lucas Theorem

$$\binom{d}{s} = \binom{d_{m-1}}{s_{m-1}}\binom{d_{m-2}}{s_{m-2}}\cdots\binom{d_0}{s_0} \quad \text{mod } 2. \tag{5.32}$$

For two binary bits a and b, $\binom{a}{b}$ is even only if $a = 0$ and $b = 1$. If 2^e is the highest power of 2 that divides d, then $d_e = 1$ and $d_i = 0$ for $0 \leq i < e$. In the case that $s = 2^e$, $s_e = 1$ and all the other bits are zero. Then all the binomial coefficients in (13) equal 1, and hence $\binom{d}{2^e}$ is odd. If $0 \leq s < 2^e$, then at least one of s_i $(0 \leq i < e)$ is 1. As a result, some of the binomial coefficients $\binom{d_i}{s_i}$ $(0 \leq i < e)$ is even, and hence $\binom{d}{s}$ is even. From these analyses, if 2^e is the highest power of 2 that divides d, then $w = 2^e$ is the minimum integer such that $\binom{d}{d-w} = \binom{d}{w}$ is odd. In the binary representation of d, e is simply the first nonzero bit position from the least significant bit. This scheme actually also applies to odd d. When d is odd, the highest power of 2 that divides d is zero, and hence $w = 2^0 = 1$. Therefore, regardless of whether d is even or odd, (5.31) can be used to compute the root.

When $\deg(\check{Q}(0, y)) > 2$, the root of $\check{Q}(0, y)$ can be derived according to the prediction using (5.31). However, even if this root is a real root, $\check{Q}(0, y)$ may have other roots. For example, if $\check{Q}_i(0, y)$ has a root of order three, then it is predicted that the corresponding $\check{Q}_{i+1}(0, y)$ in the next iteration level has a single root of order three, and the root is computed as $\theta = \check{q}_2\check{q}_3^{-1}$. Now assume that $\check{Q}_{i+1}(0, y)$ has two roots θ' and θ'', and their orders are two and one, respectively. Then

$$\check{Q}_{i+1}(0, y) = a(y + \theta')^2(y + \theta'') = a((\theta')^2\theta'' + (\theta')^2 y + \theta'' y^2 + y^3).$$

From the above equation, θ'' can be still computed as $\check{q}_2\check{q}_3^{-1}$. Nevertheless, θ' is lost by computing the root through prediction. This example tells us that it is not sufficient to check if the predicted root is indeed a root of $\check{Q}(0, y)$. Instead, the order of the predicted root also needs to be checked. The order of θ can be observed from the coefficients of $\tilde{Q}(x, y) = \check{Q}(x, y + \theta)$, which are computed from step 3 of Algorithm 8. It equals d if the coefficients of $\tilde{Q}(0,y)$ satisfy $\tilde{q}_{0,s} = 0$ for $0 \leq s < d$ and $\tilde{q}_{0,d} \neq 0$. Hence, the order check is done as a byproduct of the polynomial updating step, and does not require much extra hardware units. In the case that the real order of the root computed according to the prediction does not match the predicted order, the root is recalculated using an exhaustive search.

So far, the roots in the next iteration are computed according to the predictions made based on the roots of the current iteration level. The predictions succeed with more than 99% probability [7], and the exhaustive search is only used when the predictions fail. Hence, the average complexity of the root computations in the second and later iteration levels has been reduced to a fraction. Nevertheless, the root computation for the first iteration level is always carried out by an exhaustive search. It accounts for a significant part of the overall factorization latency after the prediction-based root computations are used for the second and later iteration levels. To speed up the root computation in the first iteration, an iterative prediction scheme was proposed in [76]. Three direct root computations are tried before the exhaustive search is activated. As a result, the average latency for the first iteration root computation is substantially reduced. Moreover, these first-iteration predictions share hardware units with the predictions for later iterations, and only bring negligible overhead.

In the first iteration level, $\check{Q}(0,y)$ may have multiple roots. It was found from simulations that, with very high probability, the maximum order of the roots is very close to the degree of $\check{Q}(0,y)$ in the first iteration level, especially when the SNR is high [76]. Assume that $\deg(\check{Q}(0,y)) = t$ in the first iteration level, and the highest order of its roots is d. It has been shown in [76] that for the KV algorithm with $m_{max} = 4$, $d \geq t - 2$ in more than 90% of the cases for high-rate codes when FER$< 10^{-2}$. This probability further increases when the SNR is higher and hence the FER is lower. This phenomenon is explained as follows. When the SNR is high, it is very likely that no error happened during the signal transmission. In this case, there are n interpolation points with multiplicity m_{max} for a RS code whose codeword length is n symbols. Then the interpolation output, $Q(x,y)$, over these points would have a factor $(y + f(x))^{m_{max}}$. Accordingly, the $\check{Q}(0,y)$ in the first iteration has degree m_{max} and a single root of order m_{max}. As the SNR decreases, more errors happened during the transmission. The chance that $Q(x,y)$ has a $y + f(x)$ factor with high order becomes smaller. The y-degree of the polynomials is decided by the maximum multiplicity, and does not change with the FER. As a result, the probability that $\check{Q}(0,y)$ has a root whose order is very close to its degree becomes smaller as the SNR decreases.

TABLE 5.2
Terms for the coefficients of $\breve{Q}(0,y)$ [76]

$\breve{q}_t$	$\breve{q}_{t-1}$	$\breve{q}_{t-2}$	$\cdots$	$\breve{q}_{t-d}$	$\breve{q}_{t-d-1}$	$\cdots$	$\breve{q}_0$
$\binom{d}{0}a_{t-d}$	$\binom{d}{1}\theta a_{t-d}$ $\binom{d}{0}a_{t-d-1}$	$\binom{d}{2}\theta^2 a_{t-d}$ $\binom{d}{1}\theta a_{t-d-1}$ $\binom{d}{0}a_{t-d-2}$	$\cdots$ $\cdots$ $\cdots$ $\ddots$ $\cdots$	$\binom{d}{d}\theta^d a_{t-d}$ $\binom{d}{d-1}\theta^{d-1}a_{t-d-1}$ $\binom{d}{d-2}\theta^{d-2}a_{t-d-2}$ $\cdots$ $\binom{d}{d-(t-d)}\theta^{d-(t-d)}a_0$	$\binom{d}{d}\theta^d a_{t-d-1}$ $\binom{d}{d-1}\theta^{d-1}a_{t-d-2}$ $\cdots$ $\binom{d}{d-(t-d-1)}\theta^{d-(t-d-1)}a_0$	$\binom{d}{d}\theta^d a_{t-d-2}$ $\cdots$ $\cdots$	$\binom{d}{d}\theta^d a_0$

TABLE 5.3
Example binomial coefficients when $t = 9$ and $d = 5$

$\check{q}_9$	$\check{q}_8$	$\check{q}_7$	$\check{q}_6$	$\check{q}_5$	$\check{q}_4$	$\check{q}_3$	$\check{q}_2$	$\check{q}_1$	$\check{q}_0$
1	1	0	0	1	1				
	1	1	0	0	1	1			
		1	1	0	0	1	1		
			1	1	0	0	1	1	
				1	1	0	0	1	1

In the case that $\check{Q}(0, y)$ has a single root whose order equals $\deg(\check{Q}(0, y)) = t$ in the first iteration level, the root can be computed by an equation similar to (5.31). This case happens less frequently when the SNR is lower or the code rate is lower. To take the advantage that $d \geq t - 2$ with high probability, root computation also needs to be done for the cases of $d = t - 1$ and $d = t - 2$. Let θ be a root of $\check{Q}(0, y)$ with order $d < t$. $\check{Q}(0, y)$ is written as

$$\check{Q}(0, y) = (y + \theta)^d a(y), \tag{5.33}$$

where $a(y)$ does not have $y + \theta$ as a factor, and $\deg(a(y)) = t - d$. Assume that $a(y) = a_0 + a_1 y + \cdots + a_{t-d} y^{t-d}$. Then (5.33) is expanded as

$$\check{Q}(0, y) = \left(\binom{d}{0} y^d + \binom{d}{1} \theta y^{d-1} + \cdots + \binom{d}{d} \theta^d \right) \\ \times (a_{t-d} y^{t-d} + a_{t-d-1} y^{t-d-1} + \cdots + a_0). \tag{5.34}$$

Multiply out the right side of the above equation, and collect the terms in Table 5.2. The terms in the same column of this table add up to be the coefficients of $\check{Q}(0, y)$ labeled in the first row. It is assumed that $d > (t - d)$ in this table. Otherwise, the terms with negative powers of θ should be replaced by zeros. For finite fields of characteristic two, the binomial coefficients are either zero or one. Hence there may exist two columns in Table 5.2 whose nonzero binomial coefficients have the same pattern. If this happens, each of the terms in one column equals the term in the same row of the other column multiplied by θ^s, where s is the difference between the indices of the two columns. In this case, θ^s can be computed by dividing the coefficients of $\check{Q}(0, y)$ corresponding to these two columns. For example, Table 5.3 lists the binomial coefficients when $t = 9$ and $d = 5$. In this table, the '1's in the column for $\check{q}_7$ have the same pattern as those in the column for $\check{q}_3$. The two terms in each row of these two columns are different by $\theta^{7-3} = \theta^4$. As a result, if $\check{q}_7 \neq 0$, then θ^4 can be computed as $\check{q}_3 / \check{q}_7$. The columns for $\check{q}_2$ and $\check{q}_6$ in Table 5.3 also have the same pattern of '1's. Therefore, θ^4 can be also computed as $\check{q}_2 / \check{q}_6$ if $\check{q}_6 \neq 0$.

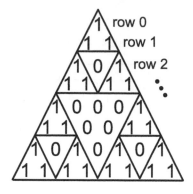

FIGURE 5.23
Binary Pascal's triangle (modified from [76])

Lemma 9 *Let d be a positive integer, and $m = \lfloor \log_2 d \rfloor$. Then the binary representation of d is $d_m d_{m-1} \ldots d_0$ with $d_m = {}'1'$. Let b be the smallest integer such that $d_i = 1$ for $b \leq i \leq m$. Then there are at least two columns in Table 5.2 that have the same pattern of nonzero binomial coefficients if $t \leq 2^{m+2} - 2^b - 2$,*

Proof: The binomial coefficients of $(y + \theta)^d$, $\binom{d}{0}, \binom{d}{1}, \cdots, \binom{d}{d}$, over finite fields of characteristic two are the numbers in the dth row of the binary Pascal's triangle as shown in Fig. 5.23. Define the 0th triangle as the single '1' on the top. The jth triangle is formed by rows 0 through $2^j - 1$. From Fig. 5.23, it can be seen that in the jth triangle, the first 2^{j-1} rows are actually the $(j-1)^{th}$ triangle, and rows 2^{j-1} through $2^j - 1$ consist of two copies of the $(j-1)^{th}$ triangles placed side by side with zeros in between. The $(j-1)^{th}$ triangle has similar structure. Rows 2^{j-2} through $2^{j-1} - 1$ are formed by two copies of the $(j-2)^{th}$ triangles placed side by side with zeros in between. By induction, rows $2^j - 2^{j-i}$ through $2^j - 1$ consist of 2^i copies of the $(j-i)^{th}$ triangles placed side by side with zeros in between.

If $\lfloor \log_2 d \rfloor = m$, then the maximum possible value of d is $2^{m+1} - 1$. Since b is the smallest integer such that $d_i = 1$ for $b \leq i \leq m$, the minimum value of d is $(2^{m+1} - 1) - (2^b - 1) = 2^{m+1} - 2^b$. In the binary Pascal's triangle, rows $2^{m+1} - 2^b$ through $2^{m+1} - 1$ consist of 2^{m+1-b} copies of the bth triangle placed side by side with zeros in between. Therefore, the dth row of the Pascal's triangle consists of 2^{m-b+1} copies of the d'^{th} row with zeros in between, where $d' = d - (2^{m+1} - 2^b)$. The numbers of digits in the dth and d'^{th} rows of the Pascal's triangle are $d + 1$ and $d' + 1$, respectively. Moreover, in the dth row, there are equal number of zeros between each copy of the d'^{th} row. Therefore, it can be computed that in the dth row, the number of zeros between adjacent

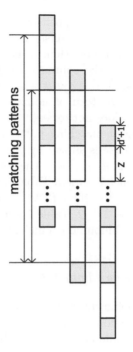

FIGURE 5.24
Matching patterns between shifted binomial coefficient sequences (modified from [76])

copies of the d'^{th} row is

$$
\begin{aligned}
z &= \frac{(d+1) - 2^{m+1-b}(d'+1)}{2^{m+1-b} - 1} \\
&= \frac{(d+1) - (2^{m+1-b} - 1)(d'+1) - (d'+1)}{2^{m+1-b} - 1} \\
&= \frac{d - d'}{2^{m+1-b} - 1} - (d'+1) \\
&= 2^b - (d'+1) \\
&= 2^{m+1} - d - 1
\end{aligned}
\tag{5.35}
$$

The binomial coefficients in each column of Table 5.2 form part or the entire sequence of $\binom{d}{0}, \binom{d}{1}, \cdots, \binom{d}{d}$. The sequence in each column is a shifted and/or truncated version of the sequences in other columns. As computed in (5.35), the binomial coefficients $\binom{d}{0}, \binom{d}{1}, \cdots, \binom{d}{d}$ are formed by copies of the $d' + 1$ binomial coefficients $\binom{d'}{0}, \binom{d'}{1}, \cdots, \binom{d'}{d'}$ with $z = 2^{m+1} - d - 1$ zeros in between. Therefore, as shown in Fig. 5.24, if the sequence $\binom{d}{0}, \binom{d}{1}, \cdots, \binom{d}{d}$

is shifted by multiple of $d' + 1 + z$ positions, then the bit patterns in the middle part match. In Table 5.2, the offset between the binomial coefficient sequences in adjacent columns is one. Hence, the nonzero binomial coefficients may have the same pattern in the columns for $\breve{q}_i$ and $\breve{q}_{i+j(d'+1+z)}$, where j is a positive integer. Since $d' + 1 + z < d + 1$, Table 5.2 would have at least two columns with the same pattern of nonzero binomial coefficients if the number of rows in the table does not exceed the maximum length of possible matching pattern. The longest matching pattern happens when the offset is the smallest possible. This case is shown by the first and second columns in Fig. 5.24. When the offset is $d' + 1 + z$, the longest matching pattern has $d - d' + z$ bits. There are $t - d + 1$ rows in Table 5.2. Therefore, $t - d + 1$ should be at most $d - d' + z = (2^{m+1} - 2^b) + (2^{m+1} - d - 1) = 2^{m+2} - 2^b - d - 1$. As a result, if $t \leq 2^{\lfloor \log_2 d \rfloor + 2} - 2^b - 2$, Table 5.2 has at least two columns with the same pattern of nonzero binomial coefficients. $\square$

Lemma 10 *When $t \leq 2^{\lfloor \log_2 d \rfloor + 2} - 2^b - 2$, θ^{2^b} can be directly computed from two coefficients of $\breve{Q}(0, y)$ using a finite field division, if the divisor is nonzero.*

Proof: The binomial coefficient sequence in each column of Table 5.2 is a shifted and/or truncated version of the sequences in other columns. For the terms in the same row, the difference between the exponents of θ equals the distance of the corresponding columns. Therefore, the power of θ that can be directly computed from the coefficients of $\breve{Q}(0, y)$ equals the distance of the corresponding columns. The offset between the sequences of adjacent columns is one. From Lemma 9, the offset between the columns that have the same pattern of nonzero binomial coefficients must be a multiple of $d' + 1 + z = d - (2^{m+1} - 2^b) + 1 + 2^{m+1} - d - 1 = 2^b$. Therefore, the exponent of the power of θ that can be directly computed must be a multiple of 2^b. Picking a nonzero coefficient of $\breve{Q}(0, y)$ to be used as the divisor, θ^{2^b} can be computed directly by a finite field element division. Similarly, θ^{j2^b} $(j \in Z^+)$ may be also computed directly if $j2^b < d$. $\square$

If the binomial coefficients $\binom{d'}{0}, \binom{d'}{1}, \cdots, \binom{d'}{d'}$ are all one, the binomial coefficients in two columns of Table 5.2 only have matching patterns when their distance is a multiple of 2^b. Depending on d', some of $\binom{d'}{0}, \binom{d'}{1}, \cdots, \binom{d'}{d'}$ may be zero. In this case, two columns in Table 5.2 may have matching patterns even if their distance is not a multiple of 2^b, and the exponent of the power of θ that can be directly computed from the coefficients of $\breve{Q}(0, y)$ does not have to be a multiple of 2^b. For example, $m = 3$ when $d = 10$. The binary representation of 10 is '1010'. Hence $b = 3$ and $d' = 2$. $[\binom{2}{0}, \binom{2}{1}, \binom{2}{2}] = [1, 0, 1]$. When $t - d = 2$, the binomial coefficients in the columns for $\breve{q}_3$ and $\breve{q}_9$ have the same pattern, and θ^6 may be computed as $\breve{q}_3 / \breve{q}_9$. The exponent 6 is not a multiple of $2^b = 8$. Nevertheless, θ^{2^b} can be always computed directly when the condition in Lemma 9 is satisfied. In the above example, θ^8 can be computed as $\breve{q}_4 / \breve{q}_{12}$. It is advantages to compute θ^{2^b} instead of other powers of θ from the coefficients of $\breve{Q}(0, y)$. When the exponent is in the format of power

TABLE 5.4
Powers of θ that can be directly computed from the coefficients of $\check{Q}(0,y)$ (data from [76])

		$t-d$														
		1	2	3	4	5	6	7	8	9	10	11	12	13	14	15
	2	2	2													
	3	1,2	1													
	4	4	4	4	4	4	4									
d	5	4	4	4	4	4										
	6	2,4,6	2,4,6	2,4	2,4	2	2									
	7	1-6	1-5	1-4	1-3	1,2	1									
	8	8	8	8	8	8	8	8	8	8	8	8	8	8	8	

of two, θ can be derived by a cyclical shift from its power if normal basis representation of finite field elements is adopted.

For practical applications, $\deg(\check{Q}(0,y))$ and hence the maximum order of the roots of $\check{Q}(0,y)$ is not high. For d in the range of 1 to 8, possible values of $t-d$ and the corresponding powers of θ that can be directly computed from the coefficients of $\check{Q}(0,y)$ are listed in Table 5.4. Take $d=6$ as an example. Table 5.4 shows that a power of θ can be computed from a pair of $\check{Q}(0,y)$ coefficients as long as $t-d = t-6 \le 6$. When $t-6 \le 2$, either θ^2, θ^4, or θ^6 can be computed directly. When $t-6 = 3$ or 4, the powers of θ that can be computed directly are θ^2 and θ^4.

As it was mentioned previously, simulation results showed that the maximum order of the roots of $\check{Q}(0,y)$ is very close to the degree of $\check{Q}(0,y)$ with high probability in the first iteration level of the factorization. Taking advantage of this feature and making use of Lemma 9 and 10, the iterative prediction scheme for the root computations in the first factorization iteration level proposed in [76] starts with predicting that $\check{Q}(0,y)$ has a single root θ with order $d=t$. In this prediction, θ is computed according to (5.31). If the first prediction fails, the second trial predicts that $d=t-1$, and θ is computed through dividing proper coefficients of $\check{Q}(0,y)$. If the second prediction fails again, d is further decreased by one. This process can be repeated until a real root is found or the condition in Lemma 9 is no longer satisfied. Similarly, to check whether a prediction is successful, the real order of the predicted root is tested. The real order is also derived from $\check{Q}(0,y+\theta)$, which is available from step 3 of Algorithm 8.

The iterative prediction process starts with predicting that the root order d equals t, and d is decreased by one each time. Hence, the root, θ, computed from each prediction has the highest order among all the roots of $\check{Q}(0,y)$. The other roots of $\check{Q}(0,y)$ also need to be computed. If

$$\check{Q}(0,y) = (y+\theta)^d a(y),$$

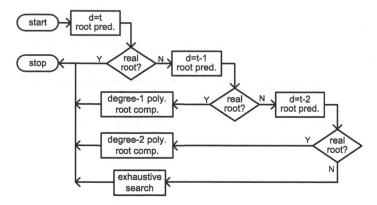

FIGURE 5.25
Iterative prediction procedure for root computation in the first factorization
iteration (modified from [76])

then the other roots are computed by solving $a(y) = 0$. The coefficients of
$(y + \theta)^d$ can be computed first. Then a polynomial division is required to
derive $a(y)$. This process brings large hardware overhead. Alternatively, the
following polynomial is derived by step 3 of Algorithm 8.

$$\check{Q}(0, y + \theta) = y^d a(y + \theta).$$

As a result, the coefficients of $a(y+\theta)$ are directly observable from the output
of step 3. Adding θ back to the roots of $a(y+\theta)$ would yield the roots of $a(y)$.

When the degree of the polynomial is higher than two, the root computation is much more involved and has longer latency if no prior information is
available. Since $d \geq t - 2$ with high probability when the FER is in practical
range, $\deg(a(y)) \leq 2$ most of the time. Moreover, Table 5.4 tells us that when
d is small or moderate, a power of the root with the highest order can be
always computed directly from $\check{Q}(0, y)$ coefficients when $t - d \leq 2$. Considering these, the iterative prediction process is limited to three trials, and the
last trial predicts that $d = t - 2$. If all three predictions fail, an exhaustive
search is activated to compute the roots. In summary, in the first iteration
level of the factorization, the root computation is done according to the iterative prediction process shown in Fig. 5.25. In this figure, the degree-1 and 2
root computation blocks find the roots of $a(y)$.

As mentioned previously, the root with the highest order may be computed directly from multiple pairs of $\check{Q}(0, y)$ coefficients. To save computations, the pairs of coefficients should be selected such that the divisor and/or
the dividend remain the same in the three predictions as much as possible.
Take $(\check{Q}(0, y))$ with degree 5 as an example. First, it is predicted that $d = 5$,
and the root is computed as $\theta^4 = \check{q}_1/\check{q}_5$. This root can be also computed as

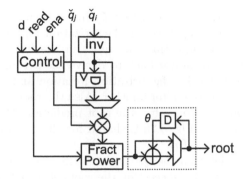

FIGURE 5.26
Root prediction (RP) architecture (modified from [76])

$\theta = \check{q}_4/\check{q}_5$. However, if the first prediction fails, then $d = 4$ is tried. Through the second prediction, θ^4 can be computed as $\check{q}_1/\check{q}_5$. Hence, if the $\check{q}_1, \check{q}_5$ pair was chosen to compute the root in the first prediction, the second prediction would require no calculation. Of course, the coefficient chosen to be the divisor should be nonzero.

When it is predicted that $d = t$, the root θ is computed using (5.31). Fig. 5.26 shows a root prediction (RP) architecture implementing the corresponding computations. In this architecture, the inputs $\check{q}_i$ and $\check{q}_j$ denote the divisor and dividend, respectively. The 'ena' input serves as the enable signal. The multiplicative inversion over finite fields can be carried out using composite field arithmetic to reduce the area requirement as explained in Chapter 1. In addition, the inverse value is stored in the register, so that it can be re-used in later predictions if needed. The 'read' signal tells whether to compute the inverse or read the inverse from the register. The input of the Fract Power block is θ^w. This block first converts the standard basis representation of θ^w to its normal basis representation, which is done by multiplying a pre-computed binary matrix. Since w is an integer power of two, θ is derived from θ^w using a cyclical shift in normal basis. After that, the normal basis representation is converted back to standard basis representation, which is also done as multiplying a pre-computed binary matrix. In the case that the predictions of $d = t - 1$ and $d = t - 2$ need to be carried out, similar computations are done to find the root θ. The root predictions for the second and later factorization iterations are also carried out in a similar way. Therefore, the RP architecture in Fig. 5.26 can be shared in all three predictions for the first factorization iteration, as well as the predictions for later iterations.

For a polynomial $a(y) = a_1 y + a_0$ with degree one, the root is $a_1^{-1} a_0$. Hence, the root computations for degree-1 polynomials are also done by using the RP architecture in Fig. 5.26. For the $a(y)$ in (5.33), which is the remaining part of $\check{Q}(0, y)$ after the root with the highest order is factored out, it is more efficient to compute the roots of $a(y + \theta)$ instead, since its coefficients are directly available after step 3 of the factorization algorithm as mentioned previously. To accommodate this, the units in the bottom right corner of Fig. 5.26 are included to add θ back to derive the actual root of $a(y)$. The three predictions for the first factorization iteration and the predictions for later iterations are done one after another. Also the root computation for the remaining $a(y)$ is always done after the root prediction. Hence, a single copy of the RP architecture in Fig. 5.26 is shared for all root predictions and root computations of degree-1 $a(y)$.

When the prediction that the $\check{Q}(0, y)$ in the first factorization iteration has a root with order $d = t - 2$ is successful, the roots of the left-over degree-2 $a(y)$ need to be computed. As discussed in Chapter 3, a degree-2 polynomial can be converted to the format of $y^2 + y + u$. The roots of such a polynomial can be derived by multiplying the basis representation of u with a pre-computed binary matrix. After that, reverse coordinate transformations are done to recover the real roots. The architecture in Fig. 3.2 can be used for the degree-2 polynomial root computation block in Fig. 5.25. Similarly, it is more efficient to compute the roots of $a(y + \theta)$ first, and then add θ back to recover the roots of $a(y)$. This architecture can be also shared to compute the roots of $\check{Q}(0, y)$ when its order is two in later factorization iterations. However, compared to the RP architecture in Fig. 5.26, the degree-2 polynomial root computation architecture has a much longer data path. In the second and later iterations of the factorization, a degree-2 $\check{Q}(0, y)$ has a single root of order two most of the time. Hence, to reduce the latency of the overall factorization, when $\check{Q}(0, y)$ has degree two in the second or a later iteration of the factorization, its coefficient of y is tested first. If it is zero, then its root satisfies $\theta^2 = \check{q}_0(\check{q}_2)^{-1}$ and is computed using the RP architecture in Fig. 5.26. Only when the coefficient of y is nonzero, the roots are computed by the left parts of the degree-2 polynomial root computation architecture in Fig. 3.2.

Let $\tilde{q}_{i,j}$ and $\check{q}_{i,j}$ be the coefficients of $x^i y^j$ in $\tilde{Q}(x, y)$ and $\check{Q}(x, y)$, respectively. Note that $\check{q}_{0,j}$ is also denoted by $\check{q}_j$ in the previous analysis. Step 3 of the factorization algorithm carries out the following computations

$$\tilde{q}_{i,j} = \sum_{s \geq j} \binom{s}{j} \check{q}_{i,s} \theta^{s-j}.$$

For an example case that t and hence the maximum y-degree of the bivariate polynomial is seven, the polynomial updating (PU) architecture in Fig. 5.27 implement these computations. If the maximum y-degree is different, units are added or deleted accordingly. In this architecture, all coefficients of the same x-degree are updated simultaneously. If high clock frequency is needed, registers are inserted as shown in Fig. 5.27 for pipelining purposes.

Using the prediction-based schemes for root computations, the factorization algorithm can be implemented by the architecture in Fig. 5.28 [76]. When the 'start' signal is asserted, the interpolation output $Q(x, y)$ is loaded into

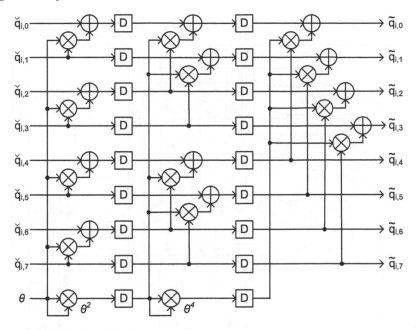

FIGURE 5.27
Polynomial updating (PU) architecture for factorization (modified from [7])

the factorization architecture. In the first iteration level, the root computations are carried out using the RP unit, degree-2 polynomial root computation engine, and/or exhaustive search according to the procedure in Fig. 5.25. The PU unit in Fig. 5.28 implements the polynomial updating in step 3 of the factorization algorithm. Through the Root & Polynomial Scheduling (RPS) block, the predicted root is sent to the first PU unit to have the corresponding polynomial updating done. At the output of this unit, the real order, d, of the predicted root and hence whether the prediction is successful is told. If the prediction succeeds, the coefficients of $a(y + \theta)$, which are also available at the output of the first PU unit, are sent back to the RP or degree-2 polynomial root computation engine to compute the remaining roots. Then the remaining roots are sent to the PU units through the RPS block for corresponding polynomial updating. The De-Scheduling block implements the reverse function of the RPS block, and the polynomial shifting (PS) units take care of step 4 and 1 of Algorithm 8. The polynomials sent back to the Polynomial & Root-order Buffers are the corresponding $\check{Q}(x, y)$ for the next iteration level. Moving polynomials in memory requires a large number of reads and writes, which cause long latency and power consumption overhead. Alternatively, the PS units employ registers to track the address displacements in step 4 and

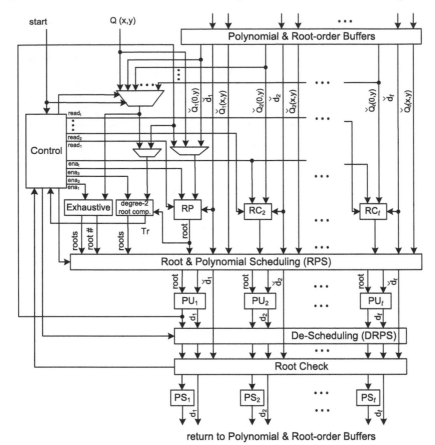

FIGURE 5.28
Factorization architecture with prediction-based root computation [76]

1. The displacements are added to the address generator of the polynomial buffers to output $\check{Q}(x, y)$ in the next iteration level.

In each of the second and later iteration levels, if there is a root with order d in the previous iteration, then it is predicted that the corresponding $\check{Q}(0, y)$ has a single root of order $\check{d} = d$. These predicted roots are computed using the RP and root computation (RC) units. Each RC unit has the same architecture as the RP unit shown in Fig. 5.26, except that it does not have the part in the bottom right corner for adding θ back. From the second iteration level, each of the RP and RC units reads one $\check{Q}(0, y)$ from the buffer and computes a root according to the predicted order $\check{d}$. Similarly, the actual orders of the predicted roots are derived from the outputs of the PU units and compared with the predicted orders. In the case of prediction failure, the failed $\check{Q}(0, y)$ is loaded into the degree-2 polynomial computation block for root re-computation if its degree is two. Otherwise, an exhaustive search is used for root re-computation.

5.5.2 Partial-parallel factorization architecture

Using the prediction-based schemes, the root computation in each factorization iteration is completed in a few clock cycles most of the time. However, using the PU architecture in Fig. 5.27, the number of clock cycles needed by the polynomial updating is dependent on the maximum x-degree of the bivariate polynomials. To reduce the latency, only the coefficients that will be involved in later factorization iterations should be updated. In this case, the number of coefficients need to be updated decreases in each iteration. Nevertheless, the polynomial updating still takes a large part of the overall factorization latency. Since the coefficients of different x-degrees are updated independently, they can be divided into groups and all groups are updated in parallel using multiple PU units. Of course, the area requirement would increase linearly with the parallel processing factor, and this approach is not efficient. One possible method to further speed up the factorization without duplicating every hardware unit is to combine the computations from adjacent iterations. However, the factorization iterations are inherently serial. The computations in the next iteration level are dependent on the results of the current iteration level. To solve this problem, a partial-parallel factorization architecture was developed in [77]. By making use of the prediction-based root computation, this architecture applies a look-ahead scheme to derive the roots of two adjacent iteration levels first. Then the polynomial updating for both iterations are combined. This partial-parallel architecture is detailed next.

The root computation for the next factorization iteration level does not have to wait until all coefficients of $\hat{Q}(x,y)$ are computed in the current iteration level. As shown in Fig. 5.22, the coefficients are represented by the symbols in a two-dimensional array. From Algorithm 8 for the factorization, the coefficients in the first nonzero row of the $\hat{Q}(x,y)$ array are the coefficients of $\check{Q}(0,y)$ for the next iteration level. Assume that a root θ with order d is computed in the current iteration level. If the prediction that the $\check{Q}(0,y)$ for the next iteration has degree d and a single root of order d is successful, then the first nonzero row of the $\hat{Q}(x,y)$ symbol array is the row for x^d. Therefore, once the coefficients in the first $d+1$ rows of the $\hat{Q}(x,y)$ array are derived, the root prediction for the next iteration level can start. As a result, the root computations in iteration levels i and $i+1$ using the prediction-based scheme can be carried out as follows:

i) Compute the root θ_i for iteration level i by using (5.31)

ii) Calculate $\hat{q}_{d,d-w}$ as

$$\hat{q}_{d,d-w} = \tilde{q}_{w,d-w} = \sum_{j \geq (d-w)} \binom{j}{d-w} \theta_i^{j-(d-w)} \check{q}_{w,j},$$

where $\hat{q}_{i,j}$ is the coefficient of $x^i y^j$ in $\hat{Q}(x,y)$.

   iii) Replace $\check{q}_{d-w}$ in (5.31) by $\hat{q}_{d,d-w}$ and compute θ_{i+1}, which is the root of $\check{Q}(0,y)$ in iteration level $i+1$. Note that the $\check{q}_d^{-1}$ in (5.31) remains the same and does not need to be recomputed if the predictions are successful.

   After θ_i and θ_{i+1} are computed, the polynomial updating for iteration levels i and $i+1$ can be combined. From steps 3 and 4 of Algorithm 8,

$$\hat{Q}_i(x,y) = \check{Q}_i(x, xy + \theta_i).$$

The purpose of step 1 in Algorithm 8 for iteration level $i+1$ is to shift the rows of the $\hat{Q}_i(x,y)$ array up, so that the first row becomes nonzero. This step can be put off to after step 4 of iteration level $i+1$ if proper coefficients are used to compute the root θ_{i+1} as was done in the three-step root computation above. If step 1 is moved to after step 4 in iteration level $i+1$, then

$$\check{Q}_{i+1}(x,y) = \hat{Q}_i(x,y) = \check{Q}_i(x, xy + \theta_i).$$

After steps 3 and 4 of iteration level $i+1$ are completed,

$$\begin{aligned}
\hat{Q}_{i+1}(x,y) &= \tilde{Q}_{i+1}(x, xy) \\
&= \check{Q}_{i+1}(x, xy + \theta_{i+1}) \\
&= \check{Q}_i(x, x^2 y + \theta_{i+1} x + \theta_i).
\end{aligned} \tag{5.36}$$

Let

$$Q'_{i+1}(x,y) = \check{Q}_i(x, y + \theta_{i+1} x + \theta_i).$$

The polynomial updating for iteration level i and $i+1$ can be jointly carried out by computing $Q'_{i+1}(x,y)$ first. Then the coefficients in the jth column of the $Q'_{i+1}(x,y)$ array are shifted down by $2j$ positions to derive $\hat{Q}_{i+1}(x,y)$.

   In steps 3 and 4 of Algorithm 8 for a single iteration level, the updating of coefficients with different x-degree is independent. However, combining the computations from adjacent iteration levels adds data dependency to the joint polynomial updating as shown in (5.36). If t is the maximum y-degree of the polynomials, then in the combined polynomial updating, a coefficient of $Q'_{i+1}(x,y)$ is affected by the coefficients of up to t different x-degrees in $\check{Q}_i(x,y)$. First consider how one row of coefficients in $\check{Q}_i(x,y)$ affect the coefficients of $Q'_{i+1}(x,y)$. Assume that the coefficients in a row of the $\check{Q}_i(x,y)$ array are the coefficients of a univariate polynomial $b(y) = b_0 + b_1 y + \cdots + b_t y^t$. Then

$$\begin{aligned}
&b(y + \theta_{i+1} x + \theta_i) = b((y + \theta_i) + \theta_{i+1} x) \\
&= b_0 + b_1((y + \theta_i) + \theta_{i+1} x) + \cdots + b_t((y + \theta_i) + \theta_{i+1} x)^t
\end{aligned}$$

t should be small to keep the hardware complexity practical. For $t < 8$, expand each term in the above equation and list them in Table 5.5.

TABLE 5.5

Expanded terms for combined polynomial updating

	b_0	$b_1(y+\theta_i)$	$b_2(y+\theta_i)^2$	$b_3(y+\theta_i)^3$	$b_4(y+\theta_i)^4$	$b_5(y+\theta_i)^5$	$b_6(y+\theta_i)^6$	$b_7(y+\theta_i)^7$
$(\theta_{i+1}x)$		b_1		$b_3(y+\theta_i)^2$		$b_5(y+\theta_i)^4$		$b_7(y+\theta_i)^6$
$(\theta_{i+1}x)^2$			b_2	$b_3(y+\theta_i)$			$b_6(y+\theta_i)^4$	$b_7(y+\theta_i)^5$
$(\theta_{i+1}x)^3$				b_3				$b_7(y+\theta_i)^4$
$(\theta_{i+1}x)^4$					b_4	$b_5(y+\theta_i)$	$b_6(y+\theta_i)^2$	$b_7(y+\theta_i)^3$
$(\theta_{i+1}x)^5$						b_5		$b_7(y+\theta_i)^2$
$(\theta_{i+1}x)^6$							b_6	$b_7(y+\theta_i)$
$(\theta_{i+1}x)^7$								b_7

TABLE 5.6

Coefficients for combined polynomial updating

	1	y	y^2	y^3	y^4	y^5	y^6	y^7
1	b_0'	b_1'	b_2'	b_3'	b_4'	b_5'	b_6'	b_7'
x	$b_1'\theta_{i+1}$		$b_3'\theta_{i+1}$		$b_5'\theta_{i+1}$		$b_7'\theta_{i+1}$	
x^2	$b_2'\theta_{i+1}^2$	$b_3'\theta_{i+1}^2$			$b_6'\theta_{i+1}^2$	$b_7'\theta_{i+1}^2$		
x^3	$b_3'\theta_{i+1}^3$				$b_7'\theta_{i+1}^3$			
x^4	$b_4'\theta_{i+1}^4$	$b_5'\theta_{i+1}^4$	$b_6'\theta_{i+1}^4$	$b_7'\theta_{i+1}^4$				
x^5	$b_5'\theta_{i+1}^5$		$b_7'\theta_{i+1}^5$					
x^6	$b_6'\theta_{i+1}^6$	$b_7'\theta_{i+1}^6$						
x^7	$b_7'\theta_{i+1}^7$							

To simplify Table 5.5, the common part of the terms in the same row are extracted and listed in the first column. For example, the actual terms of $b(y + \theta_{i+1}x + \theta_i)$ listed in the 7*th* row of Table 5.5 are $b_6(\theta_{i+1}x)^6$ and $b_7(y + \theta_i)(\theta_{i+1}x)^6$. Further expand the terms in Table 5.5 and combine the monomials with the same y-degree in each row. Then the coefficients of $b(y + \theta_{i+1}x + \theta_i)$ are those listed in Table 5.6. In this table,

$$b_0' = b_0 + b_1\theta_i + b_2\theta_i^2 + \cdots + b_7\theta_i^7$$
$$b_1' = b_1 + b_3\theta_i^2 + b_5\theta_i^4 + b_7\theta_i^6$$
$$b_2' = b_2 + b_3\theta_i + b_6\theta_i^4 + b_7\theta_i^5$$
$$b_3' = b_3 + b_7\theta_i^4$$
$$b_4' = b_4 + b_5\theta_i + b_6\theta_i^2 + b_7\theta_i^3 \qquad (5.37)$$
$$b_5' = b_5 + b_7\theta_i^2$$
$$b_6' = b_6 + b_7\theta_i$$
$$b_7' = b_7$$

As an example, the coefficient for x^2y^4 in $b(y + \theta_{i+1}x + \theta_i)$ can be computed as $b_6'\theta_{i+1}^2 = (b_6 + b_7\theta_i)\theta_{i+1}^2$. The first row in Table 5.6 actually shows the coefficients of $b(y + \theta_i)$. The coefficients in other rows are those in the first row multiplied by some powers of θ_{i+1}.

The speed of polynomial updating can be increased by processing the coefficients of $\check{Q}(x,y)$ with the same x-degree simultaneously and applying pipelining as shown in the PU architecture in Fig. 5.27. Similarly, in the combined polynomial updating, the coefficients of the same x-degree are updated in parallel, and pipelining is adopted. In each clock cycle, $b_0', b_1', \cdots, b_7'$ for a row of $\check{Q}_i(x,y)$ are first computed according to (5.37). They are loaded into registers and multiplied by powers of θ_{i+1} according to Table 5.6 in later clock cycles. The products are accumulated with those corresponding to later rows of $\check{Q}_i(x,y)$ to derive the outputs of the combined polynomial updating. The computations in (5.37) are the same as those in the original polynomial updating for a single iteration level. Hence, they can be implemented by the PU

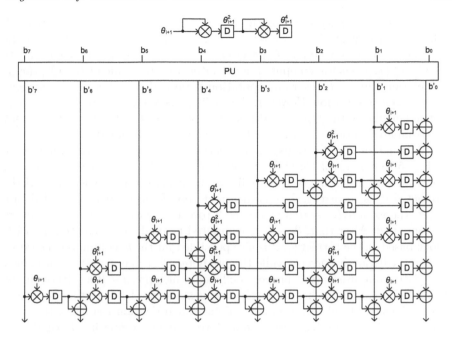

FIGURE 5.29
Combined polynomial updating (CombPU) architecture (modified from [77])

architecture in Fig. 5.27. Making use of this PU unit, the combined polynomial updating can be implemented by the CombPU architecture shown in Fig. 5.29.

When the combined polynomial updating is adopted, the roots computed through the predictions also need to be checked. As mentioned previously, θ is indeed an *dth* order root of $\check{Q}(0, y)$ if the coefficients of $1, y, y^2, \ldots, y^{d-1}$ in $\check{Q}(0, y + \theta)$ are zero. In the case that the polynomial updating from two iteration levels are combined, it is not straightforward to tell which coefficients need to be tested in order to check if the roots are real. The coefficients of $Q'_{i+1}(0, y)$ are actually the coefficients of $\check{Q}_i(0, y + \theta_i)$. Hence, whether θ_i is a real root can be checked from the coefficients of $Q'_{i+1}(0, y)$, which are the $b'_0, b'_1, b'_2, \ldots$ computed by the PU engine in the CombPU architecture. To check θ_{i+1}, the coefficients of $\check{Q}_{i+1}(0, y + \theta_{i+1})$ are needed. $Q'_{i+1}(x, xy) = \check{Q}_{i+1}(x, y + \theta_{i+1})$. However, step 1 for iteration level $i + 1$ has been put off to after the polynomial updating. Hence the coefficients to check should be those of $x^{d-j}y^j$ ($0 \le j < d$) in $Q'_{i+1}(x, y)$. Only when these coefficients are zero, the θ_{i+1} computed through prediction is a real root.

If the prediction fails, the real θ_i is re-computed using the coefficients of $\check{Q}_i(0, y)$. To re-compute θ_{i+1}, the coefficients of $\check{Q}_{i+1}(x, y)$ are required.

$\check{Q}_{i+1}(x, y) = \check{Q}_i(x, xy + \theta_i)$ when step 1 of iteration level $i + 1$ is put off. Therefore, the coefficients needed for θ_{i+1} re-computation are those in the first nonzero anti-diagonal of the $\check{Q}_i(x, y + \theta_i)$ symbol array. According to (5.37), $\check{Q}_i(x, y + \theta_i)$ is computed as an intermediate result of the $Q'_{i+1}(x, y)$ calculation. Hence, no extra computation is needed to derive the coefficients for θ_{i+1} re-computation. However, the coefficients in the upper $(d+1) \times (d+1)$ anti-triangular of the $\check{Q}_i(x, y + \theta_i)$ symbol array need to be stored when they are computed in order to find the first nonzero anti-diagonal.

With minor modifications, the prediction-based factorization architecture in Fig. 5.28 can adopt CombPU instead of PU units. First, both θ_i and θ_{i+1} are computed according to the prediction by using the RP or RC unit. If the predicted θ_i is incorrect, θ_{i+1} is incorrect too. Then $\check{Q}_i(0, y)$ is loaded into either the exhaustive search or degree-2 root computation engine to have θ_i re-computed. After that, θ_{i+1} is re-computed through the prediction in a RC block based on the correct θ_i. It is also possible that the θ_i computed according to the prediction is correct, while the predicted θ_{i+1} is not. In this case, only θ_{i+1} is re-computed using proper coefficients of $\check{Q}_i(x, y + \theta_i)$. Although the CombPU architecture has more pipelining stages and the testing of θ_i and θ_{i+1} brings overhead to the latency of each factorization iteration, the total number of iterations has been reduced to half by computing both θ_i and θ_{i+1} in a look-ahead manner and combining the polynomial updating for two iterations.

6

Interpolation-based Chase and generalized
minimum distance decoders

CONTENTS

The low-complexity Chase (LCC) [51] algebraic soft-decision decoding (ASD) algorithm for Reed-Solomon (RS) codes tests the same vectors and has the same error-correcting performance as the Chase-II algorithm [46]. It can be interpreted as the Chase-II algorithm with interpolation-based decoding. For the case that the multiplicity of each interpolation point is one, as in the LCC algorithm, additional simplification techniques have been developed for the interpolation-based decoding [78, 79, 80, 81, 82]. Moreover, the backward interpolation schemes [55, 68, 83] allow the interpolation for all test vectors to be completed in one run. The interpolation-based scheme can be also used to implement the generalized minimum distance (GMD) decoding algorithm

[45]. It starts with the erasure-only test vector and ends with error-only decoding [84, 85]. Compared to the one-pass schemes based on the Berlekamp-Massey algorithm (BMA), such as those in [61, 86], the interpolation-based method achieves higher efficiency in terms of throughput-over-area ration when adopted for the Chase and GMD algorithms. In this chapter, the simplification techniques and architectures for implementing the interpolation-based Chase and GMD decoding are presented.

6.1   Interpolation-based Chase decoder

The Chase algorithm first picks the η least reliable code positions according to the channel information. For each of these η positions, two interpolation points (α_i, β_i) and (α_i, β_i'), are assigned. α_i is the evaluation element for the ith code position. β_i and β_i' are the hard-decision and second most-likely symbol, respectively, for the ith code position. For an (n, k) RS code, there are $n - \eta$ remaining positions, and only one point (α_i, β_i) is assigned to each of them. All points have multiplicity one. One point is chosen for each code position to form a test vector of n points. Since there are two possible points for each of the η least reliable code positions, the total number of test vectors is 2^η. Decoding trials need to be carried out for each test vector.

6.1.1   Backward-forward interpolation algorithms and architectures

As mentioned in Chapter 5, there are two hardware-friendly interpolation algorithms: The Kötter's algorithm [62] and Lee-O'Sullivan (LO) algorithm [29]. The Kötter's algorithm adds constraints to the interpolation polynomials one after another. At the end of each iteration, the polynomials involved in the ötter's algorithm form a Gröbner basis of polynomials satisfying the constraints that have been covered so far with minimum weighted degree. Interpolation points can be added and multiplicities can be increased by carrying out more iterations in the Kötter's algorithm. Hence, the Kötter's algorithm is a 'forward' interpolation algorithm. On the contrary, the LO algorithm starts with a basis of polynomials that already satisfies all interpolation constraints. Then the initial basis is converted to a Gröbner basis. Once the interpolation starts in the LO algorithm, points can not be added and multiplicities can not be changed.

It is quite wasteful to start the interpolation afresh for each test vector in the Chase algorithm using either the Kötter's or the LO algorithm. The test vectors can be arranged in a Gray code manner, *i.e.* adjacent test vectors only have one pair of different points: (α_i, β_i) and (α_i, β_i'). Given the interpolation result of a test vector derived by the Kötter's algorithm, a backward interpolation scheme was proposed in [55] to eliminate the point (α_i, β_i) from the interpolation result. Then (α_i, β_i') can be added by using the Kötter's interpolation algorithm, which is a forward interpolation algorithm, to derive the interpolation result of the next test vector. Adopting such a backward-forward interpolation scheme, only the different points need to be taken care of in order to derive the interpolation result for the next test vector. Compared to repeating the interpolation for the n points in each test vector, the backward-forward interpolation leads to substantially higher throughput. In addition, a unified scheme was proposed in [68] to combine the computations in the backward and forward interpolation iterations to achieve further speedup. Later, the backward interpolation was generalized in [83] to reduce the multiplicity or eliminate a point of any multiplicity from a given interpolation result. The backward interpolation schemes and their hardware implementation architectures are introduced next.

6.1.1.1 Backward-forward interpolation

The Kötter's algorithm actually constructs a Gröbner basis of a $\mathbb{F}[x]$-module contained in $\mathbb{F}[x, y]$ if the RS code is constructed over finite field $\mathbb{F}$. Here $\mathbb{F}[x]$ is the ring of univariate polynomials over $\mathbb{F}$, and $\mathbb{F}[x, y]$ is the ring of bivariate polynomials over $\mathbb{F}$. According to the definition of Göbner basis in Chapter 5, the polynomial with the minimum weighted degree in the basis has the lowest weighted degree among all polynomials in the module. Hence, eliminating a point (α_i, β_i) from a given interpolation result can be achieved by constructing a Göbner basis that does not pass this point.

In the Chase decoding with multiplicity one for high-rate codes, the Göbner basis has two polynomials, and their maximum y-degree is one. Let the Gröbner basis be $Q^{(l)}(x, y) = q_0^{(l)}(x) + q_1^{(l)}(x)y, l \in \{0, 1\}$. The backward interpolation in [55] for eliminating (α_i, β_i) starts with computing $q_1^{(l)}(\alpha_i)$. Since $Q^{(l)}(x, y)$ passes (α_i, β_i), $Q^{(l)}(\alpha_i, \beta_i) = 0$. Hence if $q_1^{(l)}(\alpha_i) = 0$, then $q_0^{(l)}(\alpha_i) = Q^{(l)}(\alpha_i, \beta_i) + q_1^{(l)}(\alpha_i)\beta_i = 0$. Accordingly, $Q^{(l)}(x, y)$ has a factor $(x + \alpha_i)$ and $Q^{(l)}(\alpha_i, \beta_i') = q_0^{(l)}(\alpha_i) + q_1^{(l)}(\alpha_i)\beta_i' = 0$. In the case that $q_1^{(l)}(\alpha_i) \neq 0$, $Q^{(l)}(x, y)$ does not have the $(x + \alpha_i)$ factor. Since $Q^{(l)}(x, y)$ passes (α_i, β_i) and its y-degree is one, β_i is the only root of $Q^{(l)}(\alpha_i, y) = q_0^{(l)}(\alpha_i) + q_1^{(l)}(\alpha_i)y$. As a result, $Q^{(l)}(\alpha_i, \beta_i') \neq 0$ if $q_1^{(l)}(\alpha_i) \neq 0$. Denote the module of polynomials with y-degree at most one passing all interpolation points and that of polynomials passing all other points except (α_i, β_i) by $\mathcal{M}$ and $\mathcal{M}'$, respectively. Starting from a Gröbner basis $Q^{(l)}(x, y)$ ($l = 0, 1$) of $\mathcal{M}$, three categories according to whether $Q^{(l)}(x, y)$ has the $(x + \alpha_i)$ factor need to be considered to construct a Gröbner basis of $\mathcal{M}'$.

1. One of $Q^{(l)}(x, y)$ has the $(x + \alpha_i)$ factor.
   Let $u = \arg\min_l \{w_l | q_1^{(l)}(\alpha_i) \neq 0\}$, where w_l is the weighted degree of $Q^{(l)}(x, y)$. Let $v = \{0, 1\} \backslash u$. Then $Q^{(v)}(x, y)$ has the factor and

$Q^{(u)}(x,y)$ does not. $Q^{(v)}(x,y)$ can be re-written in the format of

$$Q^{(v)}(x,y) = (x+\alpha_i)F^{(v)}(x,y).$$

In this case,

$$\begin{cases} F^{(u)}(x,y) = Q^{(u)}(x,y) \\ F^{(v)}(x,y) = Q^{(v)}(x,y)/(x+\alpha_i) \end{cases}$$

form a Gröbner basis of $\mathcal{M}'$.

Proof To simplify the formulas, (x,y) is removed from bivariate polynomial notations in the following whenever no ambiguity occurs. Assume that besides (α_i, β_i), the polynomials $Q^{(l)}$ ($l = 0, 1$) in the Gröbner basis of $\mathcal{M}$ pass a set of points $(\alpha_{i'}, \beta_{i'})$ with $i' \neq i$. $(x+\alpha_i)$ does not pass any point $(\alpha_{i'}, \beta_{i'})$ with $i' \neq i$. If $F^{(v)}$ does not pass $(\alpha_{i'}, \beta_{i'})$ with $i' \neq i$, then $Q^{(v)} = F^{(v)}(x+\alpha_i)$ does not pass that point either. This is a conflict, and hence $F^{(v)}$ passes the same set of interpolation points as $Q^{(v)}$, except (α_i, β_i). As a result, both $F^{(u)}$ and $F^{(v)}$ pass all points $(\alpha_{i'}, \beta_{i'})$ with $i' \neq i$, and they belong to $\mathcal{M}'$, the module of polynomials that pass all other interpolation points except (α_i, β_i).

To prove that $F^{(u)}$ and $F^{(v)}$ form a Gröbner basis of $\mathcal{M}'$, it is sufficient to show that the leading term of any polynomial in $\mathcal{M}'$ is divisible by that of either $F^{(u)}$ or $F^{(v)}$. It should be noted that $x^a y^b$ divides $x^{a'} y^{b'}$ if $b = b'$ and $a \leq a'$. This proof is done by contradiction. If $F^{(u)}$ and $F^{(v)}$ do not form a Gröbner basis of $\mathcal{M}'$, then there exists a polynomial, A, in the module $\mathcal{M}'$ whose leading term is not divisible by those of $F^{(u)}$ or $F^{(v)}$. The polynomials in a Gröbner basis have different y-degrees in their leading terms. Since $Q^{(l)}$ ($l = 0, 1$) is a Gröbner basis and the division by $(x+\alpha_i)$ does not change the y-degree in the leading term, the y-degrees in the leading terms of $F^{(u)}$ and $F^{(v)}$ are different. There are three cases to consider based on the y-degree of $LT(A)$ and whether $A(\alpha_i, \beta_i) = 0$. Here $LT(\cdot)$ denotes the leading term.

Case i: $\deg_y(LT(A)) = \deg_y(LT(F^{(v)}))$.
Here $\deg_y(\cdot)$ denotes the y-degree of bivariate monomial. Similarly, $\deg_x(\cdot)$ is the x-degree of bivariate polynomial. Since $\deg_y(LT(A)) = \deg_y(LT(F^{(v)}))$ and $LT(A)$ is not divisible by $LT(F^{(v)})$, $\deg_x(LT(A)) < \deg_x(LT(F^{(v)}))$. Let $B = (x+\alpha_i)A$. Apparently, B passes all interpolation points including (α_i, β_i), and is a polynomial in the module $\mathcal{M}$. Since $Q^{(v)} = (x+\alpha_i)F^{(v)}$, $\deg_x(LT(B)) < \deg_x(LT(Q^{(v)}))$, and thus $LT(Q^{(v)})$ does not divide $LT(B)$. It is also true that $\deg_y(LT(Q^{(v)})) = \deg_y(LT(F^{(v)})) = \deg_y(LT(A)) = \deg_y(LT(B))$, and $\deg_y(LT(Q^{(v)})) \neq \deg_y(LT(Q^{(u)}))$. Therefore,

$LT(B)$ is not divisible by $LT(Q^{(u)})$ either. This contradicts the assumption that $Q^{(u)}$ and $Q^{(v)}$ form a Gröbner basis of $\mathcal{M}$.

Case ii: $\deg_y(LT(A)) = \deg_y(LT(F^{(u)}))$ and $A(\alpha_i, \beta_i) = 0$.
A is also a polynomial in the module $\mathcal{M}$ in this case. $LT(A)$ is not divisible by $LT(Q^{(u)})$ because it is not divisible by $LT(F^{(u)})$ and $Q^{(u)} = F^{(u)}$. Moreover, since $LT(A)$ and $LT(Q^{(v)})$ have different y-degrees, $LT(A)$ is not divisible by $LT(Q^{(v)})$ either. This contradicts that $Q^{(u)}$ and $Q^{(v)}$ form a Gröbner basis of $\mathcal{M}$.

Case iii: $\deg_y(LT(A)) = \deg_y(LT(F^{(u)}))$ and $A(\alpha_i, \beta_i) \neq 0$.
Let
$$C = F^{(v)}(\alpha_i, \beta_i)A + A(\alpha_i, \beta_i)F^{(v)}.$$

Since both $F^{(v)}(\alpha_i, \beta_i)$ and $A(\alpha_i, \beta_i)$ are nonzero and A is not a scaled version of $F^{(v)}$, C is nonzero. Apparently, C passes all points including (α_i, β_i) and is a polynomial in the module $\mathcal{M}$. The lexicographical order of $LT(A)$ according to the weighted degree can be either higher or lower than that of $LT(F^{(v)})$. In the case that the order of $LT(A)$ is higher, $LT(C) = F^{(v)}(\alpha_i, \beta_i)LT(A)$. As mentioned in Case ii, $LT(A)$ is not divisible by either $LT(Q^{(u)})$ or $LT(Q^{(v)})$. Hence, $LT(C)$ is not divisible by $LT(Q^{(u)})$ or $LT(Q^{(v)})$. If the order of $LT(A)$ is lower than that of $LT(F^{(v)})$, $LT(C) = A(\alpha_i, \beta_i)LT(F^{(v)})$. $LT(F^{(v)})$ has different y-degree from $LT(Q^{(u)})$, and has lower x-degree than $LT(Q^{(v)})$. As a result, $LT(C)$ is not divisible by $LT(Q^{(u)})$ or $LT(Q^{(v)})$. These conflict with the fact that C is in the module $\mathcal{M}$.

There are conflicts in all these cases. Hence $F^{(v)}$ and $F^{(u)}$ form a Gröbner basis of module $\mathcal{M}'$ when $Q^{(v)}$ has the $(x + \alpha_i)$ factor and $Q^{(u)}$ does not. $\square$

2. None of $Q^{(l)}(x, y)$ $(l = 0, 1)$ has the $(x + \alpha_i)$ factor.
   Again $u = \arg\min_l\{w_l | q_1^{(l)}(\alpha_i) \neq 0\}$ and $v = \{0, 1\}\backslash u$. In this case, $q_1^{(u)}(\alpha_i) \neq 0$ and $q_1^{(v)}(\alpha_i) \neq 0$. For a given module, the Gröbner basis is not unique. Consider the following set of polynomials.

$$\begin{cases} G^{(u)}(x, y) = Q^{(u)}(x, y) \\ G^{(v)}(x, y) = q_1^{(v)}(\alpha_i)Q^{(u)}(x, y) + q_1^{(u)}(\alpha_i)Q^{(v)}(x, y) \end{cases} \tag{6.1}$$

Apparently, $G^{(u)}(x, y)$ and $G^{(v)}(x, y)$ pass the same interpolation points as $Q^{(u)}(x, y)$ and $Q^{(v)}(x, y)$. $w_u \leq w_v$ since $q_1^{(u)}(\alpha_i) \neq 0$ and $q_1^{(v)}(\alpha_i) \neq 0$. Hence $LT(G^{(v)}) = LT(Q^{(v)})$ and $G^{(u)}$ and $G^{(u)}$ form an equivalent Gröbner basis of $\mathcal{M}$. Assume that $G^{(v)}(x, y)$ can be written as $g_0^{(v)}(x) + g_1^{(v)}(x)y$. From (6.1)

$$g_0^{(v)}(\alpha_i) = q_1^{(v)}(\alpha_i)q_0^{(u)}(\alpha_i) + q_1^{(u)}(\alpha_i)q_0^{(v)}(\alpha_i). \tag{6.2}$$

Since $Q^{(l)}(\alpha_i, \beta_i) = 0 = q_0^{(l)}(\alpha_i) + q_1^{(l)}(\alpha_i)\beta_i$, $q_0^{(l)}(\alpha_i) = q_1^{(l)}(\alpha_i)\beta_i$ for $l = 0, 1$. Replace $q_0^{(u)}(\alpha_i)$ and $q_0^{(v)}(\alpha_i)$ in (6.2) by $q_1^{(u)}(\alpha_i)\beta_i$ and $q_1^{(v)}(\alpha_i)\beta_i$, respectively. Then

$$g_0^{(v)}(\alpha_i) = q_1^{(v)}(\alpha_i)q_1^{(u)}(\alpha_i)\beta_i + q_1^{(u)}(\alpha_i)q_1^{(v)}(\alpha_i)\beta_i = 0.$$

In addition, from (6.1),

$$g_1^{(v)}(\alpha_i) = q_1^{(v)}(\alpha_i)q_1^{(u)}(\alpha_i) + q_1^{(u)}(\alpha_i)q_1^{(v)}(\alpha_i) = 0.$$

Therefore, $G^{(v)}(x, y)$ has the $(x + \alpha_i)$ factor because both $g_0^{(v)}(\alpha_i)$ and $g_1^{(v)}(\alpha_i)$ are zero.

Since one of the polynomials in the equivalent Gröbner basis in (6.1) has the $(x + \alpha_i)$ factor, it reduces to category 1. Dividing $(x + \alpha_i)$ from $G^{(v)}(x, y)$ will lead to a Gröbner basis of $\mathcal{M}'$.

3. Both of $Q^{(l)}(x, y)$ have the $(x + \alpha_i)$ factor.
   In this case, $Q^{(l)}(x, y)$ can be rewritten as

$$\begin{cases} Q^{(0)}(x, y) = F^{(0)}(x, y)(x + \alpha_i) \\ Q^{(1)}(x, y) = F^{(1)}(x, y)(x + \alpha_i), \end{cases}$$

where $F^{(0)}$ and $F^{(1)}$ pass all points except (α_i, β_i). Construct another polynomial

$$D = F^{(1)}(\alpha_i, \beta_i)F^{(0)} + F^{(0)}(\alpha_i, \beta_i)F^{(1)}.$$

Apparently, D is not zero, and it passes (α_i, β_i) and all the other points. Hence, it is a polynomial in the module $\mathcal{M}$. $LT(D)$ equals either $LT(F^{(0)})$ or $LT(F^{(1)})$. $\deg_x(LT(F^{(0)})) = \deg_x(LT(Q^{(0)})) - 1$ and $\deg_x(LT(F^{(1)})) = \deg_x(LT(Q^{(1)})) - 1$. Hence $LT(D)$ is not divisible by $LT(Q^{(0)})$ nor $LT(Q^{(1)})$. This contradicts with the fact that $Q^{(0)}$ and $Q^{(1)}$ form a Gröbner basis of $\mathcal{M}$. Therefore, the case that both $Q^{(l)}(x, y)$ have the $(x + \alpha_i)$ factor does not happen.

Given the interpolation result of a test vector, a point, (α_i, β_i), included in the vector can be eliminated through constructing a Gröbner basis of the module $\mathcal{M}'$ as discussed above. Then (α_i, β_i') is added by carrying out the Kötter's forward interpolation algorithm for one iteration to derive the interpolation result of the next test vector. The Gröbner basis derived through this backward-forward interpolation process may not be the same as the Gröbner basis constructed by interpolating over all the points using the Kötter's algorithm. However, they are Gröbner bases of the same module. From [72], the polynomial of the minimum weighted degree in a module must appear in any Gröbner basis of the module, although it may be different by a nonzero scaler. Scaling the interpolation output polynomial by a constant does not change its $y + f(x)$ factor. Therefore, the polynomial of minimum weighted degree from the Gröbner basis derived by the backward-forward interpolation leads to the same decoding output as using the Kötter's algorithm to interpolate over each point in the next test vector.

Algorithm 9 Backward-forward Interpolation for Chase Decoding
input: *test vectors of interpolation points with multiplicity one arranged in Gray code order*
initialization:
$Q^{(0)}(x,y) = 1, Q^{(1)}(x,y) = y$
$w_0 = 0, w_1 = k-1$ *($w_1 = -1$ if re-encoding & coordinate transformation are used)*
forward interpolation for the first test vector:
for each point (α_i, β_i) in the first test vector
    compute $Q^{(l)}(\alpha_i, \beta_i)$ for $l = 0, 1$
    $u = \arg\min_l \{w_l | Q^{(l)}(\alpha_i, \beta_i) \neq 0\}$, $v = \{0,1\} \backslash u$
    $Q^{(v)}(x,y) \Leftarrow Q^{(v)}(\alpha_i, \beta_i)Q^{(u)}(x,y) + Q^{(u)}(\alpha_i, \beta_i)Q^{(v)}(x,y)$
    $Q^{(u)}(x,y) \Leftarrow Q^{(u)}(x,y)(x + \alpha_i)$
    $w_u \Leftarrow w_u + 1$
backward-forward interpolation for each next test vector:
    backward interpolation *(eliminate (α_i, β_i)):*
1:       *compute $q_1^{(l)}(\alpha_i)$ for $l = 0, 1$*
         $u = \arg\min_l \{w_l | q_1^{(l)}(\alpha_i) \neq 0\}$, $v = \{0,1\} \backslash u$
         if $(q_1^{(v)}(\alpha_i) \neq 0)$
2:           $Q^{(v)}(x,y) \Leftarrow q_1^{(u)}(\alpha_i)Q^{(v)}(x,y) + q_1^{(v)}(\alpha_i)Q^{(u)}(x,y)$
3:       $Q^{(v)}(x,y) \Leftarrow Q^{(v)}(x,y)/(x + \alpha_i)$
         $w_v \Leftarrow w_v - 1$
    forward interpolation: *(add (α_i, β_i'))*
4:       *compute $Q^{(l)}(\alpha_i, \beta_i')$ for $l = 0, 1$*
         $u' = \arg\min_l \{w_l | Q^{(l)}(\alpha_i, \beta_i') \neq 0\}$, $v' = \{0,1\} \backslash u$
5:       $Q^{(v')}(x,y) \Leftarrow Q^{(v')}(\alpha_i, \beta_i')Q^{(u')}(x,y) + Q^{(u')}(\alpha_i, \beta_i')Q^{(v')}(x,y)$
6:       $Q^{(u')}(x,y) \Leftarrow Q^{(u')}(x,y)(x + \alpha_i)$
         $w_{u'} \Leftarrow w_{u'} + 1$

Assume that the test vectors in the Chase decoder are arranged in Gray code manner. Employing the backward-forward scheme, the interpolation results for all test vectors can be derived in one run using the process shown in Algorithm 9. In this process, the Kötter's forward interpolation in Algorithm 4, which is listed in Chapter 5, is used to derive the interpolation result of the first test vector. Then the backward-forward interpolation is adopted to compute the interpolation result for each of the next test vectors. Instead of

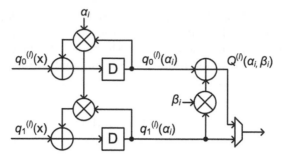

FIGURE 6.1
Polynomial evaluation architecture (modified from [55])

starting the interpolation afresh for each test vector, the backward-forward process only takes care of the differences among the test vectors. As a result, the number of iterations needed to complete the interpolation for all test vectors is substantially reduced, especially when there are a large number of test vectors. In addition, as seen from steps 1-3 and 4-6 of Algorithm 9, the backward interpolation shares many common computations with the forward interpolation. Therefore, the hardware overhead for incorporating the backward interpolation is very small.

Both of the backward and forward interpolations consist of evaluation value computation (steps 1 and 4) and polynomial updating (steps 2-3 and 4-5). The evaluation values of $Q^{(l)}(x, y)$ and $q_1^{(l)}(x)$ over a point (α_i, β_i) can be computed using the polynomial evaluation architecture shown in Fig. 6.1. In this architecture, Horner's rule is adopted, and the coefficients are input serially starting with the most significant one. $q_0^{(l)}(\alpha_i)$ and $q_1^{(l)}(\alpha_i)$ are calculated by the two feedback loops, and then $Q^{(l)}(\alpha_i, \beta_i) = q_0^{(l)}(\alpha_i) + q_1^{(l)}(\alpha_i)\beta_i$ is derived. $Q^{(l)}(\alpha_i, \beta_i)$ or $q_1^{(l)}(\alpha_i)$ is chosen as the output depending on whether the current iteration is forward or backward interpolation.

Once the evaluation value is available, the polynomial updating is carried out. The architecture in Fig. 6.2 is capable of updating the polynomials for both the backward and forward interpolations in a time-multiplexed way. This architecture updates the coefficients of $q_s^{(0)}(x)$ and $q_s^{(1)}(x)$, and two copies of this architecture can be employed to update the coefficients with $s=0$ and $s=1$ simultaneously. The coefficients of $q_s^{(0)}(x)$ and $q_s^{(1)}(x)$ are updated serially, starting with the the most significant one.

During the forward interpolation, the polynomial updating is done according to steps 5 and 6 of Algorithm 9. They are the same as those in the forward interpolation for the first test vector. Four multiplexors would be needed to choose proper polynomials and evaluation values in the computation of $Q^{(v')}(\alpha_i, \beta_i')Q^{(u')}(x, y) + Q^{(u')}(\alpha_i, \beta_i')Q^{(v')}(x, y)$. However, since

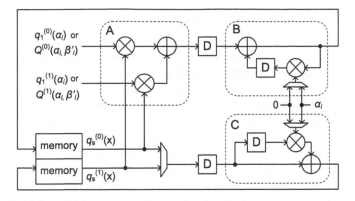

FIGURE 6.2
Polynomial updating architecture for backward-forward interpolation (modified from [55])

$u', v' \in \{0,1\}$ and $u' \neq v'$, this linear combination is done equivalently as $Q^{(0)}(\alpha_i, \beta_i')Q^{(1)}(x,y) + Q^{(1)}(\alpha_i, \beta_i')Q^{(0)}(x,y)$. Accordingly, it is implemented by block A in Fig. 6.2 without using any multiplexor. The linearly combined polynomial requires no further computation, and it is passed through block B by selecting '0' as the input to the multiplier in block B. Step 6 in Algorithm 9 for multiplying $Q^{(u')}(x,y)$ by $(x + \alpha_i)$ is implemented by the C block. After the updating, the polynomials are written back to fixed memories. Compared to writing each polynomial back to the memory from which it is read, multiplexors have been saved and the critical path is reduced at the cost of simple logic to keep track of the two polynomials. To cut the critical path, registers can be added between blocks A and B and before block C as shown in Fig. 6.1. This pipelining only leads to one clock cycle extra latency in each interpolation iteration.

During the backward interpolation, the polynomial updating architecture in Fig. 6.2 takes $q_1^{(l)}(\alpha_i)$ instead of $Q^{(l)}(\alpha_i, \beta_i')$. For the case of $q_1^{(v)}(\alpha_i) \neq 0$, the linear combination needed in step 2 of Algorithm 9 is also implemented by block A. If $q_1^{(v)}(\alpha_i) = 0$, block A outputs $q^{(u)}(\alpha_i)Q^{(v)}(x,y)$. As discussed previously, multiplying an interpolation polynomial by a nonzero scaler does not affect the decoding result. Therefore, regardless of whether $q_1^{(v)}(\alpha_i)$ is zero, the output of block A is sent to block B to be divided by $(x + \alpha_i)$. The division is achieved by passing α_i through the multiplexor in block B. In the backward interpolation, no computation is needed on $Q^{(u)}(x,y)$. Hence, it is passed through block C unchanged by selecting '0' as the multiplier input in this block. The updated coefficients of the polynomials are written back to fixed memory blocks as in the forward interpolation.

6.1.1.2 Unified backward-forward interpolation

Using the backward-forward interpolation process in Algorithm 9, it takes two iterations to derive the interpolation result for the next test vector: one backward iteration and one forward iteration. Further speedup can be achieved if the backward and forward iterations are carried out simultaneously. However, data dependency demands that the forward interpolation iteration for adding (α_i, β'_i) waits until after the backward interpolation for deleting (α_i, β_i) is completed. To solve this dilemma, a unified interpolation scheme was proposed in [68]. This scheme computes the evaluation values needed for the forward interpolation based on the input polynomials to the backward interpolation in a look-ahead manner, and then combines the polynomial updating for both iterations. As a result, one single unified backward-forward iteration is needed to derive the interpolation result for the next test vector.

According to Algorithm 9, after steps 2 and 3 for eliminating (α_i, β_i), the two involved polynomials become

$$\begin{cases} F^{(v)}(x,y) = (q_1^{(u)}(\alpha_i)Q^{(v)}(x,y) + q_1^{(v)}(\alpha_i)Q^{(u)}(x,y))/(x+\alpha_i) \\ F^{(u)}(x,y) = Q^{(u)}(x,y). \end{cases} \tag{6.3}$$

The forward interpolation iteration for adding (α_i, β'_i) starts with these polynomials. The evaluation values needed in this iteration are $F^{(l)}(\alpha_i, \beta'_i)$. From (6.3),

$$F^{(v)}(x,y)(x+\alpha_i) = q_1^{(u)}(\alpha_i)Q^{(v)}(x,y) + q_1^{(v)}(\alpha_i)Q^{(u)}(x,y).$$

Take the derivative of the above equation with respect to x.

$$F^{(v)}(x,y)+\frac{\partial(F^{(v)}(x,y))}{\partial x}(x+\alpha_i) = q_1^{(u)}(\alpha_i)\frac{\partial Q^{(v)}(x,y)}{\partial x}+q_1^{(v)}(\alpha_i)\frac{\partial(Q^{(u)}(x,y))}{\partial x}.$$

Hence,

$$F^{(v)}(\alpha_i, \beta'_i) = q_1^{(u)}(\alpha_i)\frac{\partial Q^{(v)}(x,y)}{\partial x}\bigg|_{(\alpha_i,\beta'_i)} + q_1^{(v)}(\alpha_i)\frac{\partial(Q^{(u)}(x,y))}{\partial x}\bigg|_{(\alpha_i,\beta'_i)}. \tag{6.4}$$

Also from (6.3)

$$F^{(u)}(\alpha_i, \beta'_i) = Q^{(u)}(\alpha_i, \beta'_i). \tag{6.5}$$

According to (6.4) and (6.5), $F^{(l)}(\alpha_i, \beta'_i)$ $(l = 0, 1)$ can be computed based on $Q^{(l)}(x,y)$, the input to the backward interpolation. Once the evaluation values, $q_1^{(l)}(\alpha_i)$ and $F^{(l)}(\alpha_i, \beta'_i)$, are computed and the decisions on u and u' are made, the polynomial updating in steps 2-3 and 5-6 of Algorithm 9 for both the backward and forward interpolation iterations are combined and carried out together.

Next, the VLSI architectures for implementing the unified backward-forward interpolation are presented. These architectures are also capable of

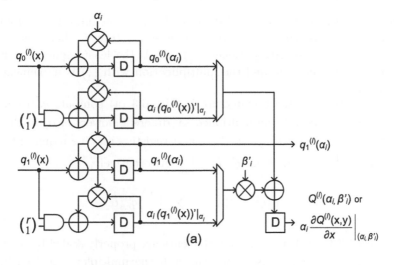

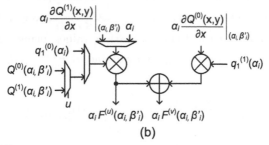

FIGURE 6.3
Polynomial evaluation architecture for unified backward-forward interpolation (modified from [68])

carrying out forward-only interpolation, which is needed to derive the result for the first test vector. In the forward-only interpolation, the evaluation values needed are $Q^{(l)}(\alpha_i, \beta_i')$. The look-ahead evaluation value computation according to (6.4) requires $\left.\frac{\partial Q^{(l)}(x,y)}{\partial x}\right|_{(\alpha_i,\beta_i')}$. These evaluation values are broken down into

$$\begin{cases} Q^{(l)}(\alpha_i, \beta_i') = q_1^{(l)}(\alpha_i)\beta_i' + q_0^{(l)}(\alpha_i) \\ \left.\frac{\partial Q^{(l)}(x,y)}{\partial x}\right|_{(\alpha_i,\beta_i')} = \left.(q_1^{(l)}(x))'\right|_{\alpha_i}\beta_i' + \left.(q_0^{(l)}(x))'\right|_{\alpha_i}. \end{cases}$$

The $q_1^{(l)}(\alpha_i)$ needed for backward interpolation are computed as an intermediate result for $Q^{(l)}(\alpha_i, \beta_i')$ calculations. $q_s^{(l)}(\alpha_i)$ and $\left.(q_s^{(l)}(x))'\right|_{\alpha_i}$ ($s = 0, 1$) can be written as

$$\begin{cases} q_s^{(l)}(\alpha_i) = \sum_{r \geq 0} q_{r,s}^{(l)}\alpha_i^r \\ \left.(q_s^{(l)}(x))'\right|_{\alpha_i} = \sum_{r \geq 1} \binom{r}{1}q_{r,s}^{(l)}\alpha_i^{r-1}, \end{cases} \tag{6.6}$$

where $q_{r,s}^{(l)}$ is the coefficient of $x^r y^s$ in $Q^{(l)}(x,y)$. Over finite fields of characteristic two, $\binom{r}{1}$ is reduced modulo 2. It is either 0 or 1 when r is an even or odd number, respectively. Hence, $\binom{r}{1}$ equals the least significant bit of r in binary representation, and the multiplication with $\binom{r}{1}$ is implemented as bit-wise AND.

The left parts of the architecture in Fig. 6.3 (a) compute $q_s^{(l)}(\alpha_i)$. (6.6) shows that $q_{r,s}^{(l)}$ needs to be multiplied by different powers of α_i in the computations of $q_s^{(l)}(\alpha_i)$ and $(q_s^{(l)}(x))'\big|_{\alpha_i}$, and registers may be required to align $q_{r,s}^{(l)}$ with α_i^{r-1} for computing $(q_s^{(l)}(x))'\big|_{\alpha_i}$. Alternatively,

$$\alpha_i (q_s^{(l)}(x))'\big|_{\alpha_i} = \sum_{r \geq 1} \binom{r}{1} q_{r,s}^{(l)} \alpha_i^r$$

can be used if the other involved coefficients are properly scaled by α_i. Either $q_s^{(l)}(\alpha_i)$ or $\alpha_i (q_s^{(l)}(x))'\big|_{\alpha_i}$ is passed through the multiplexor in Fig. 6.3 (a) depending on whether $Q^{(l)}(\alpha_i, \beta_i')$ or $\alpha_i \frac{\partial Q^{(l)}(x,y)}{\partial x}\big|_{(\alpha_i, \beta_i')}$ needs to be computed.

Fig. 6.3 (b) computes the final evaluation values using the outputs of Fig. 6.3 (a). Since $\alpha_i \frac{\partial Q^{(l)}(x,y)}{\partial x}\big|_{(\alpha_i, \beta_i')}$ is available, $\alpha_i F^{(v)}(\alpha_i, \beta_i')$ is calculated instead. From (6.4),

$$\alpha_i F^{(v)}(\alpha_i, \beta_i') = q_1^{(0)}(\alpha_i)(\alpha_i \frac{\partial Q^{(1)}(x,y)}{\partial x}\bigg|_{(\alpha_i, \beta_i')}) + q_1^{(1)}(\alpha_i)(\alpha_i \frac{\partial Q^{(0)}(x,y)}{\partial x}\bigg|_{(\alpha_i, \beta_i')})$$

because $u, v \in \{0, 1\}$ and $u \neq v$. To carry out the polynomial updating in step 5 of Algorithm 9 correctly, $F^{(u)}(\alpha_i, \beta_i') = Q^{(u)}(\alpha_i, \beta_i')$ also needs to be scaled by α_i. To save a multiplier, the architecture in Fig. 6.3 (b) is shared in a time-multiplexed manner to compute $\alpha_i F^{(v)}(\alpha_i, \beta_i')$ and $\alpha_i F^{(u)}(\alpha_i, \beta_i')$.

The polynomial updating for the unified backward-forward interpolation, which includes steps 2-3 and 5-6 of Algorithm 9, is implemented by the architecture illustrated in Fig. 6.4. During a unified backward-forward interpolation iteration, the linear combination and division by $(x + \alpha_i)$ for computing $F^{(v)}(x,y)$ according to steps 2-3 are implemented by blocks A and B in Fig. 6.4. $F^{(u)}(x,y)$ equals $Q^{(u)}(x,y)$, and hence it is chosen from $Q^{(0)}(x,y)$ and $Q^{(1)}(x,y)$ through the multiplexor whose select signal is u. The output of block C in Fig. 6.4 is

$$\alpha_i F^{(u)}(\alpha_i, \beta_i') F^{(v)}(x,y) + \alpha_i F^{(v)}(\alpha_i, \beta_i') F^{(u)}(x,y).$$

It is a scaled version of the updated $Q^{(v')}(x,y)$ in step 5 of Algorithm 9 for the forward interpolation. For step 6, $F^{(u')}(x,y)$ is multiplied by $(x + \alpha_i)$ in block D. Similarly, the updated polynomial coefficients are written back to fixed memory blocks, and registers are added for the purpose of pipelining in Fig. 6.4. During a forward-only interpolation, which is needed for the first test vector, one polynomial is updated through a linear combination and the other

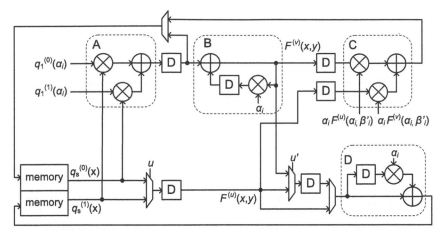

FIGURE 6.4
Polynomial updating architecture for unified backward-forward interpolation
(modified from [68])

is multiplied by $(x+\alpha_i)$. For these iterations, $Q^{(l)}(\alpha_i, \beta_i)$ instead of $q_1^{(l)}(\alpha_i)$ are
fed to the unified polynomial architecture in Fig. 6.4. The linear combination
and $(x + \alpha_i)$ multiplication are done by block A and D, respectively.

The complexities of the interpolators for the Chase decoding are summarized in Table 6.1 assuming that the re-encoding and coordinate transformation [56, 57] are employed. The maximum x-degree, dx, of the interpolation
polynomials is zero at the beginning. In the worst case, it is increased by one
in each forward interpolation iteration for the first test vector. In addition, it
remains about the same during the backward-forward interpolation iterations.
Since the polynomial coefficients with different x-degrees are updated serially,
and the updating is overlapped with the evaluation value computations for
the next iteration, the number of clock cycles needed in each interpolation
iteration is $dx + \xi$, where ξ is the pipelining latency. If the interpolation over
each test vector is done separately, an interpolator that is capable of carrying out only the forward interpolation is needed. Such an interpolator can
be implemented by the same polynomial evaluation architecture in Fig. 6.1
and the polynomial updating architecture in Fig. 6.2 with block B removed.
Hence, it has a shorter data path and requires one less pipelining stage than
the backward-forward interpolator to achieve the same critical path. As can
be seen from Fig. 6.4, the unified polynomial updating can start once $q_1^{(l)}(\alpha_i)$
is computed by part (a) of the unified polynomial evaluation architecture in
Fig. 6.3. Hence, the unified interpolator only has one more pipelining stage
than the backward-forward interpolator. When only the forward interpolator
is available, the intermediate interpolation result over the $n - k - \eta$ points
that are common to all test vectors is stored. Then η forward iterations are
needed to interpolate over the rest η points for each test vector. Employing

TABLE 6.1

Hardware complexities of interpolators for the Chase decoding with 2^η test vectors for high-rate (n, k) RS code over $GF(2^q)$

	Forward-only	Backward-forward [55]	Unified backward-forward [68]
# of iter.	$(n-k)+\eta(2^\eta-1)$	$(n-k)+2(2^\eta-1)$	$(n-k)+2^\eta-1$
# of clks/iter.	$dx + 3$	$dx + 4$	$dx + 5$
Hardware requirement			
GF mult.	12	14	24
GF add.	10	12	19
Mux	$2q$	$6q$	$15q$
Memory(bit)	$(8(n-k)-4\eta)q$	$4(n-k)q$	$4(n-k)q$
Register	$6q$	$12q$	$24q$

the backward-forward interpolation, one backward and one forward interpolation iterations are carried out to derive the result for the next test vector regardless of η. The number of iterations is further reduced to one in the unified backward-forward interpolation. As a result, the unified architecture can achieve almost η times speedup over the forward-only interpolator when η is not small. In addition, in both backward-forward schemes, no intermediate interpolation result needs to be stored. Hence, their memory requirement is much less than that of the forward-only interpolator as shown in Table 6.1.

6.1.1.3   Generalized backward interpolation

Besides the Chase decoder with multiplicity one for high-rate codes, the bit-level generalized minimum distance (BGMD) decoder [52] with $m_{max} = 2$ is also considered in the backward interpolation algorithm in [55]. However, this algorithm has limitations. It can only eliminate all the points in a code position if their multiplicities are all one and the number of points is one less than the number of polynomials in the Gröbner basis. The Chase algorithm can be generalized to have multiplicities larger than one. Besides flipping the symbols in the η least reliable code positions, the other code positions can have different multiplicities according to their reliabilities to improve the error-correcting capability. One example is the KV-LCC scheme [87], which adopts the Kötter-Vardy (KV) [47] scheme to assign multiplicities for the $n-k-\eta$ common code positions. In order to complete the interpolation for all test vectors in one-run for these cases, the backward interpolation was generalized in [83]. The generalized backward interpolation scheme is capable of iteratively reducing the multiplicity of each point in the same code position by one at a time, until all of them become zero. For example, assume that code position i has three interpolation points $(\alpha_i, \beta_{i,0}, 3)$, $(\alpha_i, \beta_{i,1}, 1)$ and $(\alpha_i, \beta_{i,2}, 2)$, where the third coordinate is the multiplicity. The generalized backward interpolation is able to reduce the multiplicities to $(\alpha_i, \beta_{i,0}, 2)$, $(\alpha_i, \beta_{i,1}, 0)$, $(\alpha_i, \beta_{i,2}, 1)$, and then to $(\alpha_i, \beta_{i,0}, 1)$, $(\alpha_i, \beta_{i,1}, 0)$, $(\alpha_i, \beta_{i,2}, 0)$, and to $(\alpha_i, \beta_{i,0}, 0)$, $(\alpha_i, \beta_{i,1}, 0)$, $(\alpha_i, \beta_{i,2}, 0)$ at the end. Unlike the backward interpolation in [55], the generalized backward interpolation is not limited by the number of points in the same code position, and the multiplicities of the points can be different. The only constraint is that the sum of the multiplicities of the points in the same code position should not exceed the maximum multiplicity of all interpolation points. This condition is satisfied in most ASD algorithms, such as the KV [47], BGMD [52], Chase [51] and KV-LCC [87] algorithms. The generalized backward interpolation can be adopted broadly to enable the sharing of computation results in the interpolation for multiple test vectors.

As mentioned before, Gröbner basis is not unique. The Gröbner bases of the same module are referred to as equivalent Gröbner bases. Interpolating over the points or constraints in different orders yields different Gröbner bases that are equivalent. They can be transformed to each other by $\mathbb{F}[x]$-linear combinations of the polynomials in the bases. $\mathbb{F}[x]$-linear combination means multiplying the basis polynomials with univariate polynomials in x over finite field $\mathbb{F}$ and then adding up the products. Assume that code position i has p interpolation points of multiplicity one. Given a Gröbner basis derived by the Kötter's forward interpolation algorithm, the backward interpolation in [55] constructs an equivalent Gröbner basis that has p polynomials with factor $(x + \alpha_i)$. Dividing $(x + \alpha_i)$ from each of these polynomials, a Gröbner basis of polynomials that pass all interpolation points except the p points in position i is derived. Note that code position i does not need to be last code position covered during the forward interpolation. The backward interpolation in [55] actually first derives a Gröbner basis equivalent to that which should have been computed if the points to be eliminated are the last ones covered during the Kötters forward interpolation. Then the $(x + \alpha_i)$ factors are divided to reverse the interpolation. The generalized backward interpolation in [83] was developed based on a similar idea. Take a point (α, β) with multiplicity m as an example. Reducing the multiplicity of this point from m to $m - 1$ is equivalent to eliminating the interpolation constraints that the coefficients of $x^{m-1}, x^{m-2}y, x^{m-3}y^2, \ldots, y^{m-1}$ in the shifted interpolation polynomial $Q^{(l)}(x + \alpha, y + \beta)$ are zero. The generalized backward interpolation constructs a Gröbner basis equivalent to the one that would have been derived by the Kötter's forward interpolation if the constraints to be eliminated are covered last. Then the computations done during the interpolation iterations for those last constraints are reversed.

To eliminate the last constraint covered by the Kötter's forward interpolation in Algorithm 4 in Chapter 5, the polynomial, $Q^{(l^*)}(x, y)$, that has been multiplied by $(x + \alpha_i)$ needs to be divided by this factor. Nevertheless, the linear combinations used to update the other polynomials do not need to be reversed, since they do not alter the weighted degree of $Q^{(l)}(x, y)$ for $l \neq l^*$. If code position i has p points of multiplicity one and they are interpolated last using the Kötter's algorithm, then there are exactly p different polynomials with the $(x + \alpha_i)$ factor at the end, as will be proved later in this section. Accordingly, the backward interpolation in [55] constructs an equivalent Gröbner basis that has p polynomials with the $(x + \alpha_i)$ factor. From Algorithm 4, when the multiplicity of a point is larger than one, multiple $(x + \alpha_i)$ factors may be multiplied to the same polynomial during the interpolation over that point. In addition, it is possible that a polynomial may lose a $(x + \alpha_i)$ factor during the linear combinations before it is multiplied by another $(x + \alpha_i)$ factor.

Definition 29 *Let $d_{a,b}^{(l)}(\alpha_i, \beta_{i,j})$ be the discrepancy coefficient defined in (5.13). In the case that $\{l | d_{a,b}^{(l)}(\alpha_i, \beta_{i,j}) \neq 0\}$ is an empty set in an iteration of the Kötter's interpolation in Algorithm 4, the corresponding interpolation constraint is said to be dependent on previously covered interpolation constraints.*

When a code position has multiple interpolation points with multiplicities larger than one, the corresponding interpolation constraints may not be all independent. If $\{l | d_{a,b}^{(l)}(\alpha_i, \beta_{i,j}) \neq 0\} = \emptyset$ in an iteration, then none of the polynomials is multiplied by $(x + \alpha_i)$ in that iteration. Therefore, the number of independent constraints during the forward interpolation over the points in position i equals the number of $(x + \alpha_i)$ factors multiplied in those interpolation iterations. Moreover, the dependency of the constraints is affected by the order of the interpolation constraints covered. For general cases, if the interpolation constraints are not covered last during the forward interpolation, the interpolation result does not tell how many $(x + \alpha_i)$ factors would have been multiplied if they were the last interpolated constraints. Due to these reasons, it is much more challenging to construct equivalent Gröbner bases and reverse the computations in the forward interpolation for general cases.

The goal of the generalized backward interpolation is to construct a Gröbner basis equivalent to that which should have been derived if certain constraints are interpolated last, so that the computations corresponding to those constraints can be reversed. To achieve this goal, we need to know what kind of equivalent Gröbner bases should be constructed and how to construct them. Before these two questions are answered, detailed analyses are given below on how the equivalent Gröbner bases computed by carrying out the Kötter's algorithm over the constraints in different orders are related, and how the multiplicities of the interpolation points in code position i are affected by the division and multiplication by $(x + \alpha_i)$.

Lemma 11 *A polynomial in a Gröbner basis can be replaced by a linear combination of itself and other basis polynomials that have lower lexicographic order in the leading term.*

Proof: According to Definition 28 for Gröbner basis in Chapter 5, a polynomial in a Gröbner basis can be replaced by any polynomial in the module with the same leading term. If one monomial has a lower weighted degree than another or has the same weighted degree but a lower y-degree, then it is said to have lower weighted lexicographic order. If a polynomial is linearly combined with other polynomials whose leading terms have lower lexicographic orders, then its leading term remains the same. Also linearly combining polynomials in a module generates a polynomial that satisfies the same constraints. Therefore, replacing a basis polynomial by a linear combination of itself (or a polynomial in the module with the same leading term) and other basis polynomials (or other polynomials in the module) that have lower lexicographic order in the leading terms yields a Gröbner basis of the same module. □

Lemma 12 *The Gröbner bases computed from carrying out the Kötter's algorithm over the same interpolation constraints in different orders can always be transformed to each other by* $\mathbb{F}[x]$*-linear combinations of the polynomials in the bases.*

   Proof: Assume that the Kötter's algorithm is carried out over the same interpolation constraints but in two different orders, and the results are two Gröbner bases $\mathbb{A}$ and $\mathbb{B}$. Since the interpolation constraints are the same, these bases correspond to the same module of polynomials. From the Kötter's algorithm, the module is actually a $\mathbb{F}[x]$-module contained in $\mathbb{F}[x,y]$. Let $\mathbb{A} = \{a_i(x,y)\}$. Any polynomial in this module can be expressed by an $\mathbb{F}[x]$-linear combination of $\{a_i(x,y)\}$ as $\sum_{a_i(x,y)\in\mathbb{A}} f_i(x)a_i(x,y)$, where $f_i(x) \in \mathbb{F}[x]$. Since a polynomial in $\mathbb{B}$ is also a polynomial in the module, it can be also expressed as an $\mathbb{F}[x]$-linear combination of $\{a_i(x,y)\}$. Therefore, the Gröbner bases computed from the Kötter's interpolation over the same constraints can be always transformed to each other through $\mathbb{F}[x]$-linear combinations.□

   From Lemma 12, a Gröbner basis computed by interpolating over the constraints in any order using the Kötter's algorithm can always be transformed to an equivalent Gröbner basis, from which a given set of interpolation constraints can be eliminated.

   Divisions by $(x + \alpha_i)$ are required to reverse the interpolation over the constraints for code position i.

Lemma 13 *After $(x + \alpha_i)$ is divided from a polynomial, the multiplicity of each point in code position i on the polynomial is reduced by one. The multiplicities of the points in other code positions on this polynomial remain the same.*

   Proof: Let $m_{i,j}(Q(x,y))$ be the multiplicity of the point $(\alpha_i, \beta_{i,j})$ on $Q(x,y)$. From the definition of multiplicity, if $Q(x,y) = F(x,y)E(x,y)$, then $m_{i,j}(Q(x,y)) = m_{i,j}(F(x,y)) + m_{i,j}(E(x,y))$. Since $Q(x,y) = (Q(x,y)/(x + \alpha_i))(x + \alpha_i)$, $m_{i,j}(Q(x,y)) = m_{i,j}(Q(x,y)/(x + \alpha_i)) + m_{i,j}(x + \alpha_i) = m_{i,j}(Q(x,y)/(x + \alpha_i)) + 1$. Accordingly, the multiplicity of $(\alpha_i, \beta_{i,j})$ on $Q(x,y)/(x+\alpha_i)$ is one less. On the other hand, for $i' \neq i$, $m_{i',j}(Q(x,y)) = m_{i',j}(Q(x,y)/(x+\alpha_i))+m_{i',j}(x+\alpha_i) = m_{i',j}(Q(x,y)/(x+\alpha_i))+0$. Therefore, the multiplicities of the interpolated points in other code positions remain the same. □

Following a similar proof, multiplying $(x + \alpha_i)$ to a polynomial increases the multiplicity of each point in code position i on this polynomial by one. Therefore, the polynomial chosen as $Q^{(l^*)}(x, y)$ and multiplied by $(x + \alpha_i)$ in an iteration of the Kötter's interpolation in Algorithm 4 for constraint (a, b) with $a + b = m_{i,j} - 1$ will have zero discrepancy coefficients in the iterations for other constraints (a', b') with $a' + b' = m_{i,j} - 1$. It will not be chosen as $Q^{(l^*)}(x, y)$ and multiplied by $(x + \alpha_i)$ again in those iterations. In other words, the $(x + \alpha_i)$ factors are multiplied to different basis polynomials during the interpolation iterations for increasing the multiplicity of each point in code position i from $m_{i,j} - 1$ to $m_{i,j}$. Reversing the interpolation iterations in which the $(x + \alpha_i)$ factors were multiplied to different polynomials is easy. However, if the interpolation constraints were covered in a different order and the $(x + \alpha_i)$ factors were not multiplied to distinct polynomials, we need to find not only which polynomials have the factors, but also how many factors each of them have. In addition, the reverse of the interpolation for different points in the same code position can not be separated, since multiplying $(x + \alpha_i)$ increases the multiplicity of each point in code position i by one. This is the reason that the generalized backward interpolation reduces the multiplicity of every point in the same code position by one at a time.

As aforementioned, there may be dependencies among the interpolation constraints. If a set of constraints is covered among all constraints in a different order during the forward interpolation, then the total number of independent constraints in this set may vary, and hence the number of $(x + \alpha_i)$ factors multiplied during the interpolation over these constraints may change. Let h be the number of independent constraints during the forward interpolation iterations for increasing the multiplicity of each $(\alpha_i, \beta_{i,j})$ from $m_{i,j} - 1$ to $m_{i,j}$, assuming all the other interpolation constraints were covered before. Note that h is unknown if the constraints were covered in a different order. Starting from a Gröbner basis derived by the Kötter's interpolation, regardless of the order of the constraints covered, Lemma 14 and 15 in the following will show that the constraints for increasing the multiplicity of each $(\alpha_i, \beta_{i,j})$ from $m_{i,j} - 1$ to $m_{i,j}$ can be always eliminated through constructing an equivalent Gröbner basis that has exactly h polynomials with the $(x + \alpha_i)$ factor. A method for constructing such an equivalent Gröbner basis will be presented in Algorithm 10, whose correctness is proved in Lemma 16 and 17. Making use of this equivalent Gröbner basis, the generalized backward interpolation is done according to Algorithm 11 [83].

Lemma 14 *There are no more than h polynomials with the $(x + \alpha_i)$ factor in any equivalent Gröbner basis.*

Proof: This is proved by contradiction. If an equivalent Gröbner basis has more than h polynomials with the $(x + \alpha_i)$ factor, then the multiplicity of $(\alpha_i, \beta_{i,j})$ on each of them is reduced from $m_{i,j}$ to $m_{i,j} - 1$ after $(x + \alpha_i)$ is divided according to Lemma 13. h is defined as the number of polynomials need to be multiplied by $(x + \alpha_i)$ in order to increase the multiplicity of each point from $m_{i,j} - 1$ to $m_{i,j}$ after the other constraints are satisfied. Multiply $(x + \alpha_i)$ back to h of the polynomials. Then at least one of the polynomials in the basis, say $Q^{(l)}(x, y)$, does not have the $(x + \alpha_i)$ factor multiplied back but still passes each $(\alpha_i, \beta_{i,j})$ with multiplicity $m_{i,j}$. This means that $Q^{(l)}(x, y)$ is a basis polynomial, and $Q^{(l)}(x, y)/(x + \alpha_i)$ is a polynomial in the corresponding module. $LT(Q^{(l)}(x, y)/(x + \alpha_i))$ is not divisible by $LT(Q^{(l)}(x, y))$ since the x-degree of $LT(Q^{(l)}(x, y)/(x + \alpha_i))$ is lower. In addition, the y-degrees of the leading terms of the polynomials in a Gröbner basis are distinct. Hence $LT(Q^{(l)}(x, y)/(x + \alpha_i))$ is not divisible by the leading term of any other basis polynomial either. This conflicts with the definition of Gröbner basis, and hence there should not be more than h polynomials with the $(x + \alpha_i)$ factor in any equivalent Gröbner basis. $\square$

In the proof for Lemma 14, no assumption has been made on the order of the constraints covered for increasing the multiplicities from 0 to $m_{i,j} - 1$, nor that for increasing the multiplicities from $m_{i,j} - 1$ to $m_{i,j}$. The orders of the constraints within each of these two groups do not affect the value of h. h is the same as long as all constraints for increasing the multiplicities from 0 to $m_{i,j} - 1$ are covered before those for increasing the multiplicities from $m_{i,j} - 1$ to $m_{i,j}$.

When a polynomial is linearly combined with another in later iterations of the Kötter's interpolation algorithm, it may lose its $(x + \alpha_i)$ factor. Therefore, the Gröbner basis output by the Kötter's algorithm may have less than h polynomials with the $(x + \alpha_i)$ factor, especially if the interpolation constraints for increasing the multiplicities of $(\alpha_i, \beta_{i,j})$ from $m_{i,j} - 1$ to $m_{i,j}$ are not covered last. Lemma 13 says that a Gröbner basis computed by the Kötter's algorithm over the same interpolation constraints can be always transformed to a basis that has h polynomials with the $(x + \alpha_i)$ factor. Although the bases that have h polynomials with the $(x + \alpha_i)$ factor are not unique either, any of them can be used to construct a Gröbner basis of polynomials passing each $(\alpha_i, \beta_{i,j})$ with multiplicity $m_{i,j} - 1$. Let $\mathbb{G}$ be a Gröbner basis that has h polynomials with the $(x + \alpha_i)$ factor. Denote the set of polynomials in $\mathbb{G}$ with the $(x + \alpha_i)$ factor by $\mathbb{G}_A$, and those without the factor by $\mathbb{G}_{\bar{A}}$. Divide each polynomial in $\mathbb{G}_A$ by $(x + \alpha_i)$ and use $\mathbb{G}'_A$ to represent the set of the quotient polynomials.

Lemma 15 $\{\mathbb{G}'_A, \mathbb{G}_{\bar{A}}\}$ *is a Gröbner basis of polynomials passing each* $(\alpha_i, \beta_{i,j})$ *with multiplicity* $m_{i,j} - 1$, *and the multiplicities of the points in other code positions remain the same.*

Proof: From Lemma 13, the multiplicities of the points in code positions other than i on each polynomial in $\{\mathbb{G}'_A, \mathbb{G}_{\bar{A}}\}$ remain the same. However, the polynomials in $\mathbb{G}'_A$ pass $(\alpha_i, \beta_{i,j})$ with multiplicity $m_{i,j} - 1$, and hence the multiplicities of these points are no longer $m_{i,j}$ on each polynomial in $\{\mathbb{G}'_A, \mathbb{G}_{\bar{A}}\}$. Let $F(x, y)$ be an arbitrary polynomial that passes each $(\alpha_i, \beta_{i,j})$ with multiplicity $m_{i,j} - 1$. It also passes the points in all other code positions with their original multiplicities. By contradiction, it will be shown next that

$LT(F(x, y))$ is always divisible by the leading term of one of the polynomials in $\{\mathbb{G}'_A, \mathbb{G}_{\bar{A}}\}$, and hence this set is a desired Gröbner basis.

Assume that $LT(F(x, y))$ is not divisible by the leading term of any polynomial in $\mathbb{G}'_A$ or $\mathbb{G}_{\bar{A}}$. Since a leading term is divisible by another only when the y-degrees of the two terms are the same, all that needs to be checked is whether the polynomial in $\{\mathbb{G}'_A, \mathbb{G}_{\bar{A}}\}$ with the same y-degree in the leading term divides $LT(F(x, y))$. This polynomial may be located at either $\mathbb{G}'_A$ or $\mathbb{G}_{\bar{A}}$.

Case 1: $\mathbb{G}'_A$ has a polynomial, $Q'^{(l)}(x, y)$, whose leading term has the same y-degree as $LT(F(x, y))$. Since $F(x, y)$ passes $(\alpha_i, \beta_{i,j})$ with multiplicity $m_{i,j} - 1$, $E(x, y) = F(x, y)(x + \alpha_i)$ passes this point with multiplicity $m_{i,j}$. In the case that $LT(Q'^{(l)}(x, y))$ does not divide $LT(F(x, y))$, $LT(Q'^{(l)}(x, y)(x + \alpha_i))$ does not divide $LT(E(x, y))$. This conflicts with the fact that $Q'^{(l)}(x, y)(x + \alpha_i)$ is a polynomial in $\mathbb{G}$, which is a Gröbner basis of polynomials passing $(\alpha_i, \beta_{i,j})$ with multiplicity $m_{i,j}$.

Case 2: $\mathbb{G}_{\bar{A}}$ has a polynomial, $Q^{(l)}(x, y)$, whose leading term has the same y-degree as $LT(F(x, y))$. There are two subcases to consider if $LT(Q^{(l)}(x, y))$ does not divide $LT(F(x, y))$.

Case 2.1: $\deg_x(LT(Q^{(l)}(x, y))) - \deg_x(LT(F(x, y))) > 1$. In this subcase, $\deg_x(LT(Q^{(l)}(x, y))) > \deg_x(LT(F(x, y))) + 1 = \deg_x(LT(E(x, y)))$. Therefore, $LT(Q^{(l)}(x, y))$ does not divide $LT(E(x, y))$. This conflicts with that $Q^{(l)}(x, y)$ is a polynomial in $\mathbb{G}_{\bar{A}}$ and hence $\mathbb{G}$, which is a Gröbner basis of polynomials passing $(\alpha_i, \beta_{i,j})$ with multiplicity $m_{i,j}$.

Case 2.2: $\deg_x(LT(Q^{(l)}(x, y))) - \deg_x(LT(F(x, y))) = 1$. In this subcase, $LT(Q^{(l)}(x, y)) = LT(E(x, y))$. Hence $\{\mathbb{G}_A, E(x, y), \mathbb{G}_{\bar{A}} \backslash Q^{(l)}(x, y)\}$ is a Gröbner basis for polynomials passing $(\alpha_i, \beta_{i,j})$ with multiplicity $m_{i,j}$. However, this set has $h + 1$ polynomials with factor $(x + \alpha_i)$, since $E(x, y)$ has this factor. This conflicts with Lemma 14. □

From any equivalent Gröbner basis $\mathbb{G}$ that has h polynomials with the $(x + \alpha_i)$ factor, Lemma 15 says that a Gröbner basis of polynomials that pass each point in code position i with the multiplicity reduced by one can be derived by dividing $(x + \alpha_i)$ from the h polynomials. Hence, the backward interpolation problem is reduced to constructing such an equivalent Gröbner basis $\mathbb{G}$ from a given Gröbner basis. This task is accomplished by Algorithm 10. Assume that the input Gröbner basis has $t + 1$ polynomials and they are written as

$$Q^{(l)}(x, y) = q_0^{(l)}(x) + q_1^{(l)}(x)y + \cdots + q_t^{(l)}(x)y^t.$$

At the beginning of Algorithm 10, $\mathbb{G}_{\bar{A}}$ is empty, and the $t + 1$ polynomials of the input Gröbner basis are put into $\mathbb{G}_A$. In each iteration, $q_{t-b}^{(l)}(\alpha_i)$ is first computed for each polynomial in $\mathbb{G}_A$. If $\mathbb{G}_A$ has polynomials with nonzero $q_{t-b}^{(l)}(\alpha_i)$, then the one with the minimum weighted degree is moved to $\mathbb{G}_{\bar{A}}$. After that, each polynomial in $\mathbb{G}_A$ is updated in step 4 of Algorithm 10. As a result, $q_{t-b}^{(l)}(\alpha_i)$ for each updated polynomial becomes zero, and hence each $q_{t-b}^{(l)}(x)$ in $\mathbb{G}_A$ has the $(x + \alpha_i)$ factor. $Q^{(l)}(x, y)$ has the

$(x + \alpha_i)$ factor if and only if $q^{(l)}_{t-b}(\alpha_i) = 0$ for $b = 0, 1, \ldots, t$. Therefore, each polynomial in $\mathbb{G}_A$ has the $(x + \alpha_i)$ factor at the end of Algorithm 10, while those in $\mathbb{G}_{\bar{A}}$ do not. In step 4, each polynomial is linearly combined with another polynomial in the basis with lower weighted degree. Hence, the output of Algorithm 10 is a Gröbner basis equivalent to the input basis.

Algorithm 10 Equivalent Gröbner Basis Transformation
input: *A Gröbner basis* $\{Q^{(l)}(x, y)\}$ *with weighted degree* w_l $(0 \leq l \leq t)$ *passing* $(\alpha_i, \beta_{i,j})$ *with multiplicity* $m_{i,j}$ *generated by Kötter's Algorithm*
initialization:
  $A = \{0, 1, \ldots, t\}$; $\bar{A} = \emptyset$;
start:
  for $b = 0$ *to* t
    for each $l \in A$
1:      compute $q^{(l)}_{t-b}(\alpha_i)$
    if $\{l | q^{(l)}_{t-b}(\alpha_i) \neq 0, l \in A\} \neq \emptyset$
2:      $l^* = \arg\min_l \{w_l | q^{(l)}_{t-b}(\alpha_i) \neq 0, l \in A\}$
3:      $A \Leftarrow A/l^*$; $\bar{A} \Leftarrow \bar{A} \cup l^*$
4:      *for each* $l \in A$
        $Q^{(l)}(x, y) \Leftarrow Q^{(l)}(x, y)q^{(l^*)}_{t-b}(\alpha_i) + Q^{(l^*)}(x, y)q^{(l)}_{t-b}(\alpha_i)$
output: $\mathbb{G} = \{Q^{(l)}(x, y)\}(0 \leq l \leq t)$; A

Even if the value of h is unknown at the input, Algorithm 10 yields an equivalent Gröbner basis that has h polynomials with the $(x + \alpha_i)$ factor. $\mathbb{G}_{\bar{A}}$ starts as an empty set. Although there are $t + 1$ iterations in Algorithm 10, no polynomial is moved to $\mathbb{G}_{\bar{A}}$ in the iterations that $\{l | q^{(l)}_{t-b}(\alpha_i) \neq 0, l \in A\} = \emptyset$. It will be shown in Lemma 16 and 17 that in the iteration that $|\mathbb{G}_{\bar{A}}|$ reaches $t + 1 - h$, the $(t + 1) - (t + 1 - h) = h$ polynomials in $\mathbb{G}_A$ all have the $(x + \alpha_i)$ factor at the end of that iteration. Therefore, in the remaining iterations of Algorithm 10, $\{l | q^{(l)}_{t-b}(\alpha_i) \neq 0, l \in A\} = \emptyset$ and no more polynomials will be moved to $\mathbb{G}_{\bar{A}}$. As a result, $|\mathbb{G}_{\bar{A}}| = t + 1 - h$ and $|\mathbb{G}_A| = h$ at the end of Algorithm 10.

In the iteration of Algorithm 10 that $|\mathbb{G}_{\bar{A}}|$ reaches $t + 1 - u$ $(u \in Z^+)$, assume that $A = \{a_0, a_1, \ldots, a_{u-1}\}$ after step 3. Let $b_0, b_1, \ldots, b_{t-u}$ $(b_0 < b_1 < \cdots < b_{t-u})$ be the indices of the iterations in which $\{l | q^{(l)}_{t-b}(\alpha_i) \neq 0, l \in A\} \neq \emptyset$. Represent the values of l^* in iterations $b_0, b_1, \ldots, b_{t-u}$ by $\bar{a}_0, \bar{a}_1, \ldots, \bar{a}_{t-u}$. Let $q^{(l)}(\alpha_i) = [q^{(l)}_t(\alpha_i), q^{(l)}_{t-1}(\alpha_i), \ldots, q^{(l)}_0(\alpha_i)]$. For the $t + 1 - u$ polynomials in $\mathbb{G}_{\bar{A}}$, define $\mathcal{Q}_{\bar{A}} = [(q^{(\bar{a}_0)}(\alpha_i))^T, (q^{(\bar{a}_1)}(\alpha_i))^T, \ldots, (q^{(\bar{a}_{t-u})}(\alpha_i))^T]^T$.

Lemma 16 *The rows of the $\mathcal{Q}_{\bar{A}}$ matrix are linearly independent.*

Proof: Form a square matrix $\mathcal{Q}'_{\bar{A}}$ by the b_0th, b_1th, $\ldots$, b_{t-u}th columns of $\mathcal{Q}_{\bar{A}}$. When a polynomial $Q^{(\bar{a}_r)}(x,y)$ $(0 \leq r \leq t-u)$ is moved from $\mathbb{G}_A$ to $\mathbb{G}_{\bar{A}}$ in an iteration of Algorithm 10, $q_{t-b}^{(\bar{a}_r)}(\alpha_i) = 0$ for $b = 0, 1, \ldots, \bar{a}_r - 1$. Hence, for all $r < s$, the entry in the rth row and sth column of $\mathcal{Q}'_{\bar{A}}$ is zero. Accordingly, $\mathcal{Q}'_{\bar{A}}$ is an upper triangular matrix, and its row rank is $t + 1 - u$. Since $\mathcal{Q}'_{\bar{A}}$ is formed by some columns of $\mathcal{Q}_{\bar{A}}$, the row rank of $\mathcal{Q}_{\bar{A}}$ is also $t+1-u$. Therefore, all rows of $\mathcal{Q}_{\bar{A}}$ are linearly independent. $\square$

Lemma 17 *In the iteration of Algorithm 10 that $|\mathbb{G}_{\bar{A}}|$ reaches $t + 1 - h$, each of the h polynomials in $\mathbb{G}_A$ has the $(x+\alpha_i)$ factor at the end of that iteration.*

Proof: This is proved by contradiction. Assume that at the end of the iteration that $|\mathbb{G}_{\bar{A}}|$ reaches $t + 1 - h$, $h' < h$ of the polynomials in $\mathbb{G}_A$ have the $(x+\alpha_i)$ factor. Then $\{l | q_{t-b}^{(l)}(\alpha_i) \neq 0, l \in A\} \neq \emptyset$ in at least one remaining iteration of Algorithm 10, and a polynomial will be moved from $\mathbb{G}_A$ to $\mathbb{G}_{\bar{A}}$. As a result, at the end of Algorithm 10, $\mathbb{G}_A$ will have fewer than h polynomials. The polynomials in $\mathbb{G}_A$ have the $(x + \alpha_i)$ factor, while those in $\mathbb{G}_{\bar{A}}$ do not. From Lemma 16, the rows of $\mathcal{Q}_{\bar{A}}$ are independent. Hence, multiplying each row of $\mathcal{Q}_{\bar{A}}$ by a nonzero finite field element leads to another matrix of full row rank. Therefore, $\mathbb{F}[x]$-linear combinations of the polynomials in $\mathbb{G}_{\bar{A}}$ do not generate any polynomial with the $(x + \alpha_i)$ factor. $\mathbb{F}[x]$-linearly combining the polynomials in $\mathbb{G}_{\bar{A}}$ with those in $\mathbb{G}_A$ does not generate any polynomial with $(x + \alpha_i)$ factor either. This is equivalent to saying that if $\mathbb{G}_A$ has any polynomial without the $(x + \alpha_i)$ factor at the end of the iteration that $|\mathbb{G}_{\bar{A}}|$ reaches $t + 1 - h$, then the Gröbner basis output by Algorithm 10 can not be converted to a basis that has h polynomials with the $(x + \alpha_i)$ factor using $\mathbb{F}[x]$-linear combinations. This conflicts with Lemma 12. $\square$

After the equivalent Gröbner basis $\mathbb{G}$ is constructed using Algorithm 10, $(x + \alpha_i)$ is divided from each polynomial in $\mathbb{G}_A$ to reduce the multiplicity of each point in code position i from $m_{i,j}$ to $m_{i,j}-1$. This process can be repeated to reduce the multiplicity of each point in code position i by one each time, until all of them become zero. Let p_i be the number of interpolation points in code position i. The generalized backward interpolation is summarized in Algorithm 11. The complexity of this algorithm is dominated by the Gröbner basis transformation using Algorithm 10. $\max\{m_{i,j}\}$ iterations of basis transformations and polynomial divisions are needed to eliminate all points in code position i from the input basis. Algorithm 10 is also iterative. In general, h is unknown if the constraints to be eliminated are not the last ones covered for that code position, and all the $t + 1$ iterations in Algorithm 10 need to be carried out. If h is known, then the iterations after $|\mathbb{G}_{\bar{A}}|$ reaches $t + 1 - h$ do not need to continue. The value of h can be pre-decided for the two special cases discussed in the following.

Algorithm 11 Generalized Backward Interpolation
input: *A Gröbner basis* $\{Q^{(l)}(x,y)\}$ *with weighted degree* w_l $(0 \leq l \leq t)$ *from the Kötter's algorithm that passes each* $(\alpha_i, \beta_{i,j})$ $(0 \leq j < p_i)$ *with multiplicity* $m_{i,j}$
start:
     while $(\max_{j=0}^{p_i-1} m_{i,j} > 0)$
         transform Gröbner basis by Algorithm 10
         for each $l \in A$
             $Q^{(l)}(x,y) \Leftarrow Q^{(l)}(x,y)/(x+\alpha_i)$
             $w_l \Leftarrow w_l - 1$
         for $j = 0$ *to* $p_i - 1$
             if $m_{i,j} > 0$
                 $m_{i,j} \Leftarrow m_{i,j} - 1$
         $\{Q^{(l)}(x,y)\}$ $(0 \leq l \leq t)$ *is a Gröbner basis that pass* $(\alpha_i, \beta_{i,j})$
         with multiplicity $m_{i,j}$ $(0 \leq j < p_i)$
output: $\{Q^{(l)}(x,y)\}$ $(0 \leq l \leq t)$, *a Gröbner basis that does not pass any* $(\alpha_i, \beta_{i,j})$ $(0 \leq j < p_i)$

Lemma 18 *If code position* i *has only one point,* (α_i, β_i), *with multiplicity* m_i, $h = m_i$ *and only the first* $t + 1 - m_i$ *iterations of Algorithm 10 need to be done.*

Proof: From the definition of multiplicity, if $Q^{(l)}(x,y)$ passes (α_i, β_i) with multiplicity m_i, then $d_{a,b}^{(l)}(\alpha_i, \beta_i) = 0$ for $a + b < m_i$. $d_{0,b}^{(l)}(\alpha_i, \beta_i) = \sum_{s=b}^{t} \binom{s}{b} q_s^{(l)}(\alpha_i) \beta_i^{s-b}$ and the constraints $d_{0,b}^{(l)}(\alpha_i, \beta_i) = 0$ $(0 \leq b < m_i)$ can be written in a matrix multiplication format as

$$\begin{bmatrix} 1 & \binom{1}{0}\beta_i & \binom{2}{0}\beta_i^2 & \cdots & \binom{t}{0}(\beta_i)^t \\ 0 & 1 & \binom{2}{1}\beta_i & \cdots & \binom{t}{1}(\beta_i)^{t-1} \\ \vdots & \vdots & \vdots & \vdots & \vdots \\ 0 & 0 & \cdots & 1 & \cdots \end{bmatrix} \begin{bmatrix} q_0^{(l)}(\alpha_i) \\ q_1^{(l)}(\alpha_i) \\ \vdots \\ q_t^{(l)}(\alpha_i) \end{bmatrix} = 0$$

There are m_i rows and $t + 1$ columns in the above two-dimensional matrix. If $q_s^{(l)}(\alpha_i) = 0$ for $m_i \leq s \leq t$, then the product of the first m_i columns of this matrix and the column vector $[q_0^{(l)}(\alpha_i), q_1^{(l)}(\alpha_i), \ldots, q_{m_i-1}^{(l)}(\alpha_i)]^T$ is zero. The first m_i columns form an upper triangular matrix. It is full-rank. Hence $q_s^{(l)}(\alpha_i)$ must be zero for $0 \leq s < m_i$ in order to have zero product in this case. In other words, if $q_s^{(l)}(\alpha_i) = 0$ for $m_i \leq s \leq t$,, then $q_s^{(l)}(\alpha_i)$ are also zero for $0 \leq s < m_i$. If there is only one point with multiplicity m_i in code position i, then m_i constraints need to be satisfied in order to increase the multiplicity of this point from $m_i - 1$ to m_i. Hence h is at most m_i. Accordingly, $|\bar{A}| \geq t + 1 - m_i$, and at least $t + 1 - m_i$ iterations should be carried out in Algorithm 10. $q_s^{(l)}(\alpha_i)$ for $s = t, t - 1, \ldots, m_i$ and $l \in A$ are forced to zero in the first $t + 1 - m_i$ iterations. This means that $q_s^{(l)}(\alpha_i)$ for $0 \leq s < m_i$ are also zero at the end of this iteration. Therefore, all polynomials in $\mathbb{G}_A$ have the $(x + \alpha_i)$ factor at the end of iteration $t + 1 - m_i$, and Algorithm 10 can stop. Since A starts with $t + 1$ polynomials and at most one polynomial is taken out of A in each iteration, $|A| \geq (t + 1) - (t + 1 - m_i) = m_i$ at this time. Moreover, according to Lemma 14, $|A|$ can not exceed h, which is at most m_i. Therefore, $|A| = m_i$ at the end of Algorithm 10 and $h = m_i$. $\square$

Lemma 19 *If all the p_i points in code position i have multiplicity one, $h = p_i$ and only the first $t + 1 - p_i$ iterations of Algorithm 10 need to be done.*

Proof: If $Q^{(l)}(x, y)$ passes each of the points, $(\alpha_i, \beta_{i,j})$ $(j = 0, 1, \ldots, p_i - 1)$, in code position i with multiplicity one, $d_{0,0}^{(l)}(\alpha_i, \beta_{i,j}) = 0$ for $j = 0, 1, \ldots, p_i - 1$. $d_{0,0}^{(l)}(\alpha_i, \beta_{i,j}) = \sum_{s=0}^{t} q_s^{(l)}(\alpha_i)\beta_{i,j}^s$ and the constraints $d_{0,0}^{(l)}(\alpha_i, \beta_{i,j}) = 0$ can be rewritten as

$$\begin{bmatrix} 1 & \beta_{i,0} & \beta_{i,0}^2 & \cdots & (\beta_{i,0})^t \\ 1 & \beta_{i,1} & \beta_{i,1}^2 & \cdots & (\beta_{i,1})^t \\ \vdots & \vdots & \vdots & \vdots & \vdots \\ 1 & \beta_{i,p_i-1} & \beta_{i,p_i-1}^2 & \cdots & (\beta_{i,p_i-1})^t \end{bmatrix} \begin{bmatrix} q_0^{(l)}(\alpha_i) \\ q_1^{(l)}(\alpha_i) \\ \vdots \\ q_t^{(l)}(\alpha_i) \end{bmatrix} = 0.$$

The first p_i columns of the above two-dimensional matrix form a square Vandermonde matrix. Since $\beta_{i,j} \neq \beta_{i,j'}$ if $j \neq j'$, this matrix has nonzero determinant, and accordingly is full-rank. Following similar analysis as that in the proof of Lemma 18, it can be derived that exactly one polynomial is moved from $\mathbb{G}_A$ to $\mathbb{G}_{\bar{A}}$ in each of the first $t + 1 - p_i$ iterations of Algorithm 10. In addition, at the end of iteration $t + 1 - p_i$, each polynomial in $\mathbb{G}_A$ has the $(x + \alpha_i)$ factor and $h = p_i$. $\square$

If $t = p_i$ in the special case of Lemma 19, Algorithm 10 only needs one iteration to generate an equivalent Gröbner basis with t polynomials having the $(x + \alpha_i)$ factor. In this case, the generalized backward interpolation reduces to the backward interpolation scheme in [55].

The computations needed in the generalized backward interpolation are also polynomial evaluations, linear combinations of polynomials, and divisions by $(x + \alpha_i)$. Therefore, many hardware units can be shared between the Kötter's forward interpolation and the generalized backward interpolation.

In both the backward and forward interpolations, fewer polynomials are involved and a smaller number of iterations is required when the multiplicities of the points are lower. The Chase algorithm with multiplicity one achieves a better performance-complexity tradeoff. Moreover, additional simplifications are enabled when the multiplicities are all one. Hence, the remainder of this chapter still focuses on the Chase algorithm with multiplicity one.

6.1.2   Eliminated factorization

Let c and e be the transmitted codeword and error vector, respectively. $r = c + e$ is the received word, and is one of the test vectors in the Chase algorithm. As explained in Chapter 5, when the re-encoding is adopted, the decoding is actually done on $\bar{r} = r + \phi$, where ϕ is a codeword that equals r in the k most reliable code positions denoted by R for an (n, k) RS code. In this case, the interpolation output polynomial would have factors $y + \bar{f}(x)$. Each $\bar{f}(x)$ is a candidate message polynomial corresponding to $\bar{c} = c + \phi$ according to evaluation map encoding. $\bar{r}_i = r_i + \phi_i = (c_i + e_i) + \phi_i = \bar{c}_i + e_i = 0$ for $i \in R$. Then $\bar{f}(\alpha_i) = \bar{c}_i = e_i$ for these code positions. If $e_i = 0$, then $\bar{f}(x)$ has a factor $(x + \alpha_i)$. Therefore, $\bar{f}(x)$ can be rewritten as

$$\bar{f}(x) = \Big(\prod_{i \in R, e_i = 0} (x + \alpha_i) \Big) \rho(x),$$

where $\rho(x)$ is a polynomial that does not have any root α_i for $i \in R$ and $e_i \neq 0$. Taking the coordinate transformation $y \Leftarrow y / \prod_{i \in R}(x + \alpha_i)$ into account, the factorization would generate factors $y + \gamma(x)$, where

$$\gamma(x) = \frac{\bar{f}(x)}{\prod_{i \in R}(x + \alpha_i)} = \frac{\rho(x)}{\prod_{i \in R, e_i \neq 0}(x + \alpha_i)}, \tag{6.7}$$

if it is applied directly to the re-encoded and coordinate transformed interpolation output. As a result, the errors in R can be corrected by using the coefficients of $\gamma(x)$ as the syndromes in the BMA. After that, another erasure decoding is needed to recover all symbols in the codeword c.

In the case of the Chase decoding with multiplicity one for high-rate codes, the interpolation output polynomial is in the format of $Q(x, y) = q_0(x) + q_1(x)y$. Hence $Q(x, y) = q_1(x)(y + q_0(x)/q_1(x))$, and $\gamma(x)$ can be rewritten as

$$\gamma(x) = \frac{q_0(x)}{q_1(x)}. \tag{6.8}$$

$\rho(x)$ and $\prod_{i \in R, e_i \neq 0}(x + \alpha_i)$ do not share any common factor. Let $p(x)$ be the possible common factor of $q_0(x)$ and $q_1(x)$. Comparing (6.7) and (6.8),

$$\begin{cases} q_0(x) = p(x)\rho(x) \\ q_1(x) = p(x) \prod_{i \in R, e_i \neq 0}(x + \alpha_i). \end{cases} \tag{6.9}$$

Lemma 20 *$p(x)$ does not contain any $(x + \alpha_i)$ factor for $i \in R$.*

Proof: When the re-encoding and coordinate transformation are applied, the interpolation generates a polynomial, $Q(x, y)$, that has the minimum weighted degree among all polynomials passing the interpolation points in code positions in $\bar{R} = \{1, 2, \ldots, n\}/R$. If $p(x)$ has a factor $(x + \alpha_i)$ for $i \in R$, $Q(x, y)$ also has this factor. From Lemma 13, $Q(x, y)/(x + \alpha_i)$ still passes the interpolation points in $\bar{R}$. However, the weighted degree of $Q(x, y)/(x + \alpha_i)$ is lower than that of $Q(x, y)$. This contradicts the fact that the interpolation output polynomial has the lowest weighted degree. Therefore, $p(x)$ does not contain any factor $(x + \alpha_i)$ for $i \in R$. $\square$.

From (6.9), the locations of the errors in R can be directly found from the roots of $q_1(x)$ since $p(x)$ does not have any factor $(x + \alpha_i)$ for $i \in R$. Also from (6.7) and (6.8),

$$\bar{f}(x) = \frac{q_0(x) \prod_{i \in R}(x + \alpha_i)}{q_1(x)}. \tag{6.10}$$

As aforementioned, $e_i = \bar{f}(\alpha_i)$. However, $q_1(\alpha_i) = 0$ for $i \in R$ and $e_i \neq 0$. L'Hôpital's rule needs to be applied to compute e_i. This rule says that for polynomials $a(x)$, $b(x)$ and $d(x)$, if $d(x) = a(x)/b(x)$ and $a(\alpha_i) = b(\alpha_i) = 0$, then $d(\alpha_i) = a'(x)/b'(x)|_{\alpha_i}$. Here $(\cdot)'$ represents the formal derivative of the polynomial. From (6.9) and Lemma 20, α_i with $i \in R$ can only be a simple root of $q_1(x)$, and hence $q_1'(\alpha_i) \neq 0$. Therefore, L'Hôpital's rule needs to be applied once to (6.10) to compute the error magnitudes as

$$e_i = \bar{f}(\alpha_i) = \frac{q_0(x)(\prod_{j \in R}(x + \alpha_j))'}{q_1'(x)}\bigg|_{x=\alpha_i}. \tag{6.11}$$

As a result, when the re-encoding and coordinate transformation are adopted for the Chase decoding of high-rate RS codes, both the error locations and error magnitudes can be directly computed from the interpolation output polynomial. Therefore, the factorization step can be eliminated [78].

6.1.3   Polynomial selection schemes

Although the factorization step is eliminated in the Chase decoder, the locations and magnitudes for the errors in R need to be computed from the interpolation output polynomial. After that, erasure decoding is needed to recover the entire codeword. The interpolation for all test vectors can be completed in one run using the backward-forward schemes. However, it is very difficult to share any intermediate result among the computations of the errors in R and erasure decoding. Repeating these computations for each of the 2^η interpolation output polynomials leads to high hardware complexity. Instead, selection criterion can be added to the interpolation outputs so that only the ones most likely leading to successful decoding are passed to the remaining steps.

6.1.3.1   Polynomial selection based on root search

Assuming the interpolation result of a test vector is $Q(x, y) = q_0(x) + q_1(x)y$, the polynomial selection method in [51] first finds the number of roots of $q_1(x)$ in the $n - k$ least reliable code positions for each interpolation output polynomial. Then one polynomial is selected based on the differences among the root number, $\deg(q_1(x))$, and $\deg(q_0(x))$, as well as the reliabilities of the symbols in these $n-k$ positions. This method has error-correcting performance loss, and the rest of the decoding steps need to wait until the interpolation for all test vectors is done. Alternatively, the polynomial selection scheme in [68] searches the roots of $q_1(x)$ over all the n code positions. If the root number of $q_1(x)$ equals its degree, the corresponding interpolation output is selected. Simulation results showed that if the vectors are tested in the order of Gray code and according to reducing reliability as much as possible, then choosing the first interpolation output satisfying this criterion does not lead to any performance loss, except when the codeword length is very short. As a result, the interpolation can stop once a polynomial has been selected, and both the interpolation latency and power consumption are reduced. Nevertheless, to find the number of roots of $q_1(x)$, the exhaustive Chien search over finite fields is needed in both of the schemes in [51] and [68]. Using the unified backward-forward architecture, it only takes one single iteration to generate the interpolation result of the next test vector. To avoid slowing down the interpolation, the root search of $q_1(x)$ for each interpolation output polynomial must be finished in the same amount of time as one unified interpolation iteration. The fast speed demands expensive highly-parallel Chien search. As a result, such polynomial selections occupy a significant part of the overall Chase decoder area, especially for RS codes constructed over large finite fields.

6.1.3.2   Polynomial selection based on evaluation value testing

If one single message symbol is sacrificed and set to a fixed value, the polynomial selection can be done by simply testing whether the evaluation value of the interpolation output polynomial over a preset point is zero [79]. This scheme is explained in detail next.

Assume that the message polynomial used in the evaluation map encoding is $f(x) = f_0 + f_1 x + \cdots + f_{k-1}x^{k-1}$. In the original ASD decoding without the re-encoding and coordinate transformation, the interpolation output polynomial, $Q(x, y)$, should have a factor $y + f(x)$, and can be written in the format of

$$Q(x, y) = A(x, y)(y - f(x)).$$

Here $A(x, y)$ is the other factor of $Q(x, y)$. It can be derived that

$$Q(0, f_0) = A(0, f_0)(f_0 + f(0)) = A(0, f_0)(f_0 + f_0) = 0.$$

Therefore, if a $Q(x, y)$ leads to successful decoding, the corresponding $Q(0, f_0)$ should be zero. Hence, if f_0 is set to a fixed value, then the $Q(x, y)$ leading to the correct codeword can be selected based on whether $Q(0, f_0) = 0$. Moreover, in the case that f_0 is set to zero, $Q(0, f_0) = Q(0, 0)$, which is the constant coefficient of $Q(x, y)$. As a result, the polynomial selection is done by choosing the $Q(x, y)$ whose constant coefficient is zero. Multiple interpolation outputs may meet this selection criteria. Also $Q(0, f_0)$ may be accidentally zero even if the corresponding $Q(x, y)$ does not lead to a correct codeword. Simulation results showed that sending only the first interpolation output with $Q(0, f_0) = 0$ to the rest decoding steps has very small coding gain loss in the Chase decoding of long RS codes. If the first two qualified interpolation outputs are chosen, the performance degradation becomes negligible [79]. The vectors are tested in the order of Gray code to enable backward-forward interpolation and according to decreasing reliability as much as possible in the simulations. Since f_0 is set to a fixed value, and is not used as a message symbol, the effective code rate is decreased from k/n to $(k-1)/n$ to allow this polynomial selection scheme. Nevertheless, this scheme leads to great hardware saving compared to the Chien-search based polynomial selections, and the code rate loss is marginal for long RS codes.

Although the polynomial selection based on $Q(0, f_0)$ testing has simple implementation, it needs to be modified to accommodate the re-encoding and coordinate transformation. These techniques decrease the number of points to be interpolated for a test vector from n to $n - k$ and reduce the length of the interpolation polynomials by $n/(n-k)$ times. They have to be adopted to reduce the complexity of the interpolation-based decoders to a practical level. However, when they are employed, the interpolation output does not have the $y + f(x)$ factor and $Q(0, f_0)$ is not directly available. In addition, although evaluation map encoding is used to explain the process of the interpolation-based decoding, systematic encoding is preferred in real systems because the message symbols can be read directly from the decoded codeword. To differentiate the notations, the input message symbols for systematic encoding are denoted by $m_0, m_1, \ldots, m_{k-1}$. Let $m(x) = m_0 + m_1 x + \cdots + m_{k-1} x^{k-1}$. As mentioned in Chapter 4, systematic encoding computes the codeword polynomial $c(x)$ as $(m(x) x^{n-k})_{g(x)} + m(x) x^{n-k}$, where $(\cdot)_{g(x)}$ denotes the remainder polynomial from the division by $g(x)$ and $g(x)$ is the generator polynomial of RS code. Since the computations involved in systematic encoding is totally different from $c_i = f(\alpha_i)$ in evaluation map encoding, f_0 can not be preset in a straightforward manner when systematic encoding is adopted. The polynomial selection based on $Q(0, f_0)$ testing has been modified in [79] to address these issues.

If a proper generator polynomial is chosen for systematic encoding, and an appropriate fixed-ordered set of distinct nonzero finite field elements are used for evaluation map encoding, then the same set of RS codewords can be generated by these two encoding approaches. One example is to use $g(x) = \prod_{i=1}^{n-k} (x + \alpha^i)$ in systematic encoding and $\{1, \alpha^1, \alpha^2, \cdots, \alpha^{n-1}\}$ in evaluation map encoding. Here α is a primitive finite field element. In this case, there is bijective mapping between the message polynomials $m(x)$ used in systematic encoding and $f(x)$ used in evaluation map encoding that lead to the same codeword.

$f(x)$ is also represent in a vector form as $f = [f_0, f_1, \ldots f_{k-1}]$. The vector representations of $c(x)$ and $m(x)$ are defined in a similar way. Then the evaluation map encoding is rewritten as

$$
\begin{aligned}
c &= f \times M_{f \to c} \\
&= f \times
\begin{bmatrix}
\alpha^{0 \times 0} & \alpha^{0 \times 1} & \cdots & \alpha^{0 \times (n-1)} \\
\alpha^{1 \times 0} & \alpha^{1 \times 1} & \cdots & \alpha^{1 \times (n-1)} \\
\vdots & \vdots & \ddots & \vdots \\
\alpha^{(k-1) \times 0} & \alpha^{(k-1) \times 1} & \cdots & \alpha^{(k-1) \times (n-1)}
\end{bmatrix},
\end{aligned}
\tag{6.12}
$$

The last k symbols of the codeword generated from systematic encoding are also the message symbols. Hence, the mapping from f to m can be done by

$$
m = f \times M_{f \to m},
$$

where $M_{f \to m}$ is a $k \times k$ matrix composed of the last k columns of $M_{f \to c}$. $M_{f \to m}$ is full-rank, and $M_{f \to m}^{-1}$ can be computed by Gaussian elimination. Therefore, the mapping from m to f is given by

$$
f = m \times M_{f \to m}^{-1}.
\tag{6.13}
$$

Let ω_i $(0 \leq i \leq k - 1)$ be the finite field elements in the first column of $M_{f \to m}^{-1}$. From (6.13),

$$
f_0 = \sum_{i=0}^{k-1} m_i \omega_i.
$$

From the above equation, the systematic encoder is modified in order to have the corresponding f_0 set to a fixed value. The modified systematic encoder takes $k - 1$ message symbols $m_1, m_2, \cdots, m_{k-1}$. m_0 is not used as a message symbol. Instead, it is computed as

$$
m_0 = \omega_0^{-1} \left(f_0 - \sum_{i=1}^{k-1} m_i \omega_i \right)
\tag{6.14}
$$

according to the value needs to be set for f_0. To simplify the computations, f_0 is set to zero. As a result, (6.14) is reduced to

$$
m_0 = \omega_0^{-1} \sum_{i=1}^{k-1} m_i \omega_i = \sum_{i=1}^{k-1} m_i (\omega_i \omega_0^{-1}).
\tag{6.15}
$$

Fig. 6.5 shows the architecture of the modified systematic encoder. Like the regular systematic encoder in Fig. 4.1, the linear feedback shift register (LFSR) computes the coefficients of the remainder polynomial

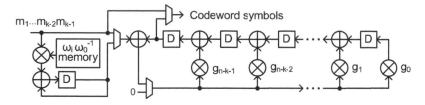

FIGURE 6.5
Modified systematic encoder (modified from [79])

$(m(x)x^{n-k})_{g(x)}$. When $m_{k-1}, m_{k-2} \ldots, m_1$ are serially input to the LFSR, they are also multiplied with $\omega_i \omega_0^{-1}$, which are pre-computed and stored in memory, to calculate m_0. After $m_{k-1}, m_{k-2} \ldots, m_1$ are input in $k-1$ clock cycles, m_0 is ready and sent to the LFSR. Then the coefficients of $(m(x)x^{n-k})_{g(x)}$ are located in the registers of the LFSR and are serially shifted out in the next $n-k$ clock cycles. Therefore, the modified encoding process also takes n clock cycles, and hence has the same latency as the original systematic encoder in Fig. 4.1.

When the re-encoding and coordinate transformation are applied, the interpolation output polynomial does not have $y + f(x)$ as a factor. It only passes $n-k$ points in code positions belonging to $\bar{R}$, and those are the points corresponding to the transformed symbols in $\bar{r} = r + \phi$. For the Chase decoding with multiplicity one for high-rate codes, the interpolation output of each test vector is in the format of $Q(x,y) = q_1(x)y + q_0(x)$. If the polynomial $\prod_{i \in R}(x + \alpha_i)$ that has been taken out of the interpolation is multiplied back and the coordinate transformation is reversed, then we get

$$\bar{Q}(x,y) = q_1(x)y + q_0(x) \prod_{i \in R}(x + \alpha_i). \qquad (6.16)$$

$\bar{r}_i = 0$ for $i \in R$. Since $\bar{Q}(\alpha_i, 0) = 0$ for $i \in R$, $\bar{Q}(x,y)$ also passes the k most reliable points corresponding to $\bar{r}$. Therefore, $\bar{Q}(x,y)$ has a factor $y + \bar{f}(x)$ if r would lead to successful decoding. Again $\bar{f}(x) = \bar{f}_0 + \bar{f}_1 x + \cdots + \bar{f}_{k-1} x^{k-1}$ is the message polynomial corresponding to $\bar{c} = c + \phi$ in the evaluation map encoding.

Assume that

$$\bar{Q}(x,y) = \bar{A}(x,y)(y + \bar{f}(x)).$$

Then

$$\bar{Q}(0, \bar{f}_0) = \bar{A}(0, \bar{f}_0)(\bar{f}_0 + \bar{f}(0)) = 0.$$

Hence, when a test vector is decodable, its corresponding $\bar{Q}(0, \bar{f}_0)$ should be zero. Therefore, the polynomial selection can be done based on whether $\bar{Q}(0, \bar{f}_0) = 0$. Nevertheless, computing $\bar{Q}(x,y)$ from $Q(x,y)$ requires the multiplication of $\prod_{i \in R}(x + \alpha_i)$, whose degree is k. Such a polynomial multiplication brings a large overhead. From (6.16),

$$\bar{Q}(0, \bar{f}_0) = q_1(0)\bar{f}_0 + q_0(0) \prod_{i \in R} \alpha_i = \left(q_1(0)\frac{\bar{f}_0}{\prod_{i \in R} \alpha_i} + q_0(0) \right) \prod_{i \in R} \alpha_i.$$

Let $\hat{f}_0 = \bar{f}_0/\prod_{i \in R} \alpha_i$. The above equation is re-written as

$$\bar{Q}(0, \bar{f}_0) = (q_1(0)\hat{f}_0 + q_0(0)) \prod_{i \in R} \alpha_i = Q(0, \hat{f}_0) \prod_{i \in R} \alpha_i.$$

$\bar{Q}(0, \bar{f}_0)$ is only different from $Q(0, \hat{f}_0)$ by a nonzero scaler $\prod_{i \in R} \alpha_i$. Therefore, the polynomial selection in the re-encoded and coordinate transformed Chase decoding can be done based on whether $Q(0, \hat{f}_0)$ is zero. Since the product of all nonzero elements in a finite field equals one, $\hat{f}_0$ is computed as $\bar{f}_0 \prod_{i \in \bar{R}} \alpha_i$ instead for high-rate primitive codes to reduce the number of field elements need to be multiplied.

Let $f^\phi(x)$ be the message polynomial corresponding to ϕ using evaluation map encoding. $\bar{f}(x) = f(x) + f^\phi(x)$ since $\bar{c} = c + \phi$ and the evaluation map encoding is linear. Accordingly, $\bar{f}_0 = f_0 + f_0^\phi$. If f_0 is set to zero, $\bar{f}_0 = f_0^\phi$. From (6.12), the evaluation map encoding for a primitive code is equivalent to a Fourier transform over finite fields. Similarly, recovering $f^\phi(x)$ from ϕ can be considered as an inverse finite Fourier transform, in which the codeword vector is multiplied by an inverse matrix. The entries in the first column of the inverse matrix are all one. Hence $f_0^\phi = \sum_{i=0}^{n-1} \phi_i$, and

$$\hat{f}_0 = \bar{f}_0 \prod_{i \in \bar{R}} \alpha_i = \left(\sum_{i=0}^{n-1} \phi_i \right)\left(\prod_{i \in \bar{R}} \alpha_i \right).$$

During the re-encoding process, ϕ_i are generated serially. Their sum can be computed by an adder-register feedback loop. Since $|\bar{R}| = n - k$, $\prod_{i \in \bar{R}} \alpha_i$ is computed by a multiplier-register loop in $n - k$ clock cycles while the re-encoding is carried out.

6.1.3.3 Reduced-complexity Chase interpolation using evaluation value testing-based polynomial selection

Although the complexity of the interpolation in the Chase decoder has been greatly reduced by the backward-forward interpolation schemes [55, 68], the interpolation result for each test vector still needs to be derived. During the backward-forward interpolation iterations, the x-degree of the interpolation polynomials is around $(n-k)/2$. If the coefficients of a polynomial are updated serially, the x-degree decides the latency of each iteration. The interpolation using the backward-forward schemes still has long latency when $n - k$ is not small or there is a large number of test vectors. Alternatively, it was proposed in [80] to first derive the $Q(0, \hat{f}_0)$ corresponding to each test vector without building the entire $Q(x, y)$. Then the interpolation is carried out for only the test vectors selected based on whether $Q(0, \hat{f}_0) = 0$.

Algorithm 12 Chase interpolation with evaluation value testing-based polynomial selection

input: (α_i, β_i) *for* $i \in \bar{R}$ *and* (α_i, β_i') *for* $i \in I_\eta$

initialize*:*

$\quad Q^{(0)}(x, y) = 1$, $Q^{(1)}(x, y) = y$

$\quad Q^{(0)}(0, \hat{f}_0) = 1$, $Q^{(1)}(0, \hat{f}_0) = \hat{f}_0$

$\quad Q^{(0)}(\alpha_i, \beta_i) = 1$, $Q^{(1)}(\alpha_i, \beta_i) = \beta_i$ *for* $i \in I_\eta$

$\quad Q^{(0)}(\alpha_i, \beta_i') = 1$, $Q^{(1)}(\alpha_i, \beta_i') = \beta_i'$ *for* $i \in I_\eta$

start*:*

1: *Interpolate over the points with code positions in $I_{\bar{\eta}}$;*

$\quad$ *Update $Q^{(l)}(0, \hat{f}_0)$, $Q^{(l)}(\alpha_i, \beta_i)$, $Q^{(l)}(\alpha_i, \beta_i')$ for $l = 0, 1$, $i \in I_\eta$*

$\quad$ *along these interpolation iterations.*

2: *Update necessary evaluation values to derive $Q(0, \hat{f}_0)$ for each test vector;*

$\quad$ *Select the first two test vectors with $Q(0, \hat{f}_0) = 0$.*

3: *Interpolate over the remaining η points for the two selected test vectors.*

The interpolation process making use of the evaluation value testing-based polynomial selection for the Chase decoding is listed in Algorithm 12. In this algorithm, I_η denotes the set of the η least reliable code positions and $I_\eta \subset \bar{R}$. $I_{\bar{\eta}}$ is the set of the rest $n - k - \eta$ code positions in $\bar{R}$. Since the initial interpolation polynomials are $Q^{(0)}(x, y) = 1$ and $Q^{(1)}(x, y) = y$, their evaluation values are derived easily. The interpolation only needs to be carried out over the points in the code positions in $\bar{R}$ when the re-encoding and coordinate transformation are employed. Forward interpolation is first done over the $n - k - \eta$ points in the code positions in $I_{\bar{\eta}}$ and the results are stored. These points are common to all test vectors. During each of these forward interpolation iterations, the evaluation values over $(0, \hat{f}_0)$, (α_i, β_i) and (α_i, β_i') for $i \in I_\eta$ are updated in the same way as the polynomials. After the $n - k - \eta$ iterations, the $Q^{(l)}(0, \hat{f}_0)$ needed for polynomial selection is derived first for each test vector without carrying out the real interpolation. This is achieved by updating $Q^{(l)}(0, \hat{f}_0)$ according to the computations that should be done in the interpolation iterations for the rest η points in I_η. From Algorithm 9, to update $Q^{(l)}(0, \hat{f}_0)$ in an iteration for interpolating over a point (α_i, β_i), $Q^{(l)}(\alpha_i, \beta_i)$ are needed. The evaluation values over (α_i, β_i) have been updated from the initial values along the polynomials in each of the first $n - k - \eta$ interpolation iterations. In the remaining η interpolation iterations for a test vector, they are also updated and stored in a similar manner until they are no longer needed to update $Q^{(l)}(0, \hat{f}_0)$ for any later iterations. As mentioned previously, selecting the two most reliable test vectors with $Q(0, \hat{f}_0) = 0$ only leads to negligible performance loss. The interpolation over the rest η points is only completed for the two vectors after they are selected based on $Q(0, \hat{f}_0)$.

During each interpolation iteration over a point in $I_{\bar{\eta}}$, $2\eta + 1$ evaluation values need to be updated for each polynomial. The remaining η points in all 2^η test vectors form a binary tree. The $Q^{(l)}(0, \hat{f}_0)$ for each test vector can be derived following a breadth-first traversal over the tree. After the level of the tree corresponding to code position $i \in I_\eta$ is gone over, $Q^{(l)}(\alpha_i, \beta_i)$ and $Q^{(l)}(\alpha_i, \beta_i')$ $(i \in I_\eta)$ are no longer needed. As a result, the number of evaluation values that need to be updated and stored gradually decreases. As a result, it takes much fewer clock cycles to carry out the necessary updating of the evaluation values compared to completing the backward-forward interpolation for each test vector.

6.1.4 Chien-search-based codeword recovery

It was shown earlier in this section that in the case of the Chase decoding for high-rate codes, the errors in the k most reliable code positions, R, can be corrected directly from the re-encoded and coordinate transformed interpolation output polynomial $Q(x, y) = q_0(x) + q_1(x)y$. The locations of those errors are the roots of $q_1(x)$ according to (6.9), and the magnitudes of those errors can be computed using (6.11). However, after these errors are corrected, an erasure decoding is required to recover the entire codeword and hence each of the message symbols.

Alternatively, (6.10) shows that

$$\bar{f}(x) = q_0(x)v(x)/q_1(x),$$

where $v(x) = \prod_{i \in R}(x + \alpha_i)$. Since $\bar{c}_i = \bar{f}(\alpha_i)$, the codeword symbols can be derived by first computing $\bar{f}(x)$, and then evaluating it over each α_i [51]. No erasure decoding is needed in this approach. Nevertheless, $\deg(v(x)) = k$. The multiplication of a polynomial with such a high degree to $q_0(x)$ and the following division by $q_1(x)$ are quite hardware-demanding. Also $\deg(\bar{f}(x)) = k - 1$. Evaluating a polynomial with a high degree requires large silicon area.

A Chien-search-based codeword recovery method was proposed in [81]. It computes $\bar{f}(\alpha_i)$ without involving polynomial multiplication or division. In addition, only evaluation values of low-degree polynomials are required. Let $w(x) = \prod_{i \in \bar{R}}(x + \alpha_i)$. Since $v(x)w(x) = \prod_{0 \leq i < n}(x + \alpha_i) = x^n - 1$ for primitive codes, $\bar{f}(x)$ is rewritten as

$$\bar{f}(x) = \frac{q_0(x)(x^n - 1)}{q_1(x)w(x)}. \tag{6.17}$$

$\deg(w(x)) = n - k$ is much lower than $\deg(v(x)) = k$ for high-rate codes, and hence $w(x)$ is much easier to calculate. Instead of computing $\bar{f}(x)$ first and then evaluating it over α_i, $\bar{f}(\alpha_i)$ is derived from $q_0(\alpha_i)$, $q_1(\alpha_i)$ and $w(\alpha_i)$ directly. The challenge here is that $q_1(\alpha_i)$ and/or $w(\alpha_i)$ may be zero. In these cases, L'Hôpital's rule needs to be applied. Depending on the possible factors of $q_1(x)$ and $w(x)$, the following four cases are considered.

Case A: $i \in R$ and $q_1(\alpha_i) = 0$. For primitive codes over $GF(2^q)$, $n = 2^q - 1$, and hence $\alpha_i^n - 1$ is always zero. For $i \in R$, $w(\alpha_i) \neq 0$. Therefore, the L'Hôpital's rule is applied to $(x^n - 1)/q_1(x)$ in this case. Since the $q_1(x)$ from the interpolation output polynomial leading to successful decoding only has simple roots, $q_1'(\alpha_i) \neq 0$. Also $n = 2^q - 1$ is an odd number for primitive codes. Hence,

$$\bar{f}(\alpha_i) = \frac{q_0(x)}{w(x)} \frac{(x^n - 1)'}{q_1'(x)}\Big|_{\alpha_i} = \frac{q_0(x)}{w(x)} \frac{x^{-1}}{q_1'(x)}\Big|_{\alpha_i}.$$

Case B: $i \in R$ and $q_1(\alpha_i) \neq 0$. $w(\alpha_i)$ is also nonzero in this case, and $\bar{f}(\alpha_i)$ is simply

$$\bar{f}(\alpha_i) = \frac{q_0(x)}{q_1(x)} \frac{(x^n - 1)}{w(x)}\Big|_{\alpha_i} = 0.$$

Case C: $i \in \bar{R}$ and $q_1(\alpha_i) \neq 0$. $w(\alpha_i) = 0$ for $i \in \bar{R}$, and all roots of $w(x)$ are simple. Therefore,

$$\bar{f}(\alpha_i) = \frac{q_0(x)}{q_1(x)} \frac{(x^n - 1)'}{w'(x)}\Big|_{\alpha_i} = \frac{q_0(x)}{q_1(x)} \frac{x^{-1}}{w'(x)}\Big|_{\alpha_i}.$$

Case D: $i \in \bar{R}$ and $q_1(\alpha_i) = 0$. $q_1(x)$ and $q_0(x)$ can be expressed as in (6.9). From this equation, if $i \in \bar{R}$ and $q_1(\alpha_i) = 0$, $p(\alpha_i)$ must be zero. Since $p(x)$ is a common factor shared by $q_1(x)$ and $q_0(x)$, $q_0(\alpha_i)$ is also zero. Hence L'Hôpital's rule is applied to $q_0(x)/q_1(x)$. Moreover, $p(x)$ should not have any $(x+\alpha_i)$ factor with order higher than one for $i \in \bar{R}$. Otherwise, $Q(x,y)/(x+\alpha_i)$ still passes each point in the code positions in $\bar{R}$. It has a lower degree than $Q(x,y)$, and this conflicts with the fact that $Q(x,y)$ has the lowest weighted degree among all polynomials passing the points in $\bar{R}$. Accordingly, $q_1'(\alpha_i) \neq 0$ for $i \in \bar{R}$. In addition, $w(\alpha_i) = 0$ for $i \in \bar{R}$, and L'Hôpital's rule is also applied to $(x^n - 1)/w(x)$. As a result,

$$\bar{f}(\alpha_i) = \frac{q_0'(x)}{q_1'(x)} \frac{(x^n - 1)'}{w'(x)}\Big|_{\alpha_i} = \frac{q_0'(x)}{q_1'(x)} \frac{x^{-1}}{w'(x)}\Big|_{\alpha_i}.$$

Denote the sum of the odd terms in a polynomial $a(x)$ by $a_{odd}(x)$. Since $xa'(x) = a_{odd}(x)$ over $GF(2^q)$, the computation of $\bar{f}(\alpha_i)$ is summarized as

$$\bar{f}(\alpha_i) = \begin{cases} \frac{q_0(\alpha_i)}{q_{1,odd}(\alpha_i)} \frac{1}{w(\alpha_i)} & i \in R \& q_1(\alpha_i) = 0 \\ 0 & i \in R \& q_1(\alpha_i) \neq 0 \\ \frac{q_0(\alpha_i)}{q_1(\alpha_i)} \frac{1}{w_{odd}(\alpha_i)} & i \in \bar{R} \& q_1(\alpha_i) \neq 0 \\ \frac{q_{0,odd}(\alpha_i)}{q_{1,odd}(\alpha_i)} \frac{1}{w_{odd}(\alpha_i)} & i \in \bar{R} \& q_1(\alpha_i) = 0 \end{cases} \qquad (6.18)$$

From (6.18), the computations of the codeword symbols in $\bar{c}$ only require the

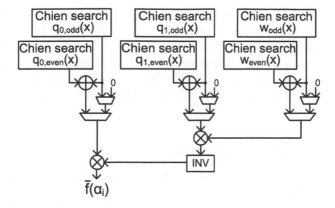

FIGURE 6.6
Full Chien-search-based codeword recovery architecture (modified from [81])

evaluation values of $q_1(x)$, $q_0(x)$ and $w(x)$ or the odd parts of these polynomials. Once they are available, $\bar{f}(\alpha_i)$ is calculated by a finite field multiplication and a division. No polynomial multiplication or division is needed. A test vector is decodable only when its error number is at most $(n-k)/2$. Therefore, the degrees of $q_0(x)$ and $q_1(x)$ in the selected interpolation output do not exceed $(n-k)/2$. Also $\deg(w(x)) = n - k$. Smaller degree means lower complexity for the polynomial evaluation value computation. Compared to evaluating $\bar{f}(x)$, whose degree is $k - 1$, the computations of the evaluation values in (6.18) are much simpler.

Whether the code position, i, belongs to R and if $q_1(\alpha_i) = 0$ decide which equation to use in (6.18). Let ω be a primitive element of $GF(2^q)$. The evaluation values of $q_1(x)$ over consecutive finite field elements, $\alpha_0 = 1, \alpha_1 = \omega, \alpha_2 = \omega^2, \cdots$ can be done by the Chien search, whose architecture is shown in Chapter 3. The Chien search architecture consists of constant multipliers. Moreover, the even and odd terms of a polynomial can be put into two groups. The Chien search is done separately on the two groups, so that the evaluation value of the odd part can be taken out to be used in (6.18). $q_1(x)$ has at most $(n - k)/2$ roots. Hence most of $\bar{f}(\alpha_i)$ is zero for high-rate codes according to the second equation in (6.18). The first equation in (6.18) needs to be used to compute at most $(n - k)/2$ $\bar{f}(\alpha_i)$. The third and fourth equations compute $(n - k)$ $\bar{f}(\alpha_i)$ with $i \in \bar{R}$ in total. These computations can start after the Chien search tells if $q_1(\alpha_i) = 0$. Although at most $3(n - k)/2$ $\bar{f}(\alpha_i)$ need to be actually computed in this case, the calculations of $q_0(\alpha_i)$ and $w(\alpha_i)$ would require regular finite field multipliers since those α_i with $i \in R$ and $q_1(\alpha_i) = 0$ and $i \in \bar{R}$ are not consecutive finite field elements.

A constant multiplier requires much fewer logic gates and has a shorter critical path than a general finite field multiplier. Therefore, it is more efficient to compute all evaluation values of $q_0(x)$ and $w(x)$ over $\alpha_0 = 1, \alpha_1 = \omega, \alpha_2 =$

$\omega^2, \ldots$ by Chien search, although most of them are not needed. Then proper evaluation values are chosen depending on whether $q_1(\alpha_i) = 0$ and $i \in R$ to compute $\bar{f}(\alpha_i)$ according to (6.18). The implementation architecture for this full Chien-search-based codeword recovery scheme is illustrated in Fig. 9.20. When $i \in R$ and $q_1(\alpha_i) \neq 0$, the first row of multiplexors in this figure pass zeros, so that the switching activities of the following hardware units are reduced.

6.1.5   Systematic re-encoding

For an (n, k) RS code, the function of the re-encoding is to find a codeword, ϕ, that equals the hard-decision received word, r, in k code positions. As a result of the re-encoding, the point with the highest multiplicity in each of those k code positions is excluded from the expensive bivariate interpolation process. For general ASD algorithms, the k most reliable code positions, R, are chosen for re-encoding. This is because the points in those code positions have higher multiplicities, and hence excluding them from the interpolation process leads to maximal complexity reduction. Such re-encoding can be implemented by erasure decoding. Simplified erasure decoder architectures have been developed in [60] as shown in Chapter 5. However, a significant proportion of the Chase decoder area is still spent on the erasure-decoder-based re-encoder.

In the case of the Chase decoding, the multiplicities of all points are the same. Hence, selecting any of the k positions for re-encoding would lead to the same complexity reduction in the interpolation. Inspired by this, it was proposed in [82] to choose the first k code positions for re-encoding in the Chase decoding. In this case, the re-encoder is implemented as a systematic encoder, whose architecture is shown in Fig. 4.1. A systematic encoder consists of constant multipliers, and has much lower complexity than an erasure decoder. Selecting different code positions to carry out the re-encoding leads to different ϕ. ϕ is just a codeword utilized to simplify the decoding process. Neither the test vectors nor the decoding results are affected by the code positions selected for re-encoding. Choosing the systematic positions for re-encoding leads to exactly the same error-correcting performance as using the k most reliable code positions. However, to accommodate systematic re-encoding, the coordinate transformation and codeword recovery need to be modified.

Without loss of generality, assume that the decoding trial is first carried out on the test vector $\{(\alpha_0, \beta_0), (\alpha_1, \beta_1), \cdots, (\alpha_{n-1}, \beta_{n-1})\}$ in the Chase algorithm. The re-encoding finds a codeword ϕ such that $\phi_i = \beta_i$ in k code positions. Previously, when the k positions in R are used for re-encoding, the interpolation needs to be carried out over $(\alpha_i, 0)$ for $i \in R$ and $(\alpha_i, \beta_i + \phi_i)$ for $i \in \bar{R}$. As discussed in Chapter 5, the interpolation result over $(\alpha_i, 0)$ $(i \in R)$ is $\{Q^{(0)}(x, y) = \prod_{i \in R}(x + \alpha_i), Q^{(1)}(x, y) = y\}$ by following the Kötter's algorithm. Then $v(x) = \prod_{i \in R}(x + \alpha_i)$ is taken out of the interpolation polynomials through coordinate transformation $y = z \prod_{i \in R}(x + \alpha_i)$. As a result, the interpolation still starts with $\{Q^{(0)}(x,z) = 1, Q^{(1)}(x,z) = z\}$, but it is only done for the transformed points $(\alpha_i, (\beta_i + \phi_i)/v(\alpha_i))$ for $i \in \bar{R}$.

Denote the systematic positions by the set S. When re-encoding is done as systematic encoding, the same aforementioned transformations apply if R is replaced by S. Nevertheless, some of the η least reliable code positions may belong to S. By coordinate transformation, the points in the re-encoding positions have been excluded from the interpolation process. Hence, the interpolation output polynomials do not pass these points, and they can not be directly eliminated from the interpolation result. In addition, since $v(\alpha_m) = 0$ for $m \in S$, the points in S can not be transformed in the same way as those in $\bar{S}$. As a result, the backward-forward interpolation can not be employed when the points in some systematic positions need to be flipped. This problem does not exist when R is used for re-encoding because none of the η least reliable code positions belongs to R.

Assume that (α_m, β_m) $(m \in S)$ is flipped to (α_m, β_m') in the next test vector. One possible approach to make (α_m, β_m) appear in the interpolation output is to adopt a modified coordinate transformation $y = \hat{z} \prod_{i \in S/m}(x+\alpha_i)$, and take out a modified polynomial $\hat{v}(x) = \prod_{i \in S/m}(x + \alpha_i)$ before the interpolation starts. Using these modifications, $\hat{v}(\alpha_m) \neq 0$ and (α_m, β_m) is transformed to $(\alpha_m, (\beta_m + \phi_m)/\hat{v}(\alpha_m)) = (\alpha_m, 0)$. In addition, the interpolation over the points in $\bar{S}$ starts with the initial basis $\{Q^{(0)}(x, \hat{z}) = (x + \alpha_m), Q^{(1)}(x, \hat{z}) = \hat{z}\}$, which is the pre-interpolation result over $(\alpha_m, 0)$. Since the Kötter's forward interpolation does not destroy any constraint that has been satisfied, the interpolation output would pass $(\alpha_m, 0)$ if this modified initial basis is used. In the worst case, S includes all the η least reliable code positions. Modifying the initial basis to accommodate the flipping of each point requires the computation of the product of η $(x + \alpha_m)$ terms. In addition, the length of the polynomials is increased by η in each interpolation iteration, which translates to longer interpolation latency.

A better approach is to modify the coordinate transformation on the fly when the points in S need to be flipped. At the beginning, the coordinate transformation $y = z \prod_{i \in R}(x+\alpha_i)$ is still used. When (α_m, β_m) $(m \in S)$ needs to be flipped, an extra coordinate transformation $z = \hat{z}/(x + \alpha_m)$ is applied. Moreover, $(x+\alpha_m)$ is taken from $v(x)$ and multiplied back to the interpolation polynomials. Assume that the interpolation result of the current test vector is $Q^{(l)}(x, z) = q_0^{(l)}(x) + q_1^{(l)}(x)z$ $(l = 0, 1)$. The following polynomials are derived as a result of the on-the-fly coordinate transformation modification and $(x + \alpha_m)$ multiply-back:

$$\begin{cases} Q^{(0)}(x, \hat{z}) = q_0^{(0)}(x)(x + \alpha_m) + q_1^{(0)}(x)\hat{z} \\ Q^{(1)}(x, \hat{z}) = q_0^{(1)}(x)(x + \alpha_m) + q_1^{(1)}(x)\hat{z} \end{cases} \tag{6.19}$$

Lemma 21 *The two polynomials in* (6.19) *form a Gröbner basis of the module* $\hat{\mathcal{M}}$ *of polynomials that pass* $(\alpha_m, 0)$ *and* $(\alpha_i, (\beta_i + \phi_i)/\hat{v}(\alpha_i))$ *for* $i \in \bar{S}$ *with* $\hat{z}$*-degree at most one.*

Proof: Since $Q^{(l)}(x,z) = q_0^{(l)}(x) + q_1^{(l)}(x)z$ $(l = 0, 1)$ are the interpolation results of the current test vector, they form a Gröbner basis of the module $\mathcal{M}$ of polynomials that pass $(\alpha_i, (\beta_i + \phi_i)/v(\alpha_i))$ for $i \in \bar{S}$ with z-degree at most one. From this and $\alpha_m + \alpha_i \neq 0$ for $i \in \bar{S}$, it can be easily derived that the two polynomials in (6.19) pass $(\alpha_m, 0)$ and $(\alpha_i, (\beta_i + \phi_i)/\hat{v}(\alpha_i))$ for $i \in \bar{S}$. Next, we prove that $Q^{(0)}(x, \hat{z})$ and $Q^{(1)}(x, \hat{z})$ in (6.19) form a Gröbner basis of $\hat{\mathcal{M}}$ by contradiction. Assume that there is a polynomial $B(x, \hat{z}) = b_0(x) + b_1(x)\hat{z}$ in the module $\hat{\mathcal{M}}$. However, $LT(B(x, \hat{z}))$ can not be divided by the leading term of either $Q^{(0)}(x, \hat{z})$ or $Q^{(1)}(x, \hat{z})$ in (6.19). Since $B(x, \hat{z})$ passes $(\alpha_m, 0)$, $b_0(\alpha_m) = 0$ and hence $b_0(x)$ has the factor $(x + \alpha_m)$. Let $F(x, \hat{z}) = B(x, \hat{z}(x + \alpha_m)) = b_0(x) + b_1(x)\hat{z}(x + \alpha_m)$. It can be derived that, for $i \in \bar{S}$,

$$F(\alpha_i, (\beta_i + \phi_i)/v(\alpha_i)) = F(\alpha_i, (\beta_i + \phi_i)/\prod_{j \in S}(\alpha_i + \alpha_j))$$

$$= b_0(\alpha_i) + \frac{b_1(\alpha_i)(\beta_i + \phi_i)}{\prod_{j \in S}(\alpha_i + \alpha_j)}(\alpha_i + \alpha_m)$$

$$= B(\alpha_i, (\beta_i + \phi_i)/\prod_{j \in S/m}(\alpha_i + \alpha_j))$$

$$= B(\alpha_i, (\beta_i + \phi_i)/\hat{v}(\alpha_i)) = 0.$$

Therefore, $F(x, \hat{z})$ passes $(\alpha_i, (\beta_i + \phi_i)/v(\alpha_i))$ for $i \in \bar{S}$. Considering that $b_0(x)$ has a factor $(x + \alpha_m)$, $F(x, \hat{z})/(x + \alpha_m) = b_0(x)/(x + \alpha_m) + b_1(x)\hat{z}$ should also pass $(\alpha_i, (\beta_i + \phi_i)/v(\alpha_i))$ for $i \in \bar{S}$. As a result, $F(x, \hat{z})/(x + \alpha_m)$ is a polynomial in $\mathcal{M}$. $\hat{z}$ is used instead of z in the notations only to track the modified coordinate transformation, and they are not differentiated in the leading term and weighted degree computations. Since $Q^{(l)}(x, z) = q_0^{(l)}(x) + q_1^{(l)}(x)z$ $(l = 0, 1)$ is a Gröbner basis of $\mathcal{M}$, $LT(F(x, \hat{z})/(x + \alpha_m))$ should be divisible by the leading term of either $Q^{(0)}(x, z)$ or $Q^{(1)}(x, z)$. However, $Q^{(l)}(x, z)$ is a Gröbner basis for polynomials with $(1, k - 1 - \deg(v(x))) = (1, -1)$ weighted degree. When the coordinate transformation is modified by taking out $\hat{v}(x)$, the degree weight should be changed to $(1, k - 1 - \deg(\hat{v}(x))) = (1, 0)$. Denote the (i, j) weight degree of a polynomial by $w_{(i,j)}(\cdot)$. $w_{(1,-1)}(b_0(x)/(x + \alpha_m)) = w_{(1,0)}(b_0(x)) - 1$ and $w_{(1,-1)}b_1(x)\hat{z} = w_{(1,0)}(b_1(x)\hat{z}) - 1$. Therefore, $LT(B(x, \hat{z}))$ and $LT(F(x, \hat{z})/(x + \alpha_m))$ have the same $\hat{z}$ degree, and the difference between their x-degree is one. Considering that dividing $(x + \alpha_m)$ does not change the $\hat{z}$-degree of the monomials, if $LT(B(x, \hat{z}))$ is not divisible by $LT(Q^{(l)}(x, \hat{z}))$, then $LT(F(x, \hat{z})/(x + \alpha_m))$ is not divisible by $LT(Q^{(l)}(x, z))$. This conflicts with the fact that $F(x, \hat{z})/(x + \alpha_m)$ is a polynomial in the module $\mathcal{M}$, for which $Q^{(l)}(x, z)$ is a Gröbner basis $\square$.

From (6.19), the backward interpolation can be applied to eliminate $(\alpha_m, 0)$. Then the forward interpolation is done over $(\alpha_m, (\beta_m' + \phi_m)/\hat{v}(\alpha_m))$ to derive the interpolation result for the next test vector. In this approach, the initial basis of the interpolation remains $\{1, z\}$, and the transformation in (6.19) is only done when a position in S is flipped for the first time.

In summary, when systematic re-encoding is applied, the interpolation is carried out by following Algorithm 13. The forward and backward interpolations in this algorithm are the same as those in Algorithm 9, except that the weighted degrees are adjusted according to the modified coordinate transformation. The equations for transforming the β coordinates are not shown in Algorithm B for the purpose of conciseness.

Algorithm 13 Systematically re-encoded Chase interpolation
initialization: $Q^{(0)}(x, z) = 1$, $Q^{(0)}(x, z) = z$
interpolation for the first test vector
    forward interpolate each transformed (α_i, β_i) $(i \in \bar{S})$
interpolation for each later test vector *(flip the point in position m)*
    if $(m \in \bar{S})$ *or* $((m \in S)$ *&* *(position m has been flipped before))*
        backward interpolate transformed (α_m, β_m)
        forward interpolate transformed (α_m, β'_m)
  else
        $Q^{(0)}(x, \hat{z}) = q_0^{(0)}(x)(x + \alpha_m) + q_1^{(0)}(x)\hat{z}$
        $Q^{(1)}(x, \hat{z}) = q_0^{(1)}(x)(x + \alpha_m) + q_1^{(1)}(x)\hat{z}$
        backward interpolate transformed (α_m, β_m)
        forward interpolate transformed (α_m, β'_m)
        multiply $(\alpha_m + \alpha_j)$ *to transformed* β_j *and* β'_j *for each of the other flipping position j*

When systematic re-encoding is adopted, the formula for codeword recovery needs to be changed from (6.17). Let S_F be the set of flipped systematic positions and $s_f(x) = \prod_{i \in S_F}(x + \alpha_i)$. Due to the modified coordinate transformation, $\bar{f}(x)$ becomes

$$\bar{f}(x) = \frac{q_0(x)}{q_1(x)} \frac{x^n - 1}{w(x) s_f(x)}, \tag{6.20}$$

where $w(x) = \prod_{i \in \bar{S}}(x + \alpha_i)$. Compared to (6.17), (6.20) only has one extra term, $s_f(x)$. However, this term is in the denominator, and its evaluation values may be zero. Moreover, at the output of the interpolation using the modified coordinate transformation, $q_1(x)$ and $q_0(x)$ have factors different from those using the original coordinate transformation. They need to be re-analyzed in order to apply L'Hôpital's rule to (6.20) to derive $\bar{f}(\alpha_i)$.

If systematic encoding has been adopted, only the symbols in S need to be recovered. Four cases are considered as follows.

    Case A: $i \in S/S_F$ and $q_1(\alpha_i) = 0$. In this case, $w(\alpha_i) \neq 0$ and $s_f(\alpha_i) \neq 0$. Since $(\alpha_i)^n - 1$ is always zero for primitive codes, L'Hôpital's rule is applied to

$x^n - 1$ and $q_1(x)$. We still need to tell if $q_1'(\alpha_i) = 0$, in which case L'Hôpital's rule needs to be applied again. Since $\bar{r} = r + \phi = \bar{c} + e$ is zero for the code positions in S/S_F, $\bar{c}_i = e_i$ for $i \in S/S_F$. Therefore, $\bar{f}(\alpha_i) = \bar{c}_i = e_i$ for these positions. If $e_i = 0$ for $i \in S/S_F$, then $\bar{f}(x)$ should have a factor $(x + \alpha_i)$. As a result, $\bar{f}(x)$ can be written as

$$\bar{f}(x) = \prod_{i \in S/S_F, e_i=0} (x + \alpha_i) \rho'(x), \tag{6.21}$$

where $\rho'(x)$ does not have any root α_i for $i \in S/S_F$ and $e_i \neq 0$. Taking the co-ordinate transformation $y = \hat{z} \prod_{i \in S/S_F}(x + \alpha_i)$ into account, the interpolation output $Q(x, \hat{z}) = q_0(x) + q_1(x)\hat{z}$ should have a factor $\hat{z} + \bar{f}(x)/\prod_{i \in S/S_F}(x + \alpha_i)$ if the corresponding test vector is decodable. From (6.21), it can be derived that

$$\frac{\bar{f}(x)}{\prod_{i \in S/S_F}(x + \alpha_i)} = \frac{\rho'(x)}{\prod_{i \in S/S_F, e_i \neq 0}(x + \alpha_i)} = \frac{q_0(x)}{q_1(x)}.$$

$\rho'(x)$ does not have any common factor with $\prod_{i \in S/S_F, e_i \neq 0}(x + \alpha_i)$. Assume that $q_1(x)$ and $q_0(x)$ have a common factor $p'(x)$. In another word, $q_1(x)$ and $q_0(x)$ can be written as

$$\begin{cases} q_0(x) = p'(x)\rho'(x) \\ q_1(x) = p'(x) \prod_{i \in S/S_F, e_i \neq 0}(x + \alpha_i) \end{cases} \tag{6.22}$$

$p'(x)$ should not have any factor $(x + \alpha_i)$ with $i \in S/S_F$. Otherwise, $Q(x, \hat{z})/(x + \alpha_i)$ is a polynomial that passes the same points in the code positions in $\bar{S} \cup S_F$ with multiplicity one as $Q(x, \hat{z})$, but with lower weighted degree. This violates that the interpolation output should have the minimum weighted degree. Accordingly, for any $i \in S/S_F$, α_i can be at most a simple root of $q_1(x)$. $q_1'(\alpha_i) \neq 0$ and L'Hôpital's rule does not need to be applied for the second time.

Case B: $i \in S/S_F$ and $q_1(\alpha_i) \neq 0$. None of the terms in the denominator evaluates to zero in this case. On the other hand, $(\alpha_i)^n - 1$ is zero for primitive codes. Hence, $\bar{f}(\alpha_i) = 0$ in this case.

Case C: $i \in S_F$ and $q_1(\alpha_i) \neq 0$. The only denominator term that evaluates to zero is $s_f(x)$ in this case. It only has simple roots. the L'Hôpital's rule is applied to $x^n - 1$ and $s_f(x)$.

Case D: $i \in S_F$ and $q_1(\alpha_i) = 0$. $w(\alpha_i) \neq 0$ in this case. Similar to Case C, L'Hôpital's rule is applied to $x^n - 1$ and $s_f(x)$. In addition, when $q_1(\alpha_i) = 0$ for $i \in S_F$, α_i must be a root of $p'(x)$ from (6.22). Hence, $q_0(x)$ has the same root, and L'Hôpital's rule is also applied to $q_0(x)$ and $q_1(x)$. Moreover, following a similar analysis as that used in Case A, it can be proved that $p'(x)$ does not have any root α_i with $i \in S_F$ and order higher than one.

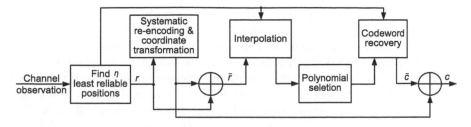

FIGURE 6.7
Interpolation-based one-pass Chase decoder (modified from [82])

Summarizing the above analysis, the codeword recovery is done as follows when systematic re-encoding is adopted [82].

$$
\bar{f}(\alpha_i) = \begin{cases}
\dfrac{q_0(x)}{w(x)s_f(x)}\dfrac{1}{q_{1,odd}(x)}\Big|_{\alpha_i} & i \in S/S_F \& q_1(\alpha_i) = 0 \\
0 & i \in S/S_F \& q_1(\alpha_i) \neq 0 \\
\dfrac{q_0(x)}{q_1(x)}\dfrac{1}{w(x)s_{f,odd}(x)}\Big|_{\alpha_i} & i \in S_F \& q_1(\alpha_i) \neq 0 \\
\dfrac{q_{0,odd}(x)}{q_{1,odd}(x)}\dfrac{1}{w(x)s_{f,odd}(x)}\Big|_{\alpha_i} & i \in S_F \& q_1(\alpha_i) = 0
\end{cases}
\tag{6.23}
$$

The codeword recovery according to (6.23) can be also implemented by Chien search architectures, whose complexities are decided by the degrees of the involved polynomials. $\deg(s_f(x)) \leq \eta$, and η is usually much smaller than $n - k$. On the other hand, the degrees of $q_0(x)$ and $q_1(x)$ are as high as $(n - k)/2$ and $\deg(w(x)) = n - k$. Therefore, having the extra term $s_f(x)$ only brings small overhead to the codeword recovery.

6.1.6 Simplified interpolation-based Chase decoder architecture

Many simplifications were enabled for the interpolation-based Chase decoder as shown in this chapter because the multiplicities of all points are one in this algorithm. Adopting available techniques, this decoder can be implemented according to Fig. 6.7. Compared to a general ASD decoder, whose data flow is shown in Fig. 5.3, the erasure re-encoder has been replaced by a systematic encoder [82], the factorization step is eliminated [78], and the iterative BMA and erasure decoder required at the end are replaced by simple Chien-search-based codeword recovery architectures [81]. In addition, the complexity of the interpolation over points with multiplicity one is much lower than those over points with larger multiplicities. Although multiple vectors are tested in the Chase algorithm, backward-forward interpolation schemes [55, 68] have been developed to share intermediate results. Through polynomial selection [68, 79], the rest of the computations only need to be carried out for one or two selected interpolation outputs without causing noticeable performance loss. As a result, the complexity of the interpolation-based Chase decoder has been reduced by more than an order of magnitude. For high-rate codes, the Chase decoder requires much lower complexity to achieve similar performance as other popular ASD algorithms, such as the KV [47] and BGMD [52] algorithms. It also achieves higher hardware efficiency than those BMA-based one-pass Chase decoders, such as that in [86]. Compared to a hard-decision BMA decoder, the interpolation-based Chase decoder with $\eta = 3$ for a (255, 239) RS code over $GF(2^8)$ requires less than twice the complexity, while the error-correcting capability is increased from 8 to $8 + 3 = 11$.

6.2   Generalized minimum distance decoder

The GMD decoding algorithm [45] carries out multiple decoding trials with different erasure patterns. Although the error-correcting performance of the GMD decoding is not as good as those of the Chase or other ASD algorithms, its implementation complexity is lower. To share intermediate results among the decoding of different erasure patterns, one-pass schemes have also been developed for the GMD algorithm in [61, 86]. In these schemes, the BMA [4] is used to build the error locator and evaluator for the error-only pattern. Then the error-and-erasure locator and evaluator for each erasure pattern are derived iteratively based on the outputs of the BMA. These schemes achieve substantial speedup over the designs that carry out each decoding trial from the beginning. Alternatively, the interpolation-based process can be adopted to achieve one-pass GMD decoding [84, 85]. Different from those based on the BMA, the interpolation-based scheme starts with erasure-only decoding. It achieves the same speed as the BMA-based approaches with lower complexity. Next, the techniques and architectures for one-pass GMD decoding are presented.

6.2.1   Kötter's one-pass GMD decoder

An erasure is a received symbol that is marked to be potentially erroneous. However, the magnitude of the symbols is unknown and needs to be found from the decoding. For an (n, k) RS code, error-and-erasure decoding can correct v errors and u erasures as long as

$$2v + u \leq n - k.$$

To increase the number of actual errors that can be corrected, the least reliable symbols in the received word are set as erasures. In the GMD decoding of an (n, k) RS code, there are $(n - k)/2$ different erasure patterns. In the ith $(0 < i \leq (n - k)/2)$ pattern, the $2i$ least reliable received symbols are erased. Error and erasure decoding is tried for each pattern.

Assume that carrying out the BMA on the hard-decision received word, r, generates an error locator polynomial $\Lambda(x)$, and an associated polynomial $B(x)$. Let the reciprocal polynomials of $\Lambda(x)$ and $B(x)$ be $\Lambda^*(x)$ and $B^*(x)$, respectively. Note that the roots of $\Lambda^*(x)$ are the error locations. Let $S(x)$ be the syndrome polynomial. The error evaluator $\Omega(x)$ and its associated polynomial $\Theta(x)$ for the error-only case are

$$\begin{cases} \Omega(x) = S(x)\Lambda^*(x) \mod x^{n-k} \\ \Theta(x) = S(x)B^*(x) \mod x^{n-k}. \end{cases} \tag{6.24}$$

In the Kötter's GMD [61], $\Omega(x), \Lambda^*(x), \Theta(x)$ and $B^*(x)$ are considered as the entries of a 2×2 matrix. To add erasures, this matrix is iteratively multiplied by a 2×2 matrix consisting of polynomials. After every two matrix multiplication iterations, the locator and evaluator for an erasure pattern are generated. Interpreting $\Omega(x), \Lambda^*(x), \Theta(x)$ and $B^*(x)$ as the coefficients of two bivariate polynomials, the Kötter's GMD algorithm is carried out according to Algorithm 14. If the output polynomial $Q^{(l^*)}(x,y)$ is written in the format of $q_0(x) + q_1(x)y$, then $q_1(x)$ and $q_0(x)$ are the error-and-erasure locator and evaluator polynomials, respectively, for the corresponding erasure pattern.

Algorithm 14 Erasure Addition in Kötter's GMD
initialization:

    $i = 0$

    $Q^{(0)}(x,y) = B^*(x)y + \Theta(x), Q^{(1)}(x,y) = \Lambda^*(x)y + \Omega(x)$

    $w_0 = deg(\Theta(x)), w_1 = deg(\Lambda^*(x)) - 1$

adding erasures:

    for each erasure code position i

        compute $q_1^{(l)}(\alpha_i)$ ($l = 0, 1$)

        $l^* = \arg\min_l\{w_l | q_1^{(l)}(\alpha_i) \neq 0\}, l' = \{0, 1\} \backslash l^*$

        $Q^{(l')}(x,y) \Leftarrow q^{(0)}(\alpha_i)Q^{(1)}(x,y) + q^{(1)}(\alpha_i)Q^{(0)}(x,y)$

        $Q^{(l^*)}(x,y) \Leftarrow Q^{(l^*)}(x,y)(x + \alpha_i)$

        $w_{l^*} \Leftarrow w_{l^*} + 1$

        $i = i + 1$

        if i is even

            output: $Q^{(l^*)}(x,y)$ ($l^* = \arg\min_l\{w_l | l = 0, 1\}$)

It takes $n - k$ iterations of Algorithm 14 to calculate the locator and evaluator polynomials for all of the $(n - k)/2$ erasure patterns. The computations involved in this algorithm are very similar to those in the Kötter's forward interpolation, except that univariate polynomial evaluation values are used

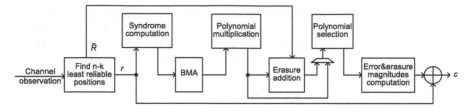

FIGURE 6.8
Kötter's one-pass GMD decoder (modified from [85])

instead of bivariate polynomial evaluation values, and the initial polynomials are the outputs of the BMA applied to the hard-decision received vector. Hence, the erasure addition process in Algorithm 14 can be implemented by architectures similar to those for the Kötter's forward interpolation. In addition, selection schemes can be added to the outputs of Algorithm 14, so that the remaining computations do not need to be repeated for every erasure pattern. It was proposed in [61] to first find the numbers of roots for each pair of the locator and evaluator polynomials. Denote these numbers by n_1 and n_0. The number of errors in the corresponding erasure pattern is estimated to be $n_1 - n_0$. If the number of erasures in the pattern is u, then the number of correctable errors does not exceed $v = \lfloor (n - k - u)/2 \rfloor$. Accordingly, the locator and evaluator polynomials are selected if $n_1 - n_0 \leq v$. The roots of the error locator and evaluator polynomials are computed by the Chien search. After the error locations are found, the magnitudes of the errors and erasures are derived by a modified Forney's formula [61]. Similar to that in the Chien-search-based codeword recovery for the interpolation-based decoding [81], the magnitudes can be also calculated through a Chien search without using Forney's algorithm. At the end, the error and erasure magnitudes are added to the received word to recover the transmitted codeword. The block diagram of Kötter's one-pass GMD decoding is depicted in Fig. 6.8.

The degrees of $\Lambda^*(x)$ and $\Theta(x)$ at the input of the erasure addition process in Algorithm 14 are usually around $(n-k)/2$. In each iteration of Algorithm 14, one of the bivariate polynomials is multiplied by $(x+\alpha_i)$. Hence, the x-degrees of the polynomials keep increasing in the $n - k$ iterations of Algorithm 14. A higher degree leads to longer latency in the erasure addition process. Also the complexity of the Chien search increases proportionally with the polynomial degree. Moreover, the BMA and polynomial multiplication require a large number of multipliers. As will be shown next, the complexity of the GMD can be reduced by using an interpolation-based approach.

6.2.2   Interpolation-based one-pass GMD decoder

Before a point is interpolated, no constraint about this point has been added to the interpolation curves. Hence, it can be considered as an erased point. Accordingly, it was proposed in [85] to adopt the interpolation-based process to achieve one-pass GMD decoding. The points that need to be interpolated are (α_i, r_i), where r_i is the hard-decision symbol in the ith code position. R is the set of the k most reliable code positions. The Kötter's interpolation algorithm [62] iteratively adds point to the interpolation curves. When the interpolation over the k points in R is completed, the resulting error locator and evaluator would be those for the test vector that has $n - k$ erasures. After every two points are interpolated, the error locator and evaluator for another erasure pattern are derived. The error locator and evaluator for the error-only vector are available after all n points are interpolated.

Similar to that in ASD decoding, re-encoding and coordinate transformation can be adopted. The re-encoding finds a codeword ϕ that equals the received word r in the code positions in R, and the interpolation is carried out on $\bar{r} = r + \phi$ instead. As discussed previously, since $\bar{r}_i = 0$ for $i \in R$, the interpolation over the points in R is pre-solved, and a coordinate transformation is used to take the common term out of the interpolation polynomials. As a result, the interpolation still starts with $\{Q^{(0)}(x, y) = 1, Q^{(1)}(x, y) = y\}$, but only needs to be done for the $n - k$ transformed points in $\bar{R}$. The interpolation result polynomials for the all-erasure case are actually these initial polynomials. Since $(1, -1)$-weighted degree should be used because of the coordinate transformation, $Q^{(1)}(x, y) = y$ is selected as the interpolation output. Hence $q_1(x) = 1$ and $q_0(x) = 0$ are the error locator and evaluator polynomials, respectively. Similar to that in the Chase decoding, $\bar{c}$, the codeword corresponding to $\bar{r}$, can be recovered according to (6.17). Substituting $q_1(x)$ and $q_0(x)$ by 1 and 0, respectively, into this equation, $\bar{c}_i = f(\alpha_i) = 0$ for every code position. When the $n - k$ points in $\bar{R}$ are set as erasures, r is decodable only if there is no error in R. In this case, ϕ is the correct codeword, which confirms that $\bar{c} = c + \phi$ should be zero as derived from the initial polynomials.

After every two transformed points in $\bar{R}$ are interpolated, the error locator and evaluator polynomials for an erasure pattern are available. To avoid repeating the rest of the computations for each pattern, selection criterion is also added to the interpolation outputs for the GMD decoding. Similarly, it can be based on testing if $\deg(q_1(x))$ equals its root number [68] or if $Q(0, f_0) = 0$ in case a message symbol can be pre-set to a fixed value and the encoder is modified [79]. Nevertheless, when more than one error locator passes the test, the one with the least number of erasures should be selected.

Although systematic re-encoding reduces the complexity of the Chase decoder, it is not suggested to be employed for the GMD decoding. The reason is that the erasures are located at the $n - k$ least reliable code positions. Picking other code positions $R' \neq R$ for re-encoding would require adding points in $\bar{R}'$ and deleting points in R' from the re-encoded and transformed initial interpolation polynomials before the error locator and evaluator for any erasure pattern is derived. Points in R' can be deleted by adopting a modified coordinate transformation similar to that [82]. However, these additions and deletions bring unnecessary overhead.

The overall process of the interpolation-based GMD decoding is similar to that of the interpolation-based Chase decoding shown in Fig. 6.7, except that the $n - k$ least reliable code positions need to be found, and the re-encoder needs to be implemented as an erasure decoder. The complexity of the erasure decoding-based re-encoder is lower than those of the syndrome computation, BMA and polynomial multiplication in the Kötter's GMD decoder shown in Fig. 6.8. The computations in the forward interpolation listed in Algorithm 9 and the erasure addition process of Algorithm 14 are very similar. However, the x-degree of the polynomials is initially zero in the re-encoded and transformed interpolation. On the other hand, it starts with $(n - k)/2$ during the erasure addition process as mentioned previously. Lower x-degree leads to not only shorter latency, but also smaller area in the following polynomial selection and codeword recovery steps. In addition, the polynomial selection for the interpolation-based decoder can be done by searching the roots of one instead of two polynomials as in the Kötter's GMD decoder. Due to these reasons, the interpolation-based GMD decoder can achieve much higher hardware efficiency than those based on the BMA [85].

7

BCH encoder & decoder architectures

CONTENTS

Binary Bose-Chaudhuri-Hocquenghem (BCH) codes [88, 89] are usually adopted in applications that have random bit errors and require low-complexity encoders and decoders. Examples of such applications are optical communications, digital video broadcasting, and Flash memories. This chapter presents the encoder and decoder architectures for binary BCH codes. Similar to Reed-Solomon (RS) [35] codes, BCH codes are linear block codes and can be defined using generator polynomials. The generator polynomial of a BCH code shares common terms with that of a RS code. However, by making use of the binary property, the en/decoding processes of BCH codes are made simpler than those for RS decoding. Besides hard-decision decoders, this chapter presents the implementation architectures for soft-decision Chase BCH decoders. Discussions are also provided for the decoder of 3-error-correcting BCH codes, which are of special interest for optical transport networks.

7.1   BCH codes

An (n, k) binary BCH code encodes k-bit messages into n-bit codewords. BCH codes are also constructed by making use of finite fields. For the codes constructed over $GF(2^q)$, n is at most $2^q - 1$. It is called a primitive code if $n = 2^q - 1$. Let the message bits $m_0, m_1, \ldots, m_{k-1}$ be the coefficients of a message polynomial $m(x)$, and the codeword bits $c_0, c_1, \ldots, c_{n-1}$ are the coefficients of a codeword polynomial $c(x)$. The BCH encoding can also be done as

$$c(x) = m(x)g(x),$$

where $g(x)$ is the generator polynomial. This encoding process is similar to that of the RS codes. However, the generator polynomials of BCH codes are constructed differently. Let α be a primitive element of $GF(2^q)$. Then the generator polynomial of a t-bit error-correcting BCH code is the product of the minimal polynomials whose corresponding conjugacy classes include $2t$ consecutive powers of α starting with α^b, *i.e.* $\alpha^b, \alpha^{b+1}, \ldots, \alpha^{b+2t-1}$. Here b is a non-negative integer, and it is called a narrow-sense BCH code when $b = 1$.

Example 33 *Consider the construction of a 2-error-correcting binary BCH code with $n = 15$ over $GF(2^4)$. The conjugacy classes and corresponding minimal polynomials for $GF(2^4)$ are listed in Table 1.1 in Chapter 1. For a narrow-sense BCH code, the four elements need to be covered by the generator polynomial are $\alpha, \alpha^2, \alpha^3$ and α^4. They are in the second and third conjugacy classes listed in Table 1.1. The minimal polynomials corresponding to these two classes are $x^4 + x + 1$ and $x^4 + x^3 + x^2 + x + 1$. Hence, the generator polynomial is*

$$g(x) = (x^4 + x + 1)(x^4 + x^3 + x^2 + x + 1) = x^8 + x^7 + x^6 + x^4 + 1.$$

Since $\deg(g(x)) = n-k = 7$, the constructed code is a (15,7) 2-error-correcting BCH code.

In the case that a different b is chosen, the $2t$ consecutive powers of α starting from α^b may be scattered in more conjugacy classes, and hence the generator polynomial equals to the product of more minimal polynomials. Accordingly, the redundancy of the code might be higher. If $b = 2$, then $\alpha^2, \alpha^3, \alpha^4, \alpha^5$ need to be covered by the generator polynomial. They belong to three conjugacy classes, whose cardinalities are 4, 4, and 2 as shown in Table 1.1. As a result, the degree of the corresponding generator polynomial is 10, and the constructed code would be a (15,5) BCH code, although the designed error-correcting capability is also $t = 2$.

Another way of BCH encoding is to multiply the message vector by a generator matrix, which can be constructed from the coefficients of $g(x)$ in the same way as that in (4.3) for RS codes. A systematic codeword is also preferred for BCH codes since the message bits can be directly read from the codeword. Similar to that for RS codes, the systematic encoding of BCH codes is achieved by

$$c(x) = m(x)x^{n-k} + (m(x)x^{n-k})_{g(x)},$$

where $(\cdot)_{g(x)}$ denotes modulo reduction by $g(x)$.

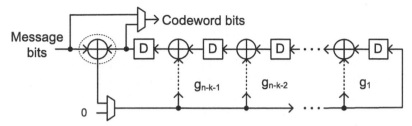

FIGURE 7.1
Serial LFSR BCH encoder architecture

7.2 BCH encoder architectures

Similar to the RS encoder in Fig. 4.1, a serial systematic BCH encoder can be implemented by a linear feedback shift register (LFSR) architecture as shown in Fig. 7.1. Since the generator polynomial, $g(x)$, of binary BCH codes is binary, each of the multipliers in the LFSR encoder for RS codes is replaced by a wire or no-connection when the corresponding coefficient of $g(x)$ is '1' or '0', respectively. Hence, the taps are represented by dashed lines in Fig. 7.1. It should be noted that g_0 and g_{n-k} always equal '1' for a BCH generator polynomial. The k message bits are input to the LFSR architecture serially in the first k clock cycles, starting with m_{k-1}. They are also sent to the output to become the systematic part of the codeword. The coefficients of the reminder $(m(x)x^{n-k})_{g(x)}$ are located in the registers after k clock cycles. Then they are shifted to the output by feeding zero through the multiplexor on the bottom. The LFSR encoder architecture can run at very high clock frequency. However, its achievable throughput is limited by the serial input and serial output. If higher throughput is desired, the unfolding technique discussed in Chapter 2 can be applied to achieve parallel processing.

BCH codes whose codeword length is thousands of bits are used in the Digital Video Broadcasting (DVB) Standard and Flash memories. In addition, very long BCH codes perform better than RS codes of similar code rate and codeword length over the AWGN channel. Because of the long codeword length, the generator polynomial is also long even if the code rate is high. For a long generator polynomial, the output of the leftmost XOR gate in the LFSR encoder shown in Fig. 7.1 needs to drive a large number of logic gates. The large fanout slows down the long BCH encoder significantly. To eliminate the fanout bottleneck, a register can be added to the input message bit. Then the retiming technique is applied to the cutset denoted by the dotted circle around the leftmost XOR gate to remove one register from each of its inputs and add one register to its output. Nevertheless, when parallel processing is applied to the encoder, registers may not always be added to the output of the leftmost XOR gate in the same way to address the large fanout issue.

In a J-unfolded architecture, each computation node in the original architecture is copied J times. If the output of node U is sent to node V after w delay elements in the original architecture, then the output of node U_i is sent to $V_{(i+w)\%J}$ after $\lfloor (i+w)/J \rfloor$ delays in the J-unfolded architecture. Here $\%$ denotes modulo reduction, and $0 \le i,j < J$. From Property 3 in Chapter 2, if $w < J$, then there are $J - w$ paths between the copies of node U and node V that do not have any delay element. The generator polynomial $g(x)$ of a BCH code can be rewritten as

$$g(x) = x^{t_0} + x^{t_1} + \cdots + x^{t_{s-2}} + 1,$$

where $t_0 > t_1 > \cdots > t_{s-2}$ are positive integers and s is the number of non-zero coefficients of $g(x)$. Accordingly, there are $t_0 - t_1$ delay elements between the two leftmost XOR gates in the LFSR encoder shown in Fig. 7.1. If $t_0 - t_1 < J$, then in the J-unfolded architecture, some paths between the copies of these two XOR gates do not have any delay element. In this case, retiming can not be applied to move one register to the output of each copy of the leftmost XOR gate in order to eliminate the large fanout effect as in the serial encoder architecture. An example of this case has been shown in Fig. 2.9 in Chapter 2. In this figure, there are two delay elements between the first and second XOR gates in the serial architecture. Therefore, when $J = 3$-unfolding is applied, one of the paths between the copies of these two gates, which is the path between the B_0 and A_2 nodes, does not have any delay element. Even though registers can be added to the inputs, retiming can not be applied to put one register at the output of the A_2 XOR gate.

To enable retiming and solve the large fanout issue in the unfolded BCH encoder, it was proposed in [90] to compute $r(x) = (m(x)x^{n-k})_{g(x)}$ in an alternative way. Rewrite $m(x)x^{n-k}$ as

$$m(x)x^{n-k} = q(x)g(x) + r(x), \tag{7.1}$$

where $q(x)$ is the quotient of dividing $m(x)x^{n-k}$ by $g(x)$. If a polynomial $p(x)$ is multiplied to both sides of (7.1), then

$$m(x)p(x)x^{n-k} = q(x)(g(x)p(x)) + r(x)p(x).$$

Since $r(x)$ is the remainder polynomial, $deg(r(x)) < deg(g(x))$. Hence $deg(r(x)p(x)) < deg(g(x)p(x))$, and $r(x)p(x)$ can be considered as the remainder of dividing $m(x)p(x)x^{n-k}$ by $g(x)p(x)$. Therefore, $r(x)$ can be derived by a three-step process:

1) multiply $m(x)$ by $p(x)$;
2) compute the remainder of dividing $m(x)p(x)x^{n-k}$ by $g(x)p(x)$;
3) compute the quotient of dividing the remainder from step 2 by $p(x)$.

The polynomial multiplication in step 1 can be done by an architecture similar to that of a finite impulse response (FIR) filter, which does not have any feedback loop. An example of such an architecture is shown in Fig. 5.6 in

Chapter 5. The second and third steps are polynomial divisions. They can be also implemented by LFSR architectures.

Assume that $g'(x) = p(x)g(x)$ is written as

$$g'(x) = x^{t'_0} + x^{t'_1} + \cdots + x^{t'_{s'-2}} + 1, \tag{7.2}$$

where $t'_0 > t'_1 > \cdots > t'_{s'-2}$ are positive integers and s' is the number of non-zero terms in $g'(x)$. A $p(x)$ can be always found by a look-ahead scheme using Algorithm 15 to make $t'_0 - t'_1 \geq J$. Accordingly retiming can be applied to avoid the large fanout in the J-unfolded LFSR implementing the division by $g'(x)$.

Algorithm 15 Computation of $p(x)$ for making $t'_0 - t'_1 \geq J$ in $p(x)g(x)$
input: *binary $g(x)$; J*
initialization:
$\tilde{p}(x) = 1, \tilde{g}(x) = g(x)$;
$a = \deg(g(x)) = t_0, b = t_1$;
start:
  while $(a - b < J)$ do
  {
      $\tilde{g}(x) \Leftarrow \tilde{g}(x) + g(x)x^{b-a}$
      $\tilde{p}(x) \Leftarrow \tilde{p}(x) + x^{b-a}$
      $num = a - b$
      $b = \deg(\tilde{g}(x) - x^a)$
  }
  $p(x) = \tilde{p}(x)x^{num}$
  $g'(x) = \tilde{g}(x)x^{num}$

The goal of Algorithm 15 is to cancel out the monomial with the second highest degree in $\tilde{g}(x)$ each time, until the difference between the degrees of the two most significant monomials is at least J. The purpose of the last two lines in this algorithm is to make the degrees of the monomials non-negative.

Example 34 *Consider that $J=4$-unfolding needs to be applied to the LFSR encoder for a (15,7) BCH code constructed over $GF(2^4)$. The generator polynomial of this code is $g(x) = x^8 + x^7 + x^6 + x^4 + 1$. Table 7.1 lists the key values at the end of each iteration of Algorithm 15 in order to derive a $p(x)$ for applying $J=4$-unfolding. At the end of this algorithm, $\tilde{g}(x)$ and $\tilde{p}(x)$ are multiplied by $x^{num} = x^3$ to get $g'(x)$ and $p(x)$ as*

$$\begin{cases} g'(x) = x^{11} + x^4 + x^3 + x^2 + 1 \\ p(x) = x^3 + x^2 + 1 \end{cases}$$

TABLE 7.1

An example for look-ahead $p(x)$ computation

iteration	$\tilde{g}(x)$	$\tilde{p}(x)$	a	b	num
-	$x^8 + x^7 + x^6 + x^4 + 1$	1	8	7	-
1	$x^8 + x^5 + x^4 + x^3 + 1 + x^{-1}$	$1 + x^{-1}$	8	5	1
2	$x^8 + x + 1 + x^{-1} + x^{-3}$	$1 + x^{-1} + x^{-3}$	8	1	3

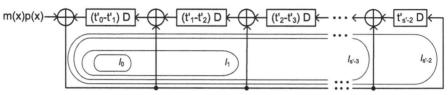

FIGURE 7.2

Serial LFSR architecture for the second step of BCH encoding (modified from [91])

The division by $p(x)$ in the third step of the encoding needs another LFSR architecture. From Algorithm 15, $\deg(p(x))$ is at most $J - 1$. Usually the unfolding factor, J, is not a very large number. Hence the LFSR for $p(x)$ division does not suffer from large fanout. Overall, the 3-step modified BCH encoding process proposed in [90] no longer has the fanout bottleneck.

As explained in Chapter 2, an architecture free of feedback loops can be pipelined to achieve higher clock frequency. However, the minimum achievable clock period of an architecture consisting of feedback loops is decided by its iteration bound, T_∞, which equals the maximum of all the loop bounds. The bound of a loop equals the computation time of the loop divided by the number of delay elements in the loop. The loop whose bound is the largest is called the critical loop. Since the first step of the three-step BCH encoding in [90] is implemented by an architecture without any feedback loop, the achievable speed of the overall encoder is determined by the iteration bounds of the two LFSRs for the second and third steps. Applying J-unfolding to an architecture with T_∞, the iteration bound of the unfolded architecture is JT_∞. Let T_∞^A and T_∞^B be the iteration bounds of the serial LFSRs for the second and third steps, respectively. Then the minimum achievable clock period of the J-unfolded three-step encoder is decided by $max\{JT_\infty^A, JT_\infty^B\}$.

Assume that $g'(x) = p(x)g(x)$ is written in the format of (7.2), the LFSR implementing the division by $g'(x)$ is shown in Fig. 7.2. Since $m(x)p(x)$ is multiplied by x^{n-k} before the division, $m(x)p(x)$ should be added to the output of the $(n-k)^{th}$ delay element from the right in the LFSR. $deg(g'(x)) \leq n-k+(J-1)$ and $t_0'-t_1' \geq J$. Hence, $m(x)p(x)$ would be added to the output of one of the $t_0'-t_1'$ consecutive delay elements in the left of the LFSR. To simplify

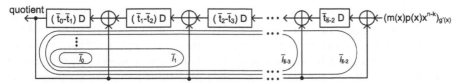

FIGURE 7.3
Serial LFSR architecture for the third step of BCH encoding (modified from [91])

the illustration, retiming is applied to move this addition to the output of the left-most delay element. As a result, the LFSR architecture in Fig. 7.2 can be used to divide $m(x)p(x)x^{n-k}$ by $g'(x)$. Note that retiming does not change the iteration bound. Since there are s' nonzero coefficients in $g'(x)$, there are $s' - 1$ loops in the LFSR in Fig. 7.2. Assume that the propagation delay of an XOR gate is T_{XOR}. Then the bound for loop l_i, $(0 \le i \le s' - 2)$ in Fig. 7.2 is $(2+i)T_{XOR}/(t'_0 - t'_{i+1})$. For long BCH codes, around half of the coefficients in the generator polynomial, $g(x)$, are nonzero. It has also been observed that in $g'(x) = p(x)g(x)$, about half of the coefficients for the terms whose degrees are lower than t'_1 are zero. Moreover, there are neither long consecutive zeros nor long consecutive ones, except for the $t'_0 - t'_1 - 1$ zeros after the most significant term $x^{t'_0}$. From these observations, there are approximately $2 + (t'_1 - t'_{i+1})/2$ XOR gates in loop l_i, and its bound is

$$\frac{(2+i)}{t'_0 - t'_{i+1}}T_{XOR} \approx \frac{4 + (t'_1 - t'_{i+1})}{2(t'_0 - t'_{i+1})}T_{XOR}.$$

Since $t'_0 - t'_1 \ge J$,

$$\frac{4 + (t'_1 - t'_{i+1})}{2(t'_0 - t'_{i+1})}T_{XOR} \le \frac{(t'_0 - t'_{i+1}) - (J - 4)}{2(t'_0 - t'_{i+1})}T_{XOR} = (\frac{1}{2} - \frac{J - 4}{2(t'_0 - t'_{i+1})})T_{XOR}.$$

To achieve high speed, $J \ge 4$ are usually needed. In this case,

$$(\frac{1}{2} - \frac{J - 4}{2(t'_0 - t'_{i+1})})T_{XOR} \le (\frac{1}{2} - \frac{J - 4}{2t'_0})T_{XOR}. \tag{7.3}$$

The iteration bound equals the maximum of all the loop bounds. Hence, (7.3) shows that T^A_∞ is at most around $T_{XOR}/2$ when $J \ge 4$. Moreover, the larger the unfolding factor J, the smaller T^A_∞.

Express $p(x)$ as

$$p(x) = x^{\bar{t}_0} + x^{\bar{t}_1} + \cdots + x^{\bar{t}_{\bar{s}-2}} + 1,$$

where $\bar{t}_0 > \bar{t}_1 > \cdots > \bar{t}_{\bar{s}-2}$ are positive integers. The third step of the BCH encoding can be implemented by the LFSR architecture illustrated in Fig. 7.3. This step computes the quotient of dividing $p(x)$ from the output of the second step, which is $(m(x)p(x)x^{n-k})_{g'(x)}$. Since $(m(x)p(x)x^{n-k})_{g'(x)}$ does not need to be multiplied by any power of x, it is directly added to the rightmost tap of the LFSR. As the inputs are taken into the LFSR, the coefficients of the quotient are available from the leftmost register, starting with the most significant one. From Fig. 7.3, the bound of loop $\bar{l}_i$ ($0 \leq i < \bar{s} - 1$) is

$$(1+i)T_{XOR}/(\bar{t}_0 - \bar{t}_{i+1}).$$

$p(x)$ does not necessarily have half of its coefficients as zero. Since $deg(p(x)) \leq J-1, \bar{t}_0 - \bar{t}_{i+1} < J$. Nevertheless, $t'_0 - t'_{i+1}$ is always larger than J and increases with i. Therefore, the loop bounds and hence the iteration bound, T_∞^B, of the LFSR for the third step are typically larger than those for the second step. As a result, the maximum clock frequency of the entire BCH encoder can be increased by reducing T_∞^B.

Actually, $t_0 - t_1 = 1$ for half of all possible generator polynomials of BCH codes. From Algorithm 15, $\bar{t}_0 - \bar{t}_1 = 1$ in this case, and hence the bound of loop $\bar{l}_0$ in the LFSR in Fig. 7.3 is T_{XOR}. Since the bound of any loop in this LFSR would not exceed T_{XOR}, $T_\infty^B = T_{XOR}$. Moreover, $\bar{t}_0 - \bar{t}_1$ does not change with the unfolding factor J in this case, and hence T_∞^B is always T_{XOR}. On the other hand, T_∞^A decreases as J increases. Therefore, reducing T_∞^B becomes more critical to speeding up the overall encoder when J is larger.

Finding a general method to reduce T_∞^B is very difficult. Alternatively, since $p(x)$ is not long, the LFSR for the third step of the BCH decoding can be replaced by architectures free of feedback loops. Pipelining can always be applied to feedback loop-free architectures, and hence the third step no longer affects the minimum achievable clock period of the overall decoder. Three approaches have been proposed in [91] to eliminate the feedback loops associated with the third step.

Approach A

Let the output of the second step be $a(x) = (m(x)p(x)x^{n-k})_{g'(x)}$. The third step needs to compute $r(x) = a(x)/p(x)$. Let a and r be the row vectors consisting of the coefficients of $a(x)$ and $r(x)$, respectively. Instead of using a second LFSR architecture, r can be computed by a matrix multiplication:

$$r^T = Qa^T.$$

From Algorithm 15, the degree of $p(x)$ should be less than J. Assume that $deg(p(x)) = z$ ($z < J$). Then $deg(a(x)) \leq deg(g'(x)) - 1 = deg(g(x)) + deg(p(x)) - 1 = n - k + z - 1$. Therefore, $a(x)$ has at most $n - k + z$ coefficients. Also $deg(r(x)) \leq deg(g(x)) - 1 = n - k - 1$, and hence $r(x)$ has at most $n - k$ coefficients. Accordingly, the dimension of Q is $(n-k) \times (n-k+z)$. Nevertheless, the monomials in $a(x)$ whose degrees are less than z do not contribute to the quotient of dividing $p(x)$. As a result, the computation of r is simplified as

$$r^T = Q'_{(n-k)\times(n-k)}(a')^T,$$

where $a' = [a_{n-k+z-1}, a_{n-k+z-2}, \cdots, a_z]$. Assume that $r = [r_{n-k-1}, r_{n-k-2}, \cdots, r_0]$ and $p(x) = p_z x^z + p_{z-1} x^{z-1} + \cdots p_0$. The entries of Q', $q'_{i,j}$ $(0 \le i, j < n-k)$, are derived using Algorithm 16. In each iteration, $\tilde{r}(x)$ and $\tilde{q}(x)$ are the remainder and quotient polynomials, respectively, of dividing $x^{n-k+z-1-j}$ by $p(x)$. By making use of these two polynomials, the contributions of $x^z, x^{z+1}, \ldots$ on the quotient are iteratively derived to form the entries of Q'.

Algorithm 16 Q' matrix construction
input: $p(x)$
initialization:
    $\tilde{r}_i = p_i$ $(0 \le i \le z - 1)$
    $\tilde{q}_0 = 1, \tilde{q}_i = 0$ $(1 \le i \le n - k - 1)$
begin construction:
    for $j = n - k - 1$ *downto* 0 *do*
        for $i = 0$ *to* $n - k - 1$ *do*
            $q'_{i,j} = \tilde{q}_{n-k-1-i}$
        $\tilde{q}(x) \Leftarrow x \cdot \tilde{q}(x) + \tilde{r}_{z-1}$
        $\tilde{r}(x) \Leftarrow (x \cdot \tilde{r}(x))_{p(x)}$

Example 35 *The generator polynomial $g(x)$ for a (15, 7) BCH code is $x^8 + x^7 + x^6 + x^4 + 1$. To apply $J = 4$ unfolding, Example 34 has showed that $p(x) = x^3 + x^2 + 1$ and $z = 3$. Using Algorithm 16, the Q' matrix is derived as*

$$Q' = \begin{bmatrix} 1 & 0 & 0 & 0 & 0 & 0 & 0 & 0 \\ 1 & 1 & 0 & 0 & 0 & 0 & 0 & 0 \\ 1 & 1 & 1 & 0 & 0 & 0 & 0 & 0 \\ 0 & 1 & 1 & 1 & 0 & 0 & 0 & 0 \\ 1 & 0 & 1 & 1 & 1 & 0 & 0 & 0 \\ 0 & 1 & 0 & 1 & 1 & 1 & 0 & 0 \\ 0 & 0 & 1 & 0 & 1 & 1 & 1 & 0 \\ 1 & 0 & 0 & 1 & 0 & 1 & 1 & 1 \end{bmatrix}.$$

The Q' matrix computed from Algorithm 16 is a lower triangular matrix of dimension $(n-k) \times (n-k)$. Hence, the complexity of multiplying this matrix is $O((n - k)^2/2)$. Since $(n - k)$ is large for long BCH codes, this matrix multiplication needs a large number of logic gates.

Approach B
Represent the quotient and remainder of dividing $m(x)x^{n-k}$ by $g(x)p(x)$ as $q'(x)$ and $r'(x)$, respectively. Then

$$m(x)x^{n-k} = q'(x)(g(x)p(x)) + r'(x), \tag{7.4}$$

Moreover, $r'(x)$ can be expressed in terms of the quotient and remainder of dividing by $g(x)$ as

$$r'(x) = q''(x)g(x) + r(x). \tag{7.5}$$

Accordingly,

$$m(x)x^{n-k} = (q'(x)p(x) + g''(x))g(x) + r(x). \tag{7.6}$$

Comparing (7.1) and (7.6), the $r(x)$ in (7.5) is the same as the remainder polynomial needs to be computed for BCH encoding. Therefore, $r(x)$ can be computed alternatively according to (7.4) and (7.5) as:

i) Compute $r'(x) = (m(x)x^{n-k})_{g(x)p(x)}$
ii) Compute $r(x) = (r'(x))_{g(x)}$

   Step i) can be implemented by a LFSR architecture similar to that in Fig. 7.2, and Algorithm 15 is also used to compute $p(x)$ so that the fanout bottleneck is eliminated from this LFSR through retiming. The division by $g(x)$ in step ii) should not be implemented by a LFSR. The large fanout bottleneck in the LFSR for $g(x)$ division was the problem to be solved in the original systematic encoder. Fortunately, the polynomial that needs to be divided by $g(x)$ in step ii), which is $r'(x)$, has a much lower degree than $m(x)x^{n-k}$ as in the original encoder. Hence, it is affordable to have the division implemented by a matrix multiplication.

Algorithm 17 H matrix construction
input: $g(x)$
initialization:
   $\hat{r}_0 = 1$, $\hat{r}_i = 0$ $(1 \le i \le n - k - 1)$
begin construction:
   for $j = n - k + z - 1$ downto 0 do
       for $i = 0$ to $n - k - 1$ do
           $h_{i,j} = \hat{r}_{n-k-1-i}$
           $\hat{r}(x) \Leftarrow (x\hat{r}(x))_{g(x)}$

   Again assume that $deg(p(x)) = z$. Since $deg(r'(x)) \le deg(g(x)p(x)) - 1 = n - k + z - 1$ and $deg(r(x)) \le n - k - 1$, the coefficients of $r(x)$ can be computed by multiplying a $(n - k) \times (n - k + z)$ matrix, H, as:

$$r^T = H_{(n-k)\times(n-k+z)}(r')^T. \tag{7.7}$$

$r = [r_{n-k-1}, r_{n-k-2} \cdots, r_0]$ and $r' = [r'_{n-k+z-1}, r'_{n-k+z-2}, \cdots, r'_0]$ are the vectors consisting of the coefficients of $r(x)$ and $r'(x)$, respectively. Algorithm

17 gives a procedure to derive the entries $h_{i,j}$ $(0 \leq i < n-k, 0 \leq j < n-k+z)$ for the H matrix. It is done through iteratively finding the remainders of dividing $1, x, x^2, \ldots, x^{n-k+z-1}$ by $g(x)$. In each iteration, $\hat{r}(x)$ is used to track the remainder polynomial.

Example 36 *Consider the same (15, 7) BCH code whose generator polynomial is* $g(x) = x^8 + x^7 + x^6 + x^4 + 1$. *To apply* $J = 4$ *unfolding,* $p(x) = x^3 + x^2 + 1$ *and* $\deg(p(x)) = z = 3$. *Hence the dimension of the* H *matrix is* 8×11. *From Algorithm 17, the* H *matrix is computed as*

$$
H = \begin{bmatrix}
1 & 0 & 1 & 1 & 0 & 0 & 0 & 0 & 0 & 0 & 0 \\
1 & 1 & 1 & 0 & 1 & 0 & 0 & 0 & 0 & 0 & 0 \\
1 & 1 & 0 & 0 & 0 & 1 & 0 & 0 & 0 & 0 & 0 \\
0 & 1 & 1 & 0 & 0 & 0 & 1 & 0 & 0 & 0 & 0 \\
0 & 0 & 0 & 0 & 0 & 0 & 0 & 1 & 0 & 0 & 0 \\
1 & 0 & 0 & 0 & 0 & 0 & 0 & 0 & 1 & 0 & 0 \\
1 & 1 & 0 & 0 & 0 & 0 & 0 & 0 & 0 & 1 & 0 \\
0 & 1 & 1 & 0 & 0 & 0 & 0 & 0 & 0 & 0 & 1
\end{bmatrix}.
$$

The rightmost $n - k$ columns of H is an identity matrix, and hence H is in the format of

$$
H = [H'_{(n-k) \times z} \; I_{(n-k) \times (n-k)}].
$$

The complexity of the H matrix multiplication is $O((n - k) \times J)$. When the unfolding factor is moderate, Approach B has lower complexity than Approach A for long BCH codes.

Approach C

As mentioned in Approach A, $r(x)$ is the quotient of dividing $a(x) = (m(x)p(x)x^{n-k})_{g'(x)}$ by $p(x)$. However, instead of dividing $p(x)$ from $a(x)$ directly, $r(x)$ can be derived alternatively as

$$
r(x) = a(x)/p(x) = a(x)b(x)/(x^v - 1),
$$

if there is a polynomial $b(x)$ such that $p(x)b(x)$ is in the format of $x^v - 1$ $(v \in Z^+)$. In this case, $r(x)$ is computed as first multiplying $a(x)$ by $b(x)$, and then dividing the product by $x^v - 1$. The multiplication with $b(x)$ can be implemented by an architecture similar to that of an FIR filter. The advantage of this approach is that the division by a polynomial in the format of $x^v - 1$ requires simple hardware to implement. Let $f(x) = a(x)b(x)$. Then $\deg(f(x)) = \deg(x^v - 1) + \deg(r(x)) = v + \deg(r(x)) \leq v + n - k - 1$. Assume that $f(x) = f_0 + f_1 x + \cdots + f_{v+n-k-1} x^{v+n-k-1}$. Let $l = \lfloor (n - k)/v \rfloor$ and $u = (n - k)\%v$. Through following the computations in long division, the coefficients of $r(x)$ can be calculated using Algorithm 18.

Algorithm 18 Computation of $r(x) = f(x)/(x^v - 1)$
input: $f(x)$
start:
    for $j = 1$ *to* v
        $r_{n-k-j} = f_{n-k+v-j}$
    for $i = 1$ *to* $l - 1$
        for $j = 1$ *to* v
            $r_{n-k-iv-j} = r_{n-k-(i-1)v-j} + f_{n-k-(i-1)v-j}$
    for $j = 1$ *to* u
        $r_{n-k-lv-j} = r_{n-k-(l-1)v-j} + f_{n-k-(l-1)v-j}$

Let $\check{r}(x)$ be the remainder of dividing $f(x)$ by $x^v - 1$. Following the long division, it can be also derived that coefficients of $\check{r}(x)$ are

$$\check{r}_i = \begin{cases} \sum_{j=0}^{l+1} f_{i+jv} & 0 \le i \le u \\ \sum_{j=0}^{l} f_{i+jv} & u < i < v. \end{cases} \tag{7.8}$$

Since it is known that $x^v - 1$ divides $f(x)$, $\check{r}_i = 0$ for $0 \le i < v$. Moreover, from Algorithm 18, $r_i + f_i = \check{r}_i$ for $0 \le i < v$. As a result, the last v coefficients of $r(x)$ are simply $r_i = f_i$ $(0 \le i < v)$. Algorithm 18 also shows that each r_i $(v \le i \le n - k - 1 - v)$ can be computed by only one additional XOR gate if substructure sharing is applied. Therefore, finding the quotient of $f(x)/(x^v-1)$ only needs $\max(n - k - 2v, 0)$ XOR gates with $l - 1$ gates in the critical path.

Example 37 *Consider again applying 4-unfolding to the encoder of the (15,7) BCH code, whose generator polynomial is* $g(x) = x^8 + x^7 + x^6 + x^4 + 1$. *It can be computed that the smallest* v *such that* $p(x) = x^3 + x^2 + 1$ *divides* $x^v - 1$ *is 6. Therefore,* $\deg(f(x)) = 13$ $l = 1$, *and* $u = 2$. *According to Algorithm 18, the coefficients of* $r(x)$ *are*

$r_7 = f_{13}$	$r_6 = f_{12}$	$r_5 = f_{11}$	$r_4 = f_{10}$
$r_3 = f_9$	$r_2 = f_8$	$r_1 = f_{13} + f_7$	$r_0 = f_{12} + f_6.$

In addition, from (7.8), $\check{r}_0 = f_0 + f_6 + f_{12} = 0$ *and* $\check{r}_1 = f_1 + f_7 + f_{13} = 0$. *Hence* $r_0 = f_0$ *and* $r_1 = f_1$. *It follows that dividing* $f(x)$ *by* $x^6 - 1$ *needs* $\max(8 - 2 \times 6, 0) = 0$ *XOR gates with 1-1=0 XOR gate in the critical path.*

The smallest v, such that the $p(x)$ computed by Algorithm 15 divides $x^v - 1$, may be large. The division of $f(x)$ by $x^v - 1$ becomes simpler for larger v. However, larger v also translates to a higher degree for $b(x)$. Additionally,

TABLE 7.2

Percentage of $p(x)$ that have extensions dividing $x^{31} - 1$, $x^{63} - 1$, $x^{127} - 1$ and $x^{255} - 1$ (data from [91])

$\deg p(x)$	8	10	16	18	20	24	32
$x^{31} - 1$	42.19%	12.89%	0.14%	0.00%	0.00%	0.00%	0.00%
$x^{63} - 1$	100%	100%	11.04%	2.87%	0.68%	0.04%	0.00%
$x^{127} - 1$	100%	100%	100%	49.07%	31.71%	3.01%	0.00%
$x^{255} - 1$	100%	100%	100%	100%	100%	100%	99.96%

the gate count of multiplying $b(x)$ becomes J times higher in a J-unfolded architecture, while the complexity of dividing $x^v - 1$ using Algorithm 18 is not affected by J. Hence, larger v leads to higher encoder complexity, especially as J increases. To reduce v, extensions of $p(x)$ can be used instead of $p(x)$ in each step of the BCH encoding. Here an extension of $p(x)$ is defined as $p'(x) = p(x)x^d + e(x)$, where d is a positive integer and $deg(e(x)) < d + deg(p(x)) - (J - 1)$. In other words, the coefficients of the J highest-degree terms in $p'(x)$ are the same as those in $p(x)$. Hence, $g'(x) = g(x)p'(x)$ still satisfies $t'_0 - t'_1 \geq J$. Although the degree of $p'(x)$ might be slightly higher than that of $p(x)$, the extra terms in $e(x)$ allow various $p'(x)$ to be examined in order to find a smaller v such that $p'(x)$ divides $x^v - 1$. For example, the minimum v such that $p(x) = x^5 + x^4 + x^2 + x + 1$ divides $x^v - 1$ is 31. On the other hand, the minimum v for the extension $p'(x) = x^3 p(x) + x + 1$ is only 15.

One method can be used to find the minimum v, such that an extension of $p(x)$ divides $x^v - 1$, is to increase v by one each time and test if $x^v - 1$ is divisible by any extension of $p(x)$. This exhaustive search takes exponential time. An easier way is to limit the possible values of v to a list. The v found from this list may be larger than the actual minimum v. However, the process for finding the extension and the $x^v - 1$ it divides becomes much simpler. A good choice of the list is $v = 2^i - 1$ ($i \in Z^+$). $x^{2^i - 1} - 1$ are factored into irreducible polynomials, which are listed in many coding books, such as [5]. If there is a combination of the irreducible factors such that the highest coefficients in their product match those in $p(x)$, then the product is an extension of $p(x)$ that divides $x^{2^i - 1} - 1$. A moderate unfolding factor leads to a $p(x)$ of moderate degree. The percentages of $p(x)$ of some moderate degrees that have extensions dividing $x^{2^i - 1}$ for $i = 5, 6, 7, 8$ are listed in Table 7.2. For example, this table shows that 42.19% of possible $p(x)$ whose degree is 8 have some extensions that divide $x^{31} - 1$. From this table, v is not large for $p(x)$ of moderate degree, and hence moderate unfolding factors. Any degree-16 $p(x)$ has some extensions that divide $x^{127} - 1$, and 99.96% of all degree-32 $p(x)$ have some extensions that divide $x^{255} - 1$.

The complexities of the above three encoding approaches are dependent on the generator polynomial, $g(x)$, and the unfolding factor J. The pattern of $g(x)$

TABLE 7.3
Complexities of Approaches A, B, C for J-unfolded architectures (data from [91])

Approach	Complexity (# of XOR gates)
A	$(J + (n - k))J/2 + (n - k)^2/4$
B	$((n - k) + (n - k))J/2$
C	$((J - 1 + d) + (n - k + d) + (v - (J - 1 + d)))J/2 + \max(n - k - 2v, 0)$

varies for different BCH codes. However, as mentioned previously, the number of nonzero coefficients in $g(x)$ is around half, and that in $g'(x)$ is about $t_1'/2$ for long BCH codes. These observations are employed to estimate the encoder hardware complexity. It is also assumed that around half of the coefficients in $p(x)$ are nonzero. This assumption is not accurate. However, since $\deg(p(x))$ is much lower than $\deg(g(x))$ for long BCH codes, the resulting estimation error is only a very small fraction.

The complexities of the three encoding approaches for the J-unfolded architecture are listed in Table. 7.3. Assume that $\deg(p(x)) = J - 1$. In Approach A, the multiplication of $m(x)$ and $p(x)$ needs $J/2 \cdot J$ XOR gates in the J-unfolded architecture. $t_1'/2 \cdot J \approx (n - k)J/2$ XOR gates are required in the LFSR implementing the division by $g'(x) = g(x)p(x)$. Assume that around half of the entries in the lower triangular part of Q' are nonzero. Then the multiplication with the Q' matrix takes roughly $(n - k)^2/4$ gates.

Approach B does not multiply any polynomial to $m(x)$. Similarly, around $(n - k)J/2$ XOR gates are required to implement the J-unfolded LFSR for dividing $m(x)x^{n-k}$ by $g'(x)$. The H matrix has two parts: an $(n - k) \times J$ H' matrix and an $(n - k) \times (n - k)$ identity matrix. Around half of the coefficients of $g(x)$ are nonzero and there are no long strings of '0's or '1's. Hence, it can be assumed that around half of the entries in H' are nonzero. As a result, multiplying H takes approximately $J/2 \cdot (n - k)$ gates.

Approach C adopts $p'(x) = x^d p(x) + e(x)$ instead of $p(x)$. If $\deg(p(x)) = J - 1$, $\deg(p'(x)) = J - 1 + d$. If around half of the coefficients in $p'(x)$ are nonzero, $m(x)p'(x)$ takes $(J - 1 + d)/2 \cdot J$ XOR gates to compute in a J-unfolded architecture. The degree of $g'(x) = p'(x)g(x)$ is $n - k + J - 1 + d$. Moreover, the coefficients of the $J - 1$ terms after $x^{n-k+J-1+d}$ in $g'(x)$ are zero. Assuming half of the remaining $n - k + d$ coefficients are nonzero, the J-unfolded LFSR implementing the division by $g'(x)$ requires $(n - k + d)/2 \cdot J$ XOR gates. $\deg(b(x)) = \deg(x^v - 1) - \deg(p'(x)) = v - (J - 1 + d)$. Hence the complexity of the J-unfolded multiplication by $b(x)$ is $(v - (J - 1 + d))/2 \cdot J$. At last, the division by $x^v - 1$ needs $\max(n - k - 2v, 0)$ XOR gates as discussed previously.

For long BCH codes, usually $n - k \geq 2J$ if J is moderate or small. In this case, it has been shown in [91] that Approach A always has higher complexity than Approach B. Moreover, if $v + d < n - k$, then the gate count of Approach B is larger than that of Approach C. From Table 7.2, $v = 255$ can be used

TABLE 7.4
Complexities of J-unfolded encoder architectures for a (8191, 7684)
BCH code (data from [91])

	$J=8$	$J=16$	$J=24$	$J=32$
$\deg(p(x))$	7	15	21	30
$\deg(p'(x))$, v for Approach C	17, 63	49, 127	71, 255	93, 255
# of clks/block ([90] & Approach A)	962	482	322	242
# of clks/block (Approach B)	961	481	321	241
# of clks/block (Approach C)	963	484	324	244
T_∞ [90] (# of T_{XOR})	8	16	24	32
T_∞ Approach A & B(# of T_{XOR})	5.1	8.0	12.8	16.0
T_∞ Approach C (# of T_{XOR})	4.2	7.8	11.1	14.0
Speedup Approach A/[90]	33%	100%	85%	100%
Speedup Approach B/[90]	33%	100%	85%	101%
Speedup Approach C/[90]	60%	99%	99%	112%
Gate count Approach A(# of XOR)	57622	67338	69230	71549
Gate count Approach B (# of XOR)	4048	8172	11981	16344
Gate count Approach C(# of XOR)	2845	5469	9432	12512

for 99.96% of all possible $p(x)$ for $J = 32$. If J is smaller, lower v would be sufficient. On the other hand, $n - k$ is much larger than v for long BCH codes even when the code rate is high. For example, when the codeword length is longer than 8000, $n - k > 500$ for code rate as high as 93.75%. Hence, $v + d < n - k$ is typically satisfied for long BCH codes and moderate unfolding factors.

Example 38 *For a (8191, 7684) BCH code constructed over $GF(2^{13})$, the complexities of different encoder architectures for various unfolding factors are listed in Table 7.4. The generator polynomial of this BCH code is $g(x) = x^{507} + x^{506} + x^{502} + \cdots + x^6 + x^2 + 1$. In the encoders, the matrix and polynomial multiplications are pipelined with the polynomial division using the LFSR architecture. The latency of the LFSR is longer and hence it decides the number of clock cycles needed to encode a block of messages. The number of clock cycles needed by a J-unfolded LFSR is $\lceil$ length of the LFSR input/$J \rceil$. For example, to apply $J = 16$ unfolding, the degree of the computed $p(x)$ is 15. The input to the LFSR in Approach A is $m(x)p(x)$, and hence its length is $7684+15=7699$-bit. Therefore, the encoding of a message block using Approach A takes $\lceil 7699/16 \rceil = 482$ clock cycles. In the encoder architecture proposed in [90], the critical loop is located at the LFSR implementing the division by $p(x)$. Since $t_0 - t_1 = 1$ for the (8191, 7684) code, the iteration bound of the J-unfolded encoder using this scheme is JT_{XOR}. In Approaches A, B and C, the LFSR dividing $p(x)$ is replaced by feedback loop-free architectures, and accordingly the iteration bounds of these encoders are substantially reduced. Since the same $p(x)$ is adopted in Approaches A and B, their iteration bounds are the same. Nevertheless, the iteration bound of Approach C is different, since it uses $p'(x)$ instead of $p(x)$. The estimated gate counts and achievable*

speedup of the three approaches over that in [90] are also listed in Table 7.4.
Approach C requires the smallest area to achieve similar or higher speed than
approaches A and B.

7.3   Hard-decision BCH decoders

Syndrome-based hard-decision BCH decoders have three steps: syndrome computation, key equation solver, and the Chien search. Let the received binary vector be $r = [r_0, r_1, \ldots, r_{n-1}]$, and the corresponding polynomial be $r(x)$. $r(x)$ is the sum of $c(x)$ and $e(x)$, where $c(x)$ and $e(x)$ are binary codeword and error polynomials, respectively. For narrow-sense BCH codes, the syndromes are defined as $S_j = r(\alpha^j)$ $(1 \le j \le 2t)$, where α is a primitive element of $GF(2^q)$ and $\alpha, \alpha^2, \ldots, \alpha^{2t}$ are roots of the generator polynomial. Assume that there are v errors and they are located at positions α^{i_l} $(1 \le l \le v)$. An error locator polynomial can be defined in the same way as that for RS codes as

$$\Lambda(x) = \prod_{l=1}^{v}(1 - X_l x) = \Lambda_0 + \Lambda_1 x + \cdots + \Lambda_v x^v,$$

where $X_l = \alpha^{i_l}$. Note that $\Lambda_0 = 1$. The key equation solver step finds $\Lambda(x)$ from the syndromes. Then the Chien search is carried out on $\Lambda(x)$ to find the error locations i_l.

Since the received vector is binary, the syndrome computation architectures for RS codes shown in Fig. 4.5 and 4.6 can be simplified for BCH decoding. In the serial architecture of Fig. 4.5, the XOR operation is carried out over a $GF(2^q)$ element and a binary input bit in the case of BCH decoding. Hence only one XOR gate is needed over the input bit and the least significant bit of the multiplier output, if standard basis representation is used. For the J-parallel architecture in Fig. 4.6, the multipliers in the first column can be replaced by hard wiring since the multiplicands, $\alpha^j, \ldots, \alpha^{j(J-1)}$, are fixed for these multipliers. For example, if $\alpha^j \in GF(2^6)$ is '001011' in basis representation, then the product of r_i and α^j is '00r_i0$r_i$$r_i$'.

The error locator polynomial in BCH decoding is over $GF(2^q)$. Hence, the Chien search for finding its roots has the same complexity as that in RS decoding. Nevertheless, the error magnitudes are all '1' in BCH decoding, and hence the BCH decoder does not have the parts for computing the error magnitude polynomial and magnitudes.

The computation of the error locator polynomial can be also simplified by making use of the binary property of BCH codes. The most broadly used BCH error locator polynomial computation algorithms are the Peterson's and Berlekamp's algorithms. The details of these algorithms are presented next.

7.3.1 Peterson's algorithm

For binary BCH codes, each error magnitude is '1'. Hence the syndromes are expressed as

$$S_j = \sum_{l=1}^{v} X_l^j, \ 1 \le j \le 2t.$$

In this case, the syndromes and coefficients of $\Lambda(x)$ are related by the following equations [5].

$$
\begin{aligned}
&S_1 + \Lambda_1 = 0 \\
&S_2 + \Lambda_1 S_1 + 2\Lambda_2 = 0 \\
&S_3 + \Lambda_1 S_2 + \Lambda_2 S_1 + 3\Lambda_3 = 0 \\
&\qquad\qquad \vdots \\
&S_v + \Lambda_1 S_{v-1} + \Lambda_2 S_{v-2} + \cdots + \Lambda_{v-1} S_1 + v\Lambda_v = 0 \\
&S_{v+1} + \Lambda_1 S_v + \Lambda_2 S_{v-1} + \cdots + \Lambda_v S_1 = 0 \\
&\qquad\qquad \vdots \\
&S_{2t} + \Lambda_1 S_{2t-1} + \Lambda_2 S_{2t-2} + \cdots + \Lambda_v S_{2t-v} = 0
\end{aligned}
\tag{7.9}
$$

Over finite field $GF(2^q)$, $S_{2j} = \sum_{l=1}^{v} X_l^{2j} = (\sum_{l=1}^{v} X_l^j)^2 = S_j^2$. $a\Lambda_j = 0$ if a is an even integer and $a\Lambda_j = \Lambda_j$ if a is odd. Therefore, if $S_1 + \Lambda_1 = 0$, then $S_2 + \Lambda_1 S_1 + 2\Lambda_2 = S_1^2 + S_1 S_1 = 0$. In other words, if the first constraint in (7.9) is satisfied, then the second constraint is also satisfied. If the third constraint $S_3 + \Lambda_1 S_2 + \Lambda_2 S_1 + 3\Lambda_3 = 0$ is also satisfied, then

$$
\begin{aligned}
S_4 + \Lambda_1 S_3 &+ \Lambda_2 S_2 + \Lambda_3 S_1 + 4\Lambda_4 \\
&= S_1^4 + S_1 S_3 + \Lambda_2 S_1^2 + S_1(S_3 + \Lambda_1 S_2 + \Lambda_2 S_1) \\
&= S_1^4 + S_1 S_3 + \Lambda_2 S_1^2 + S_1 S_3 + S_1^4 + \Lambda_2 S_1^2 \\
&= 0.
\end{aligned}
$$

Accordingly, the fourth constraint is satisfied if the first and third constraints are satisfied. Similarly, it can be proved that all even constraints in (7.9) can be eliminated. As a result, when $v = t$ errors happen, (7.9) is reduced to a set of t equations with t unknowns as specified by the following matrix multiplication format.

$$
A'\Lambda =
\begin{bmatrix}
1 & 0 & 0 & \cdots & 0 & 0 \\
S_2 & S_1 & 1 & \cdots & 0 & 0 \\
S_4 & S_3 & S_2 & \cdots & 0 & 0 \\
\vdots & \vdots & \vdots & \ddots & \vdots & \vdots \\
S_{2t-4} & S_{2t-5} & S_{2t-6} & \cdots & S_{t-2} & S_{t-3} \\
S_{2t-2} & S_{2t-3} & S_{2t-4} & \cdots & S_t & S_{t-1}
\end{bmatrix}
\begin{bmatrix}
\Lambda_1 \\
\Lambda_2 \\
\Lambda_3 \\
\vdots \\
\Lambda_{t-1} \\
\Lambda_t
\end{bmatrix}
=
\begin{bmatrix}
S_1 \\
S_3 \\
S_5 \\
\vdots \\
S_{2t-3} \\
S_{2t-1}
\end{bmatrix}
\tag{7.10}
$$

The set of equations in (7.10) have a unique solution if and only if A' is nonsingular. Due to the property that $S_{2j} = S_j^2$ for narrow-sense binary BCH codes, A' is nonsingular if there are t or $t-1$ errors. If the matrix is singular, the two rightmost columns and two bottom rows are removed iteratively until a nonsingular matrix is derived. Then the set of linear equations corresponding to the nonsingular matrix is solved to find the error locator polynomial. The Peterson's binary BCH decoding algorithm is shown in Algorithm 19.

Algorithm 19 Peterson's Algorithm
1. *Compute the syndromes S_j ($1 \leq j < 2t$)*
2. *Construct the syndrome matrix A'*
3. *Compute the determinant of A'*
4. *If the determinant is nonzero, go to step 6*
5. *Reduce A' by deleting the two rightmost columns and two bottom rows;*
   Go back to step 3
6. *Solve for Λ_i ($0 < i \leq v$)*
7. *Compute the roots of $\Lambda(x)$*
8. *If the number of distinct roots of $\Lambda(x)$ in $GF(2^q)$ does not equal*
   to $\deg(\Lambda(x))$, declare decoding failure

When t is small, formulas for directly computing the error locator polynomial coefficients can be derived from (7.10). For example, the formulas for the cases of $t = 1, 2, 3, 4$ are listed below.

- $t = 1$

$$\Lambda_1 = S_1$$

- $t = 2$

$$\Lambda_1 = S_1$$
$$\Lambda_2 = \frac{S_3 + S_1^3}{S_1}$$

- $t = 3$

$$\Lambda_1 = S_1$$
$$\Lambda_2 = \frac{S_1^2 S_3 + S_5}{S_1^3 + S_3}$$
$$\Lambda_3 = (S_1^3 + S_3) + S_1 \Lambda_2$$

- $t = 4$

$$\Lambda_1 = S_1$$

$$\Lambda_2 = \frac{S_1(S_7 + S_1^7) + S_3(S_1^5 + S_5)}{S_3(S_1^3 + S_3) + S_1(S_1^5 + S_5)}$$

$$\Lambda_3 = (S_1^3 + S_3) + S_1\Lambda_2$$

$$\Lambda_4 = \frac{(S_5 + S_1^2 S_3) + (S_1^3 + S_3)\Lambda_2}{S_1}$$

From these formulas, the error locator polynomial can be calculated without an iterative process when t is small. Nevertheless, the complexity of the Peterson's algorithm increases significantly for larger t. In those cases, the Berlekamp's algorithm is more efficient and has been widely adopted.

7.3.2 The Berlekamp's algorithm and implementation architectures

The error locator polynomial for binary BCH decoding can also be found using the Berlekamp's algorithm [4]. Compared to the the Berlekamp-Massey algorithm (BMA) for RS codes, the number of iterations needed in the Berlekamp's algorithm is reduced to half. The reason is that the even constraints in (7.9) can be eliminated in the case of binary BCH codes. The Berlekamp's algorithm is given in Algorithm 20 without proof. Similar to the original BMA in Algorithm 2, this algorithm also suffers from the long latency caused by the computation of the discrepancy $\delta^{(2r)}$.

Algorithm 20 Berlekamp's Algorithm
inputs: S_j *(1 ≤ j ≤ 2t)*
initialization: $r = 0$, $\Lambda^{(0)}(x) = 1$, $B^{(0)}(x) = 1$
begin:
1. $\delta^{(2r)} =$ *the coefficient of* x^{2r+1} *in* $\Lambda^{(2r)}(x)(1 + S_1x + S_2x^2 + \cdots + S_{2t}x^{2t})$
2. $\Lambda^{(2r+2)}(x) = \Lambda^{(2r)}(x) + \delta^{(2r)}xB^{(2r)}(x)$
3.

$$B^{(2r+2)}(x) = \begin{cases} x^2 B^{(2r)}(x) & \text{if } \delta^{(2r)} = 0 \text{ or } \deg(\Lambda^{(2r)}(x)) > r \\ x\Lambda^{(2r)}(x)/\delta^{(2r)} & \text{if } \delta^{(2r)} \neq 0 \text{ and } \deg(\Lambda^{(2r)}(x)) \leq r \end{cases}$$

4. $r \Leftarrow r + 1$
5. *if* $r < t$, *go to step 1*
output: $\Lambda(x) = \Lambda^{(2t)}(x)$

Example 39 *Consider the 2-error-correcting (15,7) BCH code over $GF(2^4)$ constructed in Example 33. The generator polynomial is $g(x) = x^8 + x^7 + x^6 + x^4 + 1$, and $p(x) = x^4 + x + 1$ is the irreducible polynomial used to construct $GF(2^4)$. For a message polynomial $m(x) = x^2 + 1$, the corresponding codeword polynomial is $m(x)g(x) = x^{10} + x^9 + x^7 + x^4 + x^2 + 1$. Assume that the received polynomial is $x^{11} + x^{10} + x^9 + x^7 + x^4 + 1$. It can be computed that the syndromes are $S_1 = \alpha^9$, $S_2 = \alpha^3$, $S_3 = \alpha^2$ and $S_4 = \alpha^6$, where α is a root of $p(x)$. Applying Algorithm 20, the results are listed in Table 7.5. From this table, $\Lambda(x) = \Lambda^{(4)}(x) = 1 + \alpha^9 x + \alpha^{13} x^2$. The roots of this polynomial are α^4 and α^{13}, and hence it is a valid error locator polynomial. The corresponding error locations are $\alpha^{-4} = \alpha^{11}$ and $\alpha^{-13} = \alpha^2$.*

TABLE 7.5

Example values in the Berlekamp's algorithm for binary BCH decoding

r	$\Lambda^{(2r)}(x)$	$B^{(2r)}(x)$	$\delta^{(2r)}$
0	1	1	α^9
1	$1 + \alpha^9 x$	$\alpha^6 x$	α^7
2	$1 + \alpha^9 x + \alpha^{13} x^2$	-	-

The reformulated inversionless Berlekamp-Massey (riBM) and RiBM algorithms [43] discussed in Chapter 4 can be simplified for decoding binary BCH codes, and their architectures are more efficient than those based on the Berlekamp's algorithm. Since the RiBM algorithm is just a reorganized version of the riBM algorithm for the purpose of making all the involved processing elements (PEs) identical, the simplifications are explained next using the riBM algorithm.

The riBM algorithm is listed in Algorithm 3 in Chapter 4. To avoid the long latency associated with the discrepancy computation, the riBM algorithm updates a discrepancy polynomial $\hat{\Delta}^{(r+1)}(x)$ in iteration r in a similar way as updating the error locator polynomial $\Lambda^{(r+1)}(x)$. The constant coefficient of $\hat{\Delta}^{(r+1)}(x)$, denoted by $\hat{\Delta}_0^{(r+1)}$, is the actual discrepancy for iteration $r+1$. For binary BCH codes, the even constraints in (7.9) are satisfied if the odd constraints are satisfied. Hence, the discrepancies, $\hat{\Delta}_0^{(r)}$, are zero for odd r in the riBM algorithm. Making use of this property, those iterations of the riBM algorithm can be skipped by applying a look-ahead scheme. Assume that r is an odd integer. Then following the computations in the riBM algorithm in

Algorithm 3 and $\hat{\Delta}_0^{(r)} = 0$, the updating needs to be done in iteration r is

$$\Lambda^{(r+1)}(x) = \gamma^{(r)}\Lambda^{(r)}(x)$$
$$\hat{\Delta}^{(r+1)}(x) = \gamma^{(r)}\hat{\Delta}^{(r)}(x)/x$$
$$B^{(r+1)}(x) = xB^{(r)}(x)$$
$$\hat{\Theta}^{(r+1)}(x) = \hat{\Theta}^{(r)}(x)$$
$$\gamma^{(r+1)} = \gamma^{(r)}$$
$$k^{(r+1)} = k^{(r)} + 1$$

Accordingly, $\hat{\Delta}_0^{(r+1)} = \gamma^{(r)}\hat{\Delta}_1^{(r)}$. Substituting these equations into those for iteration $r + 1$, $\Lambda^{(r+2)}(x)$ and $\hat{\Delta}^{(r+2)}(x)$ are expressed in terms of $\Lambda^{(r)}(x)$, $\hat{\Delta}^{(r)}(x)$, $B^{(r)}(x)$, $\hat{\Theta}^{(r)}(x)$, $\gamma^{(r)}$, and $k^{(r)}$, and hence iteration $r + 1$ can be skipped. In addition, multiplying the same nonzero scaler to both $\Lambda(x)$ and $\hat{\Delta}(x)$ does not affect the error locations nor the error magnitudes. As a result, the riBM algorithm can be simplified for BCH decoding as in Algorithm 21.

Algorithm 21 Look-ahead riBM algorithm for binary BCH decoding

input: S_j $(1 \leq j \leq 2t)$
initialization: $\Lambda^{(-1)}(x) = 1$, $B^{(-1)}(x) = x^{-1}$
$\qquad\qquad k^{(-1)} = -1$, $\gamma^{(-1)} = 1$
$\qquad\qquad \hat{\Delta}^{(-1)}(x) = S_1 x + S_2 x^2 + \cdots + S_{2t} x^{2t}$
$\qquad\qquad \hat{\Theta}^{(-1)}(x) = S_1 + S_2 x + \cdots + S_{2t} x^{2t-1}$
begin:
$\qquad$ *for* $r = -1, 1, 3, \ldots, 2t - 3$
$\qquad\qquad \Lambda^{(r+2)}(x) = \gamma^{(r)}\Lambda^{(r)}(x) + \hat{\Delta}_1^{(r)} x^2 B^{(r)}(x)$
$\qquad\qquad \hat{\Delta}^{(r+2)}(x) = \gamma^{(r)}\hat{\Delta}^{(r)}(x)/x^2 + \hat{\Delta}_1^{(r)}\hat{\Theta}^{(r)}(x)$
$\qquad\qquad$ *if* $\hat{\Delta}_1^{(r)} \neq 0$ *and* $k^{(r)} \geq -1$
$\qquad\qquad\qquad B^{(r+2)}(x) = \Lambda^{(r)}(x)$
$\qquad\qquad\qquad \hat{\Theta}^{(r+2)}(x) = \hat{\Delta}^{(r)}(x)/x^2$
$\qquad\qquad\qquad \gamma^{(r+2)} = \hat{\Delta}_1^{(r)}$
$\qquad\qquad\qquad k^{(r+1)} = -k^{(r)} - 2$
$\qquad$ *else*
$\qquad\qquad\qquad B^{(r+2)}(x) = x^2 B^{(r)}(x)$
$\qquad\qquad\qquad \hat{\Theta}^{(r+2)}(x) = \hat{\Theta}^{(r)}(x)$
$\qquad\qquad\qquad \gamma^{(r+2)} = \gamma^{(r)}$
$\qquad\qquad\qquad k^{(r+2)} = k^{(r)} + 2$
output: $\Lambda(x) = \Lambda^{(2t-1)}(x)$

Following the original riBM algorithm in Algorithm 3, the updating for iteration $r = 0$ still needs to be done based on the initial polynomials and variables. To make the computations for every iteration identical, which simplifies the hardware units and control logic, Algorithm 21 starts with iteration $r = -1$, and the initial values are modified accordingly to accommodate this change. Originally, whether $\hat{\Delta}_0^{(r+1)} = \gamma^{(r)}\hat{\Delta}_1^{(r)}$ is zero needs to be tested in the look-ahead computations to decide how $B^{(r+2)}(x)$, $\hat{\Theta}^{(r+2)}(x)$, etc. should be updated. Since $\gamma^{(r)}$ is always nonzero, the testing is applied to $\hat{\Delta}_1^{(r)}$ in Algorithm 21. It can be seen from this algorithm that only t iterations are needed to find the error locator polynomial for binary BCH codes. The error magnitudes are always '1' in BCH decoding, and hence the error evaluator polynomial is not needed. However, $\hat{\Delta}(x)$ should not be removed from Algorithm 21 since it is used to generate the discrepancy coefficient $\hat{\Delta}_1^{(r)}$ for each iteration.

Example 40 *Consider the same 2-error-correcting BCH code and received word as in Example 39. Table 7.6 lists the results of applying Algorithm 21 for the decoding. From this table, $\Lambda(x) = \Lambda^{(3)}(x) = \alpha^9 + \alpha^3 x + \alpha^7 x^2$. It is the error locator polynomial found in Example 39 scaled by α^9, and hence has the same roots, α^4 and α^{13}, and same error locations.*

TABLE 7.6
Example values in the look-ahead riBM algorithm for binary BCH decoding

r	$\Lambda^{(r)}(x)$	$B^{(r)}(x)$	$k^{(r)}$	$\gamma^{(r)}$	$\hat{\Delta}^{(r)}(x)$, $\hat{\Theta}^{(r)}(x)$
-1	1	x^{-1}	-1	1	$\alpha^9 x + \alpha^3 x^2 + \alpha^2 x^3 + \alpha^6 x^4$, $\alpha^9 + \alpha^3 x + \alpha^2 x^2 + \alpha^6 x^3$
1	$1 + \alpha^9 x$	1	-1	α^9	$\alpha^7 x + \alpha x^2 + x^3$, $\alpha^3 + \alpha^2 x + \alpha^6 x^2$
3	$\alpha^9 + \alpha^3 x + \alpha^7 x^2$	-	-	-	-

In Algorithm 21, the polynomial updating involved in each iteration is the same as that in the original riBM algorithm for RS codes, except that multiplications and divisions by x are replaced with those by x^2. Hence the same PE1 architecture shown in Fig. 4.2 can be employed to implement a look-ahead RiBM architecture for BCH codes. However, to accommodate the multiplications and divisions by x^2, the $\hat{\Delta}(x)$ or $\Lambda(x)$ input coefficient of PE1$_i$ should come from the output of PE1$_{i+2}$ as shown in Fig. 7.4. The degree of the modified initial $\hat{\Delta}^{(-1)}(x)$ has been increased by one. Nevertheless, the discrepancy coefficients used in Algorithm 21 are $\hat{\Delta}_1^{(r)}$, and hence the constant coefficients of $\hat{\Delta}(x)$ and $\hat{\Theta}(x)$ do not need to be computed or stored. As a result, the number of PEs in Fig. 7.4 remains at $3t + 1$. In the beginning, the registers in PE1$_0$-PE1$_{2t-1}$ are initialized with the non-constant coefficients of

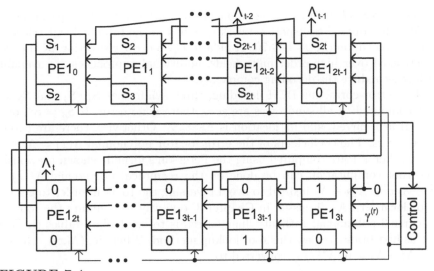

FIGURE 7.4
Look-ahead RiBM architecture for binary BCH codes

$\hat{\Delta}^{(-1)}(x)$ and $\hat{\Theta}^{(-1)}(x)$, and those in $\text{PE1}_{2t}\text{-PE1}_{3t}$ are initialized according to $\Lambda^{(-1)}(x)$ and $B^{(-1)}(x)$. After t iterations, the coefficients of $\Lambda(x) = \Lambda^{(2t-1)}(x)$ are located at PE1_t through PE1_{2t} as shown in Fig. 7.4.

Originally, in the RiBM algorithm for RS codes, $\Lambda(x) = \Lambda^{(2t)}(x)$ and $\hat{\Omega}(x) = \hat{\Delta}^{(2t)}(x) \mod x^t$. The look-ahead RiBM algorithm for BCH codes stops at $\Lambda^{(2t-1)}(x)$ and $\hat{\Delta}^{(2t-1)}(x)$. Since $\hat{\Delta}_1^{(2t-1)} = 0$, $\Lambda^{(2t)}(x)$ would be $\Lambda^{(2t-1)}(x)$ scaled by $\gamma^{(2t-1)}$, which is nonzero. Hence $\Lambda^{(2t-1)}(x)$ can be also used as the error locator polynomial. In addition, $\hat{\Delta}^{(2t)}(x) = \gamma^{(2t-1)}\hat{\Delta}^{(2t-1)}(x)/x$. Since the constant coefficient of $\hat{\Delta}(x)$ is not kept in Fig. 7.4, PE1_0 through PE1_{t-1} have $\hat{\Delta}^{(2t-1)}(x)/x \mod t$ at the end of Algorithm 21. To verify the error magnitudes, these two polynomials are plugged into (4.14), and the computed error magnitudes are indeed '1'.

7.3.3 3-error-correcting BCH decoder architectures

When the error-correcting capability of a BCH code is 2 or 3, the error locator polynomial can be easily computed by the Peterson's algorithm. In addition, the roots of the error locator polynomial can be derived directly using the methods in Chapter 3 without an exhaustive search. As a result, such BCH codes can achieve very high throughput with simple hardware implementations, and hence have been considered for applications such as computer memories [92, 93] and optical communications [94]. In particular, it was found in [94] that product codes with a 3-error-correcting BCH code constructed over $GF(2^{10})$ as the component code achieve the best performance for 100Gbps optical transport network specified in the International Telecommunication Union (ITU)-T G.709 standard. For 3-error-correcting BCH codes constructed over high-order finite fields, additional approaches are available to further reduce the decoding complexity.

In 3-error-correcting BCH decoding, three syndromes S_1, S_3 and S_5 are first computed. If all syndromes are zero, there is no error. If $S_3 = S_1^3$, only one error occurred, and its location is $\Lambda_1 = S_1$. Otherwise, there are two or three errors. The coefficients of the error locator polynomial $\Lambda(x) = \Lambda_3 x^3 + \Lambda_2 x^2 + \Lambda_1 x + 1$ are computed using the Peterson's formula shown previously. The root computation can be done on the reciprocal polynomial $\Lambda^*(x) = x^3 + \Lambda_1 x^2 + \Lambda_2 x + \Lambda_3$ instead to avoid reversing the roots in order to get the error locations. As discussed in Chapter 3, one way to compute the roots of a degree-3 polynomial without Chien search is to use lookup tables (LUTs). Let us first briefly review this root lookup method. When $\deg(\Lambda(x)) = 3$, and $a = \Lambda_1^2 + \Lambda_2 \neq 0$, $\Lambda^*(x)$ is converted to

$$z^3 + z + c = 0 \qquad (7.11)$$

employing coordinate transformations $y = x + \Lambda_1$ and $z = ya^{-1/2}$. Here $c = ba^{-3/2}$ and $b = \Lambda_3 + \Lambda_1 \Lambda_2$. Accordingly, for high-order finite fields, the roots can be pre-computed and stored in a LUT whose access address is c. Only one root, β, is stored for each c to reduce the LUT size. The other two roots are computed by solving a degree-2 polynomial $(z^3 + z + c)/(z + \beta) = z^2 + \beta z + (1 + \beta^2)$, which is done as a binary constant matrix multiplication as explained in Chapter 3. In the case that $\deg(\Lambda(x)) = 2$, the same root computation process is used, except that one root is zero and is discarded.

As it was found in [32], if β is a root of $z^3 + z + c = 0$, then β^2 is a root of $z^3 + z + c^2 = 0$. Adopting normal basis, the root corresponding to c can be derived by cyclically shifting the roots corresponding to the conjugates of c. Nevertheless, the patterns of the normal basis representations of the field elements in different conjugacy classes are irregular. It was proposed in [32] to store the roots of all field elements whose normal basis bit-vectors start with '10' or '01'. Then the access address of the LUT is the remaining $q - 2$ bits of a cyclically shifted version of c that starts with such a prefix. Using a 2-bit prefix, the LUT size is reduced to $q2^{q-2}$. Nevertheless, it is still large when q is not small.

The above methods handle the cases where c can be any element of $GF(2^q)$. However, when a degree-3 polynomial is a valid error locator polynomial, it must have three distinct roots. Considering this, the number of possible values of c for which $z^3 + z + c$ is a valid error locator polynomial is quite limited. Lemma 22 and its proof have been given in [34].

Lemma 22 *The number of elements c in $GF(2^q)$ for which $z^3 + z + c$ has*

TABLE 7.7
Number of conjugacy classes formed by good c in $GF(2^q)$ (data from [34])

q	# of good c	# of conjugacy classes	conjugacy class sizes
4	2	1	2
5	5	1	5
6	10	3	1, 3, 6
7	21	3	7(3)
8	42	7	2, 4(2), 8(4)
9	85	11	1, 3, 9(9)
10	170	19	5(4), 10(15)
11	341	31	11(31)
12	682	63	1, 2, 3, 4, 6(6), 12(53)

three distinct roots in $GF(2^q)$ is

$$\begin{cases} (2^{p-1} - 1)/3 & \textit{if } p \textit{ is odd} \\ (2^{p-1} - 2)/3 & \textit{if } p \textit{ is even} \end{cases}$$

From Lemma 22, roughly $1/6$ of the field elements can be valid c for an error locator polynomial. These field elements are referred to as 'good c'. From [32], $z^3 + z = c^2$ has a solution iff $z^3 + z = c$ has a solution. Therefore, the set of good c can be partitioned into conjugacy classes. A conjugacy class of $GF(2^q)$ consists of $\{\omega, \omega^2, \omega^4, \cdots\}$ for some $\omega \in GF(2^q)$. The cardinality of each class is a divisor of q and there are q elements in most classes. Therefore, the good c form a very small number of conjugacy classes. The numbers of good c and the conjugacy classes they form are listed in Table 7.7 for some finite fields. In the last column, the digits in parentheses indicate the numbers of classes of good c with certain cardinality if there is more than one of them. For example, when $q = 8$, the good c are partitioned into one class of size 2, 2 classes of size 4, and 4 classes of size 8.

The LUT size can be further reduced by storing a root for only one entry in each conjugacy class consisting of good c. This is made possible due to the small number of classes formed by good c. The elements picked from all classes should have a common prefix in the corresponding binary strings, and the remaining bits should be as distinct as possible. In this case, a given c can be shifted to have the prefix, then the remaining bits are used to access the LUT after small adjustment if needed. The purpose of shifting the bit strings to start with a certain prefix is to reduce the number of bits to be examined in order to decide which conjugacy class the given c belongs to, and hence which line of the LUT to access. A given finite field has multiple normal bases, and they lead to different bit patterns in the representation of each field element. All normal bases are searched in order to find the one that minimizes the complexity of generating the LUT access address. Two examples are given next to help explain the address generation process.

Example 41 *As it can be observed from Table 7.7, the good c in $GF(2^7)$ form three conjugacy classes of size 7. The leader of a conjugacy class is the one whose exponent in power representation is the smallest. Let α be a root of the irreducible polynomial $x^7 + x^6 + x^5 + x^2 + 1$. Then α and its 6 conjugates form a normal basis. The leaders of the conjugacy classes for the good c are α^{11}, α^{19} and α^{43}. Using the normal basis $\{\alpha, \alpha^2, \ldots, \alpha^{2^6}\}$, the leading elements are represented as '1010011', '0001010', and '1001010'. The other elements in the conjugacy classes are cyclical shifts of these bit strings. '1010011', '1010000' and '1010010' are selected to represent the three classes. They all start with '10100', and the last two bits in these strings are distinct. It should be also noted that each of these strings only has one occurrence of the '10100' pattern. Accordingly, the LUT stores only three roots, one for each class representative. Given c, it is first shifted so that its binary pattern starts with '10100'. Then the last two bits are used as the LUT address to read the root out. Of course, the root from the LUT needs to be reversely shifted. Comparatively, $2^{7-2} = 32$ roots need to be stored in the LUT using the approach in [32].*

For larger finite fields, the good c form more conjugacy classes. Although there are also more choices of normal bases, it may not always happen that a representative can be found for each conjugacy class of good c, so that each of them starts with a common pattern and s of the remaining bits are distinct. Here $s = \lceil \log_2(\# \text{ of conjugacy classes for good } c \text{ represented in LUT}) \rceil$. From Table 7.7, the majority of the cosets are of size q. If the size of a class is $q' < q$, the bit string for each element in the class has q/q' repeated q'-bit patterns. Hence, these classes can be easily differentiated from those with size q, and their root lookups are handled separately. Simulation codes can be written to help finding a normal basis leading to simple LUT address generation. For every normal basis, the bit strings of the leading elements for those good c conjugacy classes with size q are cyclically shifted to find the longest matching pattern. Each leader may have been shifted by a different number of bit positions. Then the remaining bits are analyzed to find s bits that are as distinct as possible. If multiple leaders are the same in these bits, additional logic is added to map those leaders to different addresses of the LUT. Although a longer matching pattern does not necessarily lead to higher probability of having s remaining bits that are distinct, it helps to reduce the possible combinations of remaining bits that need to examined.

Example 42 *Construct $GF(2^{10})$ by using the irreducible polynomial $p(x) = x^{10} + x^6 + x^5 + x^3 + x^2 + x + 1$. Let α be a root of $p(x)$. Among all possible normal bases, it can be found that the one consists of α^{511} and its conjugates leads to the longest matching pattern, which is 4-bit, in the bit strings of the leaders for those size-10 good c conjugacy classes. Using this normal basis, the bit strings for the class leaders are listed in the second column of Table 7.8.*

TABLE 7.8

Bit strings for the class leaders of good c in $GF(2^{10})$ (data from [34])

class #	leaders	shifted leaders	LUT address
1	0110001100	01100	$(0 \oplus 1)0 = 10$
2	0100001000	01000	$(0 \oplus 0)0 = 00$
3	1100111001	01110	$(0 \oplus 1)1 = 11$
4	1111011110	01111	$(1 \oplus 1)1 = 01$
5	1111111100	0111111110	1110
6	0001000111	0111000100	$0100 \oplus 0000 = 0100$
7	0100101110	0111001001	1001
8	1000100111	0111100010	0010
9	1001101011	0111001101	$1101 \oplus 0000 = 1101$
10	0110001111	0111101100	1100
11	1101111100	0111110011	0011
12	1111010011	0111111010	1010
13	0011100011	0111000110	0110
14	1011110011	0111100111	0111
15	1110010110	0111001011	1011
16	1000101011	0111000101	0101
17	1110010001	0111100100	$0100 \oplus 0100 = 0000$
18	1101011110	0111101101	$1101 \oplus 0010 = 1111$
19	1010001111	0111110100	$0100 \oplus 0101 = 0001$

The sizes of the first 4 classes are 5, and the sizes of the remaining 15 classes are 10. For the size-10 classes, shifted strings of the leaders are listed in the third column. All of them start with '0111'. The LUT stores 15 roots, and each corresponds to one of these shifted leaders. Hence, a 4-bit address is needed to read the root from the LUT. Among the remaining 6 bits of the binary strings, the 15 shifted leaders are almost distinct in the right-most four bits. The only exceptions are that classes 6, 17, and 19 are all '0100', and classes 9 and 18 are both '1101' in these bits. These classes need to be differentiated by using the other bits in the strings. Let l_i be the ith bit in the string of the shifted leader, and l_0 is the leftmost bit. For classes 6, 17 and 19, if their common rightmost four bits, '0100', are XORed with their respective '$0l_4 0l_5$', then they are mapped to distinct addresses as listed in the last column of Table 7.8. The same goal can be achieved for classes 9 and 18 by XORing their common rightmost four bits, '0100', with their respective '$00l_4l_5$'.

Another issue to consider is whether the matching pattern appears more than once in any of the leaders. If each string is shifted to move the first matching pattern to the left, as is done by the architecture in Fig. 3.3, then the shifting results of the bit strings in the same conjugacy class might be different if they consist of more than one matching pattern. For example, in the bit string of class 14 leader, '1011110011', the matching pattern '0111' *appears twice. This pattern also appears twice in the bit string of every other element in this class. The elements in this class whose bit strings have '111' before '1111' will be shifted to '0111011110', and the last four bits of this string conflict with those of class 5. In this case, extra logic needs to be added to adjust the last four bits before they are used as the LUT address. Fortunately, the l_4 bits in this string and that of class 5 are different. If the last four bits are '1110' and $l_4 = '0'$, then '1110' is changed back to '0111' to be used as the LUT address.*

In summary, the address of the LUT storing the roots for the size-10 classes can be generated as follows. Given c, it is first shifted to start with '0111'. If the last four bits do not equal '0100', '1101' or '1110', they are directly used as the LUT address. Otherwise, they are adjusted as aforementioned by utilizing other bits in the shifted string. The good c in those size-5 classes can be easily identified by XORing the first 5 bits with the last 5 bits. If c is an element of those classes, the first 5 bits of c are sufficient to tell which class c belongs to. The first 5 bits are shifted to start with '10'. Then $(l_4 \oplus l_2)l_3$ are distinct for the size-5 classes and they can be used as the address of the 4-entry LUT storing the four roots of these classes.

Two LUTs storing a total of 19 roots, one for each class of good c, are needed in this approach. Comparatively, the method in [32] needs a LUT with $2^{10-2} = 256$ entries trying to recover the root for every $c \in GF(2^q)$. The adjustment on the bit strings in this approach can be implemented with very simple logic. Accordingly, the hardware overhead of the address generation is minimal despite the large reduction achieved on the LUT size.

This root lookup method for error locator polynomials in the format of $z^3 + z + c$ can be implemented by an architecture similar to that in Fig. 3.3, with small modifications for testing the exceptional bit strings.

When $\Lambda_1^2 + \Lambda_2 = 0$, the degree-3 error locator is transformed to the format of $y^3 + b$. The direct root computation for polynomials in such a format has been discussed in Chapter 3. When q is odd, this polynomial has a unique cube root. Since the roots for an error locator polynomial should be distinct, the error locator polynomial would not be in this format when q is odd.

7.4   Chase BCH decoder based on Berlekamp's algorithm

Analogous to RS codes, the Chase decoding scheme can be applied to BCH codes. In a Chase decoder, the η least reliable bits are flipped, and the decoding is carried out on each of the 2^η test vectors. The decoding for the first test vector can be done by using the Berlekamp's algorithm. To avoid repeating the computations for each test vector, one-pass Chase schemes have been proposed in [95, 96] to build the error locators for the other test vectors based on the result of the Berlekamp's algorithm for the first test vector. In addition, modifications have been proposed in [97] to reduce the hardware complexity of the one-pass scheme in [96].

Algorithm 22 One-pass Chase BCH decoding based on Berlekamp's algorithm

input: $r^{(0)}, r^{(1)}, \ldots, r^{(2^\eta - 1)}$

start:

> *compute* $S_j = \sum_{i=0}^{n-1} r_i^{(0)} \alpha^{ji}$ $(1 \le j \le 2t)$
>
> *compute* $\Lambda(x)$ *and* $B(x)$ *using the Berlekamp's algorithm*
>
> $\Lambda^{(0)}(x) = \Lambda(x),\ B^{(0)}(x) = x^2 B(x)$
>
> $L_\Lambda^{(0)} = \deg(\Lambda(x)),\ L_B^{(0)} = \deg(B(x)) + 2$
>
> *for* $l = 1$ *to* $2^\eta - 1$
>
> > *if* $r^{(l)}$ *is different from* $r^{(p)}(p < l)$ *at code position* i
> >
> > > *compute* $a = \Lambda^{(p)}(\alpha^{-i}),\ b = B^{(p)}(\alpha^{-i})$
> > >
> > > **case 1:** $a = 0 \mid (a \ne 0\ \&\ b \ne 0\ \&\ L_\Lambda^{(p)} \ge L_B^{(p)})$
> > >
> > > > $\Lambda^{(l)}(x) = b \cdot \Lambda^{(p)}(x) + a \cdot B^{(p)}(x)$
> > > >
> > > > $B^{(l)}(x) = (x^2 + \alpha^{-2i}) \cdot B^{(p)}(x)$
> > > >
> > > > $L_\Lambda^{(l)} = L_\Lambda^{(p)},\ L_B^{(l)} = L_B^{(p)} + 2$
> > >
> > > **case 2:** $b = 0 \mid (a \ne 0\ \&\ b \ne 0\ \&\ L_\Lambda^{(p)} < L_B^{(p)} - 1)$
> > >
> > > > $\Lambda^{(l)}(x) = (x^2 + \alpha^{-2i}) \cdot \Lambda^{(p)}(x)$
> > > >
> > > > $B^{(l)}(x) = b \cdot x^2 \Lambda^{(p)}(x) + \alpha^{-2i} \cdot a \cdot B^{(p)}(x)$
> > > >
> > > > $L_\Lambda^{(l)} = L_\Lambda^{(p)} + 2,\ L_B^{(l)} = L_B^{(p)}$
> > >
> > > **case 3:** $a \ne 0\ \&\ b \ne 0\ \&\ L_\Lambda^{(p)} = L_B^{(p)} - 1$
> > >
> > > > $\Lambda^{(l)}(x) = b \cdot \Lambda^{(p)}(x) + a \cdot B^{(p)}(x)$
> > > >
> > > > $B^{(l)}(x) = b \cdot x^2 \Lambda^{(p)}(x) + \alpha^{-2i} \cdot a \cdot B^{(p)}(x)$
> > > >
> > > > $L_\Lambda^{(l)} = L_\Lambda^{(p)} + 1,\ L_B^{(l)} = L_B^{(p)} + 1$
>
> **output:** $\Lambda^{(l)}(x)$

Consider an (n, k) t-bit error-correcting BCH code constructed over $GF(2^q)$. Let the hard-decision received bit-vector be $r^{(0)} = [r_0^{(0)}, r_1^{(0)}, \cdots, r_{n-1}^{(0)}]$. Represent the other test vectors in the Chase decoding by $r^{(1)}, r^{(2)}, \cdots$. The one-pass Chase BCH decoding scheme developed in [96] is carried out according to Algorithm 22. Assuming α is a primitive element of $GF(2^q)$, the syndromes S_j are computed for the hard-decision vector at the beginning. From these syndromes, the Berlekamp's algorithm computes the error locator polynomial $\Lambda(x)$ and its associated polynomial $B(x)$ for the hard-decision vector. Building on these polynomials, the error locator polynomials for the other test vectors are derived one after another. In the iterative process, $L_\Lambda^{(l)}$ and $L_B^{(l)}$ are the degrees of $\Lambda^{(l)}(x)$ and $B^{(l)}(x)$, respectively. To facilitate the iterations, the test vectors are arranged in a tree structure, so that each child vector has only one bit different from its parent vector [96]. If $r^{(l)}$ is different from its parent vector $r^{(p)}$ in code position i, then $\Lambda^{(l)}(x)$ and $B^{(l)}(x)$ are computed according to one of the three cases depending on the relative degrees and evaluation values of $\Lambda^{(p)}(x)$ and $B^{(p)}(x)$ over α^i. The calculations involved are linear combinations and multiplications with $(x + \alpha^{-2i})$. The computed $\Lambda^{(l)}(x)$ is the error locator polynomial for $r^{(l)}$.

It can be observed from Algorithm 22 that $L_\Lambda^{(l)} + L_B^{(l)}$ is increased by two in each iteration. In an architecture that updates the coefficients of the polynomials serially, longer polynomials means more clock cycles in each iteration. As the iterations in Algorithm 22 are carried out, the polynomial degrees become higher and higher. The increasing polynomial degrees lead to long latency in the decoding, especially when η is not small. In addition, $r^{(l)}$ needs to be different from $r^{(p)}$ in a bit that has not been flipped before in Algorithm 22. The reason is that $\Lambda^{(p)}(x)$ and $B^{(p)}(x)$ would have the factor $(x + \alpha^{-i})$ if $r^{(p)}$ is flipped in code position i. If the bit in this code position needs to be flipped again to derive $r^{(l)}$, then $a = b = 0$ and hence the polynomials for $r^{(l)}$ can not be computed according to any of the cases listed in Algorithm 22. As a result, in this algorithm, the vectors need to be ordered in a tree such that none of the nodes in a path from the root to a leaf share the same flipped bits. Testing the vectors according to the order of such a tree requires multiple pairs of polynomials to be recorded.

Example 43 *A tree for ordering the test vectors in the one-pass Chase decoding in Algorithm 22 for $\eta = 4$ is shown in Fig. 7.5. In this figure, each of the 4-bit tuple represents a test vector. The hard-decision vector is denoted by '0000', and a '1' in a vector means the bit in the corresponding code position has been flipped from the hard-decision vector. Starting from the root, the test vectors are covered in a depth-first manner to reduce the number of intermediate polynomials need to be stored. Nevertheless, the polynomials for those vectors that have more than one child must be be stored. These vectors are represented by the nodes with circles in Fig. 7.5.*

It has been proved in [96] that the $\Lambda^{(l)}(x)$ and $B^{(l)}(x)$ computed from Algorithm 22 always have the factor $(x + \alpha^{-i})$. The goal of the decoding trial over $r^{(l)}$ is to find the error locations, whose inverse are the roots of $\Lambda^{(l)}(x)$, and flip the bits in those positions. Therefore, $\Lambda^{(l)}(x)/(x + \alpha^{-i})$ can be used as the error locator instead, if the bit at code position i is flipped in the corresponding vector at the end [97]. Moreover, if $(x + \alpha^{-i})$ is also taken out of $B^{(l)}(x)$, then $\Lambda^{(l)}(x)$ and $B^{(l)}(x)$ have the same relative degrees and their evaluation values are scaled by the same nonzero factor. As a result, for all the three cases in Algorithm 22, the testing conditions and involved computations can remain the same even if the following step is added to the end of each iteration:

$$\Lambda^{(l)}(x) \Leftarrow \Lambda^{(l)}(x)/(x + \alpha^{-i})$$
$$B^{(l)}(x) \Leftarrow B^{(l)}(x)/(x + \alpha^{-i}) \qquad (7.12)$$
$$L_\Lambda^{(l)} = L_\Lambda^{(l)} - 1, L_B^{(l)} = L_B^{(l)} - 1.$$

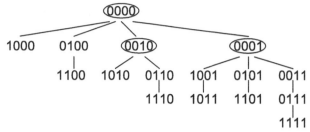

FIGURE 7.5
Test vector ordering tree for one-pass Chase BCH decoding with $\eta = 4$ based on the Berlekamp's algorithm

After this step is added, $L_\Lambda^{(l)} + L_B^{(l)}$ does not change with the iterations. Although extra hardware is needed to implement this step, the degrees of the polynomials and hence the decoding latency is greatly reduced.

The extra step in (7.12) also helps to simplify the ordering of the test vectors. Since the factor $(x + \alpha^{-i})$ has been taken out of the polynomials, a bit can be flipped again and again. As a result, the test vectors can be ordered according to Gray code, so that adjacent vectors are different in only one bit. In this case, only one pair of error locator and associate polynomials need to be stored at any time, and the memory requirement of the decoder is reduced. Moreover, the $\Lambda^{(l)}(x)$ and $B^{(l)}(x)$ computed from the current iteration will be used as the $\Lambda^{(p)}(x)$ and $B^{(p)}(x)$ in the next iteration. Hence, once the coefficients of $\Lambda^{(l)}(x)$ and $B^{(l)}(x)$ are derived, they are sent to the evaluation value computation engine to compute the variables a and b to be used in the next iteration. As a result, the computations from adjacent iterations are overlapped to increase the throughput.

The syndrome computation and Berlekamp's algorithm can be implemented as discussed earlier in this chapter. From the outputs of the Berlekamp's algorithm, the error locator polynomials for the second and later test vectors are derived through evaluation value computations, linear combinations, multiplications by $(x + \alpha^{-2i})$, and divisions by $(x + \alpha^{-i})$. They can be implemented by architectures similar to those for the evaluation value computation and polynomial updating shown in Chapter 6 for the Chase RS decoding. One of the 2^η error locator polynomials needs to be picked to generate the final decoding output. Similar to that for RS codes, the error locator polynomial whose degree equals its root number can be chosen. Such a selection scheme does not lead to noticeable error-correcting performance degradation unless the code is short or the error-correcting capability of the code is very small. To match the fast error locator polynomial generation, highly-parallel Chien search is needed for the polynomial selection if the code is not short.

7.5 Interpolation-based Chase BCH decoder architectures

The interpolation-based Chase RS decoders achieve higher hardware efficiency than those based on the BMA. Inspired by the interpolation-based RS decoder, interpolation-based Chase BCH decoder was proposed in [98]. It also has much lower hardware complexity than those based on the Berlekamp's algorithm.

The interpolation-based decoding was originally developed based on the interpretation that the codeword symbols are evaluation values of the message polynomial over finite field elements. Binary BCH codes can not be encoded by this evaluation map method, since evaluating a binary message polynomial over elements of $GF(2^q)$ generally does not lead to binary codeword symbols. The generator polynomial, $g(x)$, of a narrow-sense t-bit error-correcting (n, k) BCH code over $GF(2^q)$ is the product of the minimal polynomials that contain $\alpha, \alpha^2, \ldots, \alpha^{2t}$ as roots. Usually, $n - k$ equals or is slightly less than qt, depending on the cardinalities of the conjugacy classes containing these $2t$ elements. On the other hand, the generator polynomial, $g'(x)$, of a narrow-sense t-symbol error-correcting (n, k') RS code over $GF(2^q)$ is $\prod_{i=1}^{2t}(x + \alpha^i)$, and $n - k' = 2t$. $g(x)$ is divisible by $g'(x)$. Hence, all the codewords of a t-bit error-correcting (n, k) BCH code over $GF(2^q)$ are also codewords of a t-symbol error-correcting (n, k') RS code over the same finite field. As a result, BCH codewords are considered as RS codewords to apply the interpolation-based decoding.

The same systematic re-encoding [82], coordinate transformation, backward-forward interpolation [55, 68] can be adopted, and the factorization can be eliminated [78] for interpolation-based BCH decoding. It should be noted that the re-encoding is applied to the first k' instead of k positions since the BCH codewords are considered as RS codewords in the decoding. Nevertheless, by utilizing the binary property, further simplifications were made on the polynomial selection and codeword recovery for binary BCH codes in [98].

An (n, k') RS code over $GF(2^q)$ has $2^{qk'}$ codewords. However, only 2^k of them are also codewords of the (n, k) binary BCH code. Since $k' = n - 2t$ and k is around $n - qt$, k is much smaller than k' and the BCH codewords only form a tiny proportion of the RS codewords. Therefore, it is extremely unlikely that the decoding will generate a binary BCH codeword if the test vector is undecodable. Hence, for BCH decoding, the interpolation output polynomials can be selected according to whether they will generate binary codewords. Testing if each symbol in the recovered codeword is binary would require the entire decoding process to be finished first, and render the purpose of having a polynomial selection scheme invalid. Instead, by making use of the property that the codeword symbols are evaluation values of the message polynomial, a few codeword symbols easy to compute can be recovered first for the purpose of polynomial selection.

Let r be a test vector. When systematic re-encoding is applied, the decoding is carried out on $\bar{r} = r + \phi$, where ϕ is a codeword that equals r in the first k' code positions denoted by the set S. Assume that the codeword corresponding to $\bar{r}$ is $\bar{c}$, and the message polynomial is $\bar{f}(x)$. Let the interpolation output be $Q(x, y) = q_0(x) + q_1(x)y$ and S_F is the set of systematic bits that have been flipped in r from the hard-decision vector. It has been shown in Chapter 6 that

$$\bar{f}(x) = \frac{q_0(x)}{q_1(x)} \frac{v(x)}{s_f(x)}, \tag{7.13}$$

where $v(x) = \prod_{i \in S}(x + \alpha_i)$ and $s_f(x) = \prod_{i \in S_F}(x + \alpha_i)$. Here $\{\alpha_0, \alpha_1, \ldots, \alpha_{n-1}\}$ is the set of distinct field elements used for evaluation map encoding. From analyzing the possible factors of the polynomials in (7.13) and applying L'Hôpital rule when necessary, the systematic symbols in $\bar{c}$ for a primitive RS code can be computed as

$$\bar{c}_i = \bar{f}(\alpha_i) = \begin{cases} \frac{q_0(x)}{w(x)s_f(x)} \frac{1}{q_{1,odd}(x)}\Big|_{\alpha_i} & i \in S/S_F \,\&\, q_1(\alpha_i) = 0 \\ 0 & i \in S/S_F \,\&\, q_1(\alpha_i) \neq 0 \\ \frac{q_0(x)}{q_1(x)} \frac{1}{w(x)s_{f,odd}(x)}\Big|_{\alpha_i} & i \in S_F \,\&\, q_1(\alpha_i) \neq 0 \\ \frac{q_{0,odd}(x)}{q_{1,odd}(x)} \frac{1}{w(x)s_{f,odd}(x)}\Big|_{\alpha_i} & i \in S_F \,\&\, q_1(\alpha_i) = 0 \end{cases} \tag{7.14}$$

where $w(x) = (x^n - 1)/v(x)$. $q_1(x)$ is considered as the error locator polynomial. Even if the test vector is undecodable, the degree and hence the root number of $q_1(x)$ should not exceed $t = (n - k')/2$. $t << k'$ for high-rate codes. Hence, most of the systematic bits are zero according to (7.14). ϕ_i is binary for $i \in S$. Therefore, if $\bar{c}$ is computed using (7.14), the recovered codeword $c = \bar{c} + \phi$ would be binary in the systematic positions even if the test vector is undecodable. On the other hand, when the test vector is undecodable, $\bar{c}_i$ and ϕ_i are mostly non-binary for $i \in \bar{S}$. The probability that $\bar{c}_i + \phi_i$ ($i \in \bar{S}$) is binary is extremely small. Accordingly, it was proposed in [98] to test two symbols of c in non-systematic code positions for polynomial selection. Simulations showed that such a polynomial selection scheme does not lead to any performance loss for BCH codes with moderate or long codeword length. Any two symbols in $\bar{S}$ can be used for testing. However, testing only one symbol would lead to very small performance degradation, since a symbol in $\bar{S}$ can be accidentally zero in those undecodable cases. $\bar{c}_i$ for $i \in \bar{S}$ can be also derived through evaluating (7.13) over α_i.

If systematic BCH encoding is employed, only the codeword symbols in the systematic positions need to be recovered. Since $\bar{c}$ is binary when the test vector is decodable, all needs to be derived is whether $\bar{c}_i = \bar{f}(\alpha_i)$ is zero or nonzero for the k systematic positions. Note that these systematic positions are a subset of S, the systematic positions for the corresponding RS code. Making use of this binary feature, the codeword recovery in the interpolation-based BCH decoding is simplified.

$\bar{f}(\alpha_i)$ can be computed from (7.13). Whether $\bar{f}(\alpha_i) = 0$ is dependent on if $(x + \alpha_i)$ is a factor of $q_0(x)$ and $q_1(x)$. It was shown in Chapter 6 that

$$\begin{cases} q_0(x) = p(x)\rho(x) \\ q_1(x) = p(x) \prod_{i \in S/S_F, e_i \neq 0}(x + \alpha_i), \end{cases} \tag{7.15}$$

where e_i is the error in code position i, and $\rho(x)$ does not have $(x + \alpha_i)$ as a factor for $i \in S/S_F$ and $e_i \neq 0$. In addition, $p(x)$ does not have any $(x + \alpha_i)$ factor for $i \in S/S_F$ nor any $(x + \alpha_i)$ factor with order higher than one for $i \in \bar{S} \cup S_F$.

Five cases are considered for recovering the message bits in S.

Case 1: $i \in S/S_F$ and $q_1(\alpha_i) \neq 0$. In this case, $(x + \alpha_i)$ is a factor of $v(x)$, but is not a factor of the denominators in (7.13). Hence, $\bar{f}(\alpha_i) = 0$.

Case 2: $i \in S/S_F$ and $q_1(\alpha_i) = 0$. $p(x)$ does not have any factor $(x + \alpha_i)$ with $i \in S/S_F$. Hence $(x + \alpha_i)$ must be a simple factor of the second term of $q_1(x)$ in (7.15). Since it can not be a factor of $\rho(x)$, it is not a factor of $q_0(x)$. As a result, $(x + \alpha_i)$ is canceled out from $v(x)$ and $q_1(x)$, and $\bar{f}(\alpha_i)$ is '1' in this case.

Case 3: $i \in S \cap S_F$ and $q_0(\alpha_i) \neq 0$. $v(x)$ and $s_f(x)$ have the $(x + \alpha_i)$ factor in this case, and it is canceled out from these two polynomials. In addition, neither $p(x)$ nor $\rho(x)$ has this factor, and hence $q_1(x)$ does not have it as a factor. Therefore, $\bar{f}(\alpha_i)$ is '1'.

Case 4: $i \in S \cap S_F$, $q_0(\alpha_i) = 0$ and $q_1(\alpha_i) \neq 0$. $(x + \alpha_i)$ is canceled out from $v(x)$ and $s_f(x)$ as in Case 3. However, it is also a factor of $\rho(x)$, and hence $\bar{f}(\alpha_i) = '0'$.

Case 5: $i \in S \cap S_F$, $q_0(\alpha_i) = 0$ and $q_1(\alpha_i) = 0$. $(x + \alpha_i)$ is canceled out from $v(x)$ and $s_f(x)$. In addition, it is a factor of $p(x)$, and is canceled out from $q_0(x)$ and $q_1(x)$ as well. It may also be a factor of $\rho(x)$. Since $(x + \alpha_i)$ can not be a factor of $p(x)$ with order higher than one, whether it is a factor of $\rho(x)$ can be told from $q_0'(\alpha_i)$.

Summarizing the above cases, the codeword $\bar{c}$ for BCH decoding is recovered by

$$\bar{c}_i = \bar{f}(\alpha_i) = \begin{cases} 0, & i \in S/S_F \& q_1(\alpha_i) \neq 0 \\ 1, & i \in S/S_F \& q_1(\alpha_i) = 0 \\ 1, & i \in S \cap S_F \& q_0(\alpha_i) \neq 0 \\ 0, & i \in S \cap S_F \& q_0(\alpha_i) = 0 \& q_1(\alpha_i) \neq 0 \\ 0, & i \in S \cap S_F \& q_0'(\alpha_i) = 0 \& q_1(\alpha_i) = 0 \\ 1, & i \in S \cap S_F \& q_0'(\alpha_i) \neq 0 \& q_1(\alpha_i) = 0 \end{cases} \tag{7.16}$$

After that, the message bits are computed as $c_i = \bar{c}_i + \phi_i$.

It has been shown in [98] that the interpolation-based Chase BCH decoder achieves much higher hardware efficiency than those based on the Berlekamp's algorithm [97]. The major reason is the low-complexity polynomial selection. Testing whether two of the recovered symbols are binary can be implemented by simple hardware for the interpolation-based decoder. Nevertheless, to select the error locator polynomial in the Chase decoder based on the Berlekamp's algorithm, highly parallel Chien search is needed as was discussed in the previous section The simple polynomial selection according to whether a couple of the recovered codeword symbols are binary can not be applied to those decoders based on the Berlekamp's algorithm. This is because the codewords are not interpreted as evaluation values of the message polynomial in those approaches, and no information about the codeword can be derived before the Chien search for the error locator polynomial is completed.

8

Binary LDPC codes & decoder architectures

CONTENTS

Low-density parity-check (LDPC) codes are a class of linear block codes that can approach the Shannon limit. They were first introduced in 1962 [99]. However, in those early years, the capacity of integrated circuits was very limited, and the LDPC decoding algorithms were too complicated to be implemented. As a result, these codes have been forgotten for many years. The development of VLSI technology in the past two decades rejuvenated these codes. Binary LDPC codes are currently adopted in many applications and standards, such as 10GBase-T Ethernet, WiMAX wireless communications, digital video broadcasting, and magnetic and solid-state drives.

This chapter first introduces LDPC codes. LDPC encoding can be done as a generator matrix multiplication. For quasi-cyclic (QC)-LDPC codes [100], which are a type of LDPC codes that enable efficient hardware implementations, the generator matrix consists of circulant submatrices [101]. In this case,

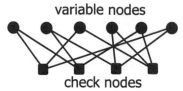

FIGURE 8.1
Example Tanner graph

the encoder can be also implemented by shift-register architectures. Comparatively, LDPC decoders are much more complicated, and hence are the focus of this chapter. Different types of decoding algorithms and scheduling schemes are introduced before the hardware architectures are presented. The emphasis will be given to the Min-sum decoder for QC-LDPC codes, which are usually employed in practical systems.

There are also non-binary LDPC codes. Their decoding algorithms and implementations will be discussed in the next chapter.

8.1   LDPC codes

Like Reed-Solomon (RS) and Bose-Chaudhuri-Hocquenghem (BCH) codes, LDPC codes are linear block codes. They can be specified by the parity check matrix H. A sequence $x = [x_0, x_1, x_2, \ldots]$ is a codeword iff $Hx^T = 0$. However, for LDPC codes, only a small proportion of the entries in H are nonzero. The H matrix is also associated with a Tanner graph, which is a bipartite graph. Each row of H corresponds to a check node in the Tanner graph, and each column is represented by a variable node. Also a check node and a variable node are connected by an edge if the corresponding entry in H is nonzero. For example, if

$$H = \begin{bmatrix} 1 & 0 & 1 & 0 & 0 & 1 \\ 1 & 1 & 0 & 0 & 1 & 0 \\ 0 & 1 & 0 & 1 & 0 & 1 \\ 0 & 0 & 1 & 1 & 1 & 0 \end{bmatrix}, \tag{8.1}$$

then the corresponding Tanner is the one shown in Fig. 8.1. For real LDPC codes, the H matrix is very sparse, and the codeword length should be at least a few thousands of bits to achieve good error-correcting performance.

The number of '1's in a row of H is called the row weight, and the number of '1's in a column is referred to as the column weight. When each row (or column) has the same weight, the code is said to be regular. The locations of the nonzero entries in H can appear either random or structured. Particularly,

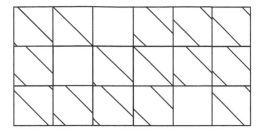

FIGURE 8.2
The H matrix for an example QC-LDPC code

QC-LDPC codes [100] are of great interest for hardware implementations. The H matrix of a QC-LDPC code consists of square sub-matrixes. Each sub-matrix is either zero or a cyclically shifted identity matrix. A QC H matrix example is shown in Fig. 8.2. The diagonal lines denote the locations of the '1's. Compared to random codes, QC-LDPC codes do not necessarily have inferior error-correcting performance.

The row and column weights, the length of the shortest cycle in the corresponding Tanner graph, which is also called the girth, as well as the distribution of the cycles, affect the error-correcting performance in the waterfall region and the error-floor of LDPC codes. The error-floor is a phenomenon that the error-correcting performance curve in high signal-to-ratio (SNR) region flattens out. It happens due to the presence of cycles in the Tanner graph. Apparently, the Tanner graph should not have 4-cycles, which are two variable nodes and two check nodes connected by four edges. This means that two rows of H should not have '1's in the same two columns. Further discussions on the construction of LDPC codes for good waterfall region performance and low error floor is beyond the scope of this book. The interested reader is referred to [102, 103].

8.2 LDPC decoding algorithms

Given the channel information, the goal of the LDPC decoding is to find a sequence x such that $Hx^T = 0$. Each row of H specifies a check equation that needs to be satisfied. For example, assume that $x = [x_0, x_1, x_2, x_3, x_4, x_5]$, then the check equation specified by the first and second rows of H in (8.1) are $x_0 + x_2 + x_5 = 0$ and $x_0 + x_1 + x_4 = 0$, respectively. During the decoding process, messages with regard to whether each received bit should be '1' or '0' are passed iteratively between the connected check and variable nodes. Hard-decision of each bit is made at the end of each iteration. If the hard decisions satisfy all the check equations, then a codeword is found and the decoding can be stopped. Otherwise, the iteration is repeated until a pre-set maximum iteration number I_{max}.

8.2.1 Belief propagation algorithm

Traditionally, LDPC codes are decoded by the belief propagation (BP) [104], also called the sum-product algorithm. Let y_n be the observation from the channel for the *nth* received bit. Then $\gamma_n = Pr(x_n = 1|y_n)$ is used as the input to the decoder. Since each received symbol can be only '0' or '1', $Pr(x_n = 0|y_n) = 1 - Pr(x_n = 1|y_n)$ and there is no need to track both of them. Denote the message from variable node n to check node m by $u_{m,n}$, and that from check node m to variable node n by $v_{m,n}$. Similar to γ_n, these messages are associated with the probabilities that $x_n=$'1'. Let $S_c(n)$ $(S_v(m))$ be the set of check (variable) nodes connected to variable (check) node n (m). Use $\mathcal{L}(m|n)$ to denote the set of sequences x_j $(j \in S_v(m), j \neq n)$ such that the *mth* check equation is satisfied with $x_n=$'1'.

γ_n is used as variable-to-check (v2c) messages $u_{m,n}$ for each $m \in S_c(n)$ in the first iteration. Check node m computes the probability that $x_n=$'1' for each $n \in S_v(m)$ given that the *mth* check equation is satisfied. In the BP algorithm, this is done as

$$v_{m,n} = \sum_{\mathcal{L}(m|n)} \prod_{j\in S_v(m),j\neq n} (x_j u_{m,j} + x'_j(1 - u_{m,j})). \tag{8.2}$$

In the above equation, x'_j is the logic NOT of x_j, and hence only one of x_j and x'_j is '1'. When $x_j=$'1', $u_{m,j}$ is multiplied up. Otherwise, $(1 - u_{m,j})$, which is the probability associated with $x_j =$'0' is included in the product. From the check-to-variable (c2v) messages, $v_{m,n}$, the v2c messages are updated as

$$u_{m,n} = \gamma_n \prod_{i\in S_c(n),i\neq m} v_{i,n}. \tag{8.3}$$

Besides, the *a posteriori* information is derived by

$$\tilde{\gamma}_n = \gamma_n \prod_{i\in S_c(n)} v_{i,n}.$$

The hard-decision of x_n is '1' if $\tilde{\gamma}_n \geq 0.5$, and is '0' otherwise. If all check equations are satisfied by the hard decisions, the decoding can be stopped. If not, the c2v and v2c message computations are repeated iteratively.

Multiplications are costly to implement in hardware. Multiplications are converted to additions in log domain. Moreover, log-domain decoders are less sensitive to quantization noise. Let $\gamma_n = \log(Pr(x_n = 1|y_n)/Pr(x_n = 0|y_n))$, and define $u_{m,n}$ and $v_{m,n}$ in a similar way as log-likelihood ratios (LLRs). Then the probability-domain variable node processing in (8.3) becomes

$$u_{m,n} = \gamma_n + \sum_{i\in S_c(n),i\neq m} v_{i,n}, \tag{8.4}$$

and the check node processing in (8.2) is converted to

$$v_{m,n} = 2\tanh^{-1}\left(\prod_{j \in S_v(m), j \neq n} \tanh(u_{m,j}/2)\right). \tag{8.5}$$

The log-domain variable node processing in (8.4) only requires additions. Nevertheless, hyperbolic tangent function is involved in the log-domain check node processing in (8.5). This function requires lookup tables (LUTs) to implement in hardware. Although each LUT is small if the word length of the messages is kept short, a large number of LUTs are needed when the processing of multiple check nodes is carried out simultaneously to achieve high throughput. As a result, the computations in (8.5) lead to high hardware complexity.

8.2.2   Min-sum algorithm

To enable efficient and high-speed LDPC decoding, the Min-sum algorithm was proposed in [105]. In this algorithm, the check node processing in (8.5) is approximated as

$$v_{m,n} = \left(\prod_{j \in S_v(m), j \neq n} sign(u_{m,j})\right) \min_{j \in S_v(m), j \neq n} |u_{m,j}|, \tag{8.6}$$

while the variable node processing is the same as that in (8.4). The product of signs can be computed by XOR gates. Since only $u_{m,n}$ is excluded from all input v2c messages in the computation of $v_{m,n}$, many computations can be shared to derive $v_{m,n}$ with different $n \in S_v(m)$. Particularly, if $u_{m,n}$ has the smallest magnitude among all input v2c messages, then from (8.6), the magnitudes of $v_{m,n'}$ with $n' \neq n$ are the same, and they all equal $|u_{m,n}|$. On the other hand, the magnitude of $v_{m,n}$ should be equal to the smallest of all $|u_{m,j}|$ with $j \neq n$, which is actually the second smallest among all $|u_{m,j}|$. As a result, the Min-sum algorithm for LDPC decoding can be carried out according to Algorithm 23. In this algorithm, z_n denotes the hard-decision of the nth bit.

The approximations adopted in the Min-sum algorithm result in over-estimation of LLRs that causes performance degradation. To compensate for the over-estimation, the c2v messages can be offset or scaled. In the offset and normalized Min-sum algorithms, the c2v message computations are modified as

$$v_{m,n} = \left(\prod_{j \in S_c(m), j \neq n} sign(u_{m,j})\right) \max\left(\min_{j \in S_v(m), j \neq n} |u_{m,j}| - \lambda, 0\right) \tag{8.7}$$

and

$$v_{m,n} = \alpha\left(\prod_{j \in S_c(m), j \neq n} sign(u_{m,j})\right) \min_{j \in S_v(m), j \neq n} |u_{m,j}|,$$

respectively. λ is a positive real number, and α is a scaler smaller than 1. Their optimum values can be derived from simulations. Moreover, α should be chosen so that its binary representation only has a few nonzero bits. In this case, the scaler multiplication is implemented by simple additions. If proper offset or scaler is used, the Min-sum algorithm only has very small performance loss compared to the BP algorithm. Algorithm 23 can be easily modified to take into account the offset subtraction and scaler multiplication. In addition, these two compensation schemes can be jointly adopted and made adaptive to the channel condition in order to further improve the performance.

Algorithm 23 The Min-sum Decoding Algorithm
initialization: $u_{m,n} = \gamma_n$; $z_n = sign(\gamma_n)$
for $k = 1$ *to* I_{max}
     Compute zH^T; *Stop if* $zH^T = 0$

    check node processing
    $min1_m = \min_{j \in S_v(m)} |u_{m,j}|$
    $idx_m = \arg\min_{j \in S_v(m)} |u_{m,j}|$
    $min2_m = \min_{j \in S_v(m), j \neq idx_m} |u_{m,j}|$
    $s_m = \prod_{j \in S_v(m)} sign(u_{m,j})$
    for each $n \in S_v(m)$
$$|v_{m,n}| = \begin{cases} min1_m & if \ n \neq idx_m \\ min2_m & if \ n = idx_m \end{cases}$$
        $sign(v_{m,n}) = s_m sign(u_{m,n})$

    variable node processing
    $u_{m,n} = \gamma_n + \sum_{i \in S_c(n), i \neq m} v_{i,n}$

    a posteriori information computation & tentative decision
    $\tilde{\gamma}_n = \gamma_n + \sum_{i \in S_c(n)} v_{i,n}$
    $z_n = sign(\tilde{\gamma}_n)$

8.2.3   Majority-logic and bit-flipping algorithms

Majority-logic and bit-flipping algorithms are another two types of LDPC decoding algorithms. Although their error-correcting performance is not as good as those of the sum-product algorithm or its approximations, they enjoy low-complexity and high-speed decoder implementations.

     To simplify the notations, assume that the LDPC code has constant column weight d_v. The one-step majority-logic decoding (OSMLGD) [106] takes as input the hard decision of each received bit from the channel. Denote the hard-decision vector by $z = [z_0, z_1, z_2, \dots]$. Assume that the *ith* row of the parity check matrix is $h_i = [h_{i,0}, h_{i,1}, h_{i,2}, \dots]$. $s_i = z h_i$ is called the *ith* check sum or syndrome. If $s_i = 0$, then the *ith* check equation is satisfied. In addition, those s_i for $i \in S_c(j)$ are referred to as the check sums orthogonal to the variable node j. In the OSMGLD, z_j is flipped if the majority of the check sums orthogonal to variable node j are nonzero. Since two rows of the H matrix share at most one nonzero common entry in any two columns, the OSMLGD can correct at most $\lfloor d_v/2 \rfloor$ errors. If d_v is low, the decision on whether to flip a bit may not be very reliable. Therefore, the OSMLGD is only effective when d_v is not small.

By adopting an iterative process to flip the hard-decision bits, the bit-flipping algorithms can correct more errors than the OSMLGD. In each iteration of the bit-flipping algorithms, hard decisions of the received bits are flipped based on some criteria applied to the number of nonzero orthogonal check sums (NNOCS). The Gallager A bit-flipping algorithm is listed in Algorithm 24 [99].

Algorithm 24 Gallager A Bit-flipping LDPC Decoding Algorithm
initialization: $z = [z_0, z_1, z_2, \dots]$
for $k = 1 : I_{max}$
    Compute zH^T; Stop if $zH^T = 0$
    For each variable node j
        if $|\{s_i | i \in S_c(j) \text{ and } s_i \neq 0\}| \geq d_v - 1$
            flip z_j

The Gallager B algorithm is the same as the Gallager A algorithm except that the threshold of the NNOCS for flipping is changed from $d_v - 1$ to a lower number. Improvements of the bit-flipping algorithm employ various rules utilizing the NNOCS. For example, in [107], a hard-decision bit is flipped with a probability $p < 1$ when its corresponding NNOCS is larger than a threshold. In [106], the hard decisions whose corresponding NNOCS are the largest are flipped.

When soft information is available from the channel, the performance of the majority-logic and bit-flipping algorithms can be improved. Assume that the reliability of the *jth* received symbol is r_j. The weight associated with check sum s_i is defined in [106] as

$$w_i = \min\{r_j | j \in S_v(i)\}.$$

Then the weighted check sums are added up to derive

$$\sum_{i \in S_c(j)} s_i w_i,$$

which is used in place of the NNOCS to decide whether z_j should be flipped. Other methods for computing the weighted check sums are available in literature. If this process is adopted iteratively, then it becomes a weighted bit-flipping algorithm.

In the weighted majority logic and bit-flipping algorithms described above, the soft information available in each decoding iteration is the same. By iteratively updating the soft information, more errors can be corrected by the iterative soft reliability-based majority-logic decoding (ISRB-MLGD) [108]. In this algorithm, a multi-bit reliability measure, $R_j^{(k)}$, is adopted for the *jth* received bit. The superscript (k) is added when necessary to denote the values in the *kth* decoding iteration. Define the extrinsic information for variable node j as

$$E_j = \sum_{i \in S_c(j)} (2\sigma_{i,j} - 1),$$

where

$$\sigma_{i,j} = \sum_{l \in S_v(i) \setminus j} z_l h_{i,l}.$$

The sum in the above equation is done as XOR, and $\sigma_{i,j}$ is either '0' or '1'. Then the reliability of the *jth* received bit in decoding iteration k is given by

$$R_j^{(k)} = R_j^{(k-1)} + E_j^{(k-1)}.$$

The value of $E_j^{(k-1)}$ is in the range of $[-d_v, d_v]$, and $R_j^{(k-1)}$ are LLRs. The sign of $R_j^{(k)}$ is an indication of whether $z_j^{(k)}$ should be '1' or '0'.

Assume that the reliability from the channel for the *jth* bit is γ_j, the pseudo codes of the ISRB-MLGD algorithm are listed in Algorithm 25. Similar to that in the Min-sum algorithm, computation results are shared to derive E_j for different j. $\sum_{l \in S_v(i)} z_l h_{i,l}$ is calculated first. Then each $\sigma_{i,j}$ only takes one more XOR operation to compute.

When only hard decisions are available from the channel, the iterative process in Algorithm 25 can still be utilized to achieve better performance than those hard-decision bit-flipping algorithms. However, in this case, $R_j^{(0)}$ should be initialized to γ and $-\gamma$ when the hard decision is '1' and '0', respectively, where γ is a constant. Such a hard-input decoder is referred to as iterative hard-reliability based (IHRB)-MLGD [108].

Algorithm 25 ISRB-MLGD Algorithm
initialization: $z^{(0)} = [z_0, z_1, z_2, \dots]$
$$R_j^{(0)} = \gamma_j \text{ for variable node } j$$
for $k = 1 : I_{max}$
  compute $z^{(k-1)}H^T$; *Stop if* $z^{(k-1)}H^T = 0$
  for each variable node j
    for each check node i
$$\sigma_{i,j} = \sum_{l \in S_v(i) \backslash j} z_l^{(k-1)} h_{i,l}$$
$$E_j = \sum_{i \in S_c(j)} (2\sigma_{i,j} - 1)$$
$$R_j^{(k)} = R_j^{(k-1)} + E_j$$
    if $R_j^{(k)} > 0$
$$z_j^{(k)} = 1$$
    else
$$z_j^{(k)} = 0$$

8.2.4 Finite alphabet iterative decoding algorithm

For RS and BCH codes, the error rate curve goes down smoothly following the same slope as the SNR increases. Although LDPC codes can correct more errors, except for short or very high-rate codes, their error-correcting performance curve may flatten out in the high SNR region. As mentioned previously, this phenomenon is called the error floor. It happens because the decoding may converge to trapping sets [109, 110], which are harmful loops of variable and check nodes in the Tanner graph. Even though there are not many errors, the decoding may fail to correct them no matter how many decoding iterations are carried out. The behavior of trapping sets is dependent on the algorithm and word length used for decoding. LDPC codes can be designed so that there is no cycle of length four or six in the Tanner graph. However, cycles of other lengths and trapping sets are unavoidable. The lengths and distributions of the cycles affect the error-correcting performance in the waterfall region and the error floor of LDPC codes. Codes with lower column weight usually have better performance in the waterfall region and lead to simpler hardware implementation. On the other hand, they may have a higher error floor. The error-floor is an important issue to address for those applications requiring a very low error rate, such as Flash memory and optical communications.

Besides constructing better LDPC codes, which may not always be possible for given system constraints, such as code rate and codeword length, much research has been devoted to lowering the error floor through modifying the

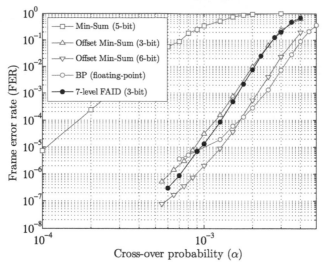

FIGURE 8.3

FERs of decoding algorithms for a (7807, 7177) QC-LDPC code with $d_c = 37$ and $d_v = 3$ over a binary symmetric channel [115]

decoding algorithm and/or post-processing. The error floor can be lowered by setting variable nodes in trapping sets as erasures, applying outer algebraic codes, such as BCH codes to correct residual errors, or combining problematic check node processors with generalized-constraint processors [111]. These methods require the trapping sets to be known, lead to extra complexity, and/or lower the code rate. In the backtracking scheme [112], decoding is retried by flipping a variable node connected to unsatisfied check nodes each time. Although this method does not need to pre-analyze the trapping sets, recording and selecting the decoding state of unsatisfied check nodes, and carrying out a large number of decoding trials result in high complexity overhead and long latency. Another method is to bias the LLR from the variable nodes connected to unsatisfied check nodes and then carry out more iterations [113]. This scheme does not require pre-knowledge about the trapping sets either. However, it still leads to a significant area and power consumption increase.

The finite alphabet iterative decoders (FAIDs) [114] tackle the error-floor problem by introducing randomness through the variable node processing. The check node processing in these decoders is the same as that in the Min-sum decoders. However, in the variable node processing, the values of the input c2v messages are considered as one of the limited number of alphabets. Instead of adding up the c2v messages and channel information, the v2c messages are derived by using a pre-defined Boolean map, which has the c2v messages and channel information as the inputs. Through comparing a certain cost function with regard to trapping sets that usually occur in various LDPC codes, the

Boolean maps that lead to good performance in the error-floor region are selected. Moreover, multiple Boolean maps can be used to address different trapping sets and further improve the performance [114]. Fig. 8.3 shows the frame error rates (FERs) of several decoding algorithms for a (7807, 7177) regular QC-LDPC code over the binary symmetric channel (BSC). This code has row weight $d_c = 37$ and column weight $d_v = 3$. The size of the submatrices is 211×211. It can be seen from this figure that the floating-point BP has an error floor show up as early as FER=10^{-4}. With only 7 possible levels in the alphabets, which translates to 3-bit word length, the FAID decoder has a much lower error floor than the BP. This figure also shows the performance of the original Min-sum, and offset Min-sum with optimized offset using different word length. The FAID decoder achieves better performance than the offset Min-sum with the same data precision in the waterfall region. Next, details about the FAID algorithm are given.

In a FAID, a message equals one of the alphabets $\{-L_s, \ldots, -L_1, 0, L_1, \ldots, L_s\}$, where L_i is a positive integer and $L_i > L_j$ for any $i > j$. Similar to that in the BP and Min-sum decoders, the magnitude of the alphabet tells how reliable the message is, and the sign is an estimation of whether the corresponding received bit should be '0' or '1'. For BSC, the channel information for a received bit takes one of the two values. Denote the channel information by $\pm C$, where C is a positive integer. The check node processing in the FAID is done according to (8.6), the same as that in the Min-sum decoder. Denote the c2v messages to a variable node n with degree d_v by $v_{0,n}, v_{1,n}, \ldots v_{d_v-1,n}$. The v2c message to check node $d_v - 1$ in the FAID is derived as

$$u_{d_v-1,n} = \Phi(y_n, v_{0,n}, v_{1,n}, \ldots, v_{d_v-2,n}) = Q\left(\sum_{i=0}^{d_v-2} v_{i,n} + w_n \cdot y_n\right), \quad (8.8)$$

where y_n is the channel information for the nth received bit. The v2c message to check node m $(0 \le m < d_v - 1)$ can be computed by an equation similar to (8.8) with $v_{m,n}$ excluded from the input c2v messages. The function Q is defined as

$$Q(x) = \begin{cases} sign(x)L_i, & \text{if } T_i \le |x| < T_{i+1} \\ 0, & \text{otherwise} \end{cases}$$

The thresholds T_i $(1 \le i \le s + 1)$ in the function Q are real numbers, and they satisfy $T_i > T_j$ if $i > j$ and $T_{s+1} = \infty$. The major difference between the FAID and Min-sum decoders results from the weight, w_n, multiplied to the channel information, y_n, in (8.8). It is computed by a symmetric function, Ω, that takes the $d_v - 1$ c2v messages involved in (8.8) as the inputs. If Ω is a constant regardless of the values of the input c2v messages, then the function Φ is a linear threshold function and a FAID with such a Φ is a linear FAID. Otherwise, Φ is a non-linear threshold function and a FAID with a non-linear Φ is called a non-linear FAID. Actually, the Min-sum algorithm is a linear FAID with $\Omega = 1$. In the FAID, the *a posteriori* information is generated as $y_n + \sum_{i=0}^{d_v-1} v_{i,n}$, from which hard decisions are made.

TABLE 8.1

Variable node processing LUT for $y_n = -C$ of a 7-level FAID when $d_v = 3$

$v_{0,n}/v_{1,n}$	$-L_3$	$-L_2$	$-L_1$	0	$+L_1$	$+L_2$	$+L_3$
$-L_3$	$-L_3$	$-L_3$	$-L_3$	$-L_3$	$-L_3$	$-L_3$	0
$-L_2$	$-L_3$	$-L_3$	$-L_3$	$-L_2$	$-L_2$	$-L_1$	L_1
$-L_1$	$-L_3$	$-L_3$	$-L_2$	$-L_2$	$-L_1$	0	L_2
0	$-L_3$	$-L_2$	$-L_2$	$-L_1$	0	L_1	L_2
L_1	$-L_3$	$-L_2$	$-L_1$	0	0	L_1	L_2
L_2	$-L_3$	$-L_1$	0	L_1	L_1	L_2	L_3
L_3	0	L_1	L_2	L_2	L_2	L_3	L_3

Note that the function Φ is symmetric and monotonic. In other words,

$$\Phi(y_n, v_{0,n}, v_{1,n}, \ldots, v_{d_v-2,n}) = -\Phi(-y_n, -v_{0,n}, -v_{1,n}, \ldots, -v_{d_v-2,n})$$

and

$$\Phi(y_n, v_{0,n}, v_{1,n}, \ldots, v_{d_v-2,n}) \geq \Phi(y_n, v'_{0,n}, v'_{1,n}, \ldots, v'_{d_v-2,n})$$

if $v_{i,n} \geq v'_{i,n}$ for any $i = 0, 1, \ldots, d_v - 2$.

When the column weight of the code is three, the Φ function has only two c2v messages and y_n as the inputs. Also $y_n = \pm C$ for BSC. Hence, the Φ function can be expressed as two LUTs, one for the case that $y_n = C$, and the other for $y_n = -C$. Due to the symmetry of the Φ function, the two tables are symmetric to each other. Table 8.1 shows an example LUT for the FAID when the messages have 7 different alphabets [115]. The two c2v input messages are denoted by $v_{0,n}$ and $v_{1,n}$ in this table. From (8.8), the order of the c2v messages at the input of the Φ function can be changed without affecting the output of the Φ function. Hence, the entries in the LUT for the FAID are symmetric with respect to the diagonal.

As mentioned previously, the Min-sum algorithm is as a linear FAID with $w_n = 1$ for any n. In the LUT of the variable node processing for the Min-sum algorithm, the entries whose corresponding $v_{0,n}$ and $v_{1,n}$ have the same sum are equal. For example, all of the entries in the diagonal equal y_n. For the offset Min-sum, the offset subtraction in (8.7) can be moved to the variable node processing step. Accordingly, the offset Min-sum is also an instance of the FAID. Table 8.2 lists the LUT for the variable node processing in a 3-bit offset Min-sum decoding [115]. Although this table is also very regular, the sequences of three zeros show some nonlinearity. This nonlinearity contributes to the better performance of the offset Min-sum over the original Min-sum decoder. Comparing Table 8.1 and 8.2, the LUT for the FAID is even more nonlinear. The differences between the magnitudes of two adjacent entries, such as the last two entries in the first row, can be more than one level in

TABLE 8.2

Variable node processing LUT for $y_n = -C$ of a 3-bit offset Min-sum decoder represented as a FAID when $d_v = 3$

$v_{0,n}/v_{1,n}$	$-L_3$	$-L_2$	$-L_1$	0	$+L_1$	$+L_2$	$+L_3$
$-L_3$	$-L_3$	$-L_3$	$-L_3$	$-L_3$	$-L_3$	$-L_2$	$-L_1$
$-L_2$	$-L_3$	$-L_3$	$-L_3$	$-L_3$	$-L_2$	$-L_1$	0
$-L_1$	$-L_3$	$-L_3$	$-L_3$	$-L_2$	$-L_1$	0	0
0	$-L_3$	$-L_3$	$-L_2$	$-L_1$	0	0	0
L_1	$-L_3$	$-L_2$	$-L_1$	0	0	0	L_1
L_2	$-L_2$	$-L_1$	0	0	0	L_1	L_2
L_3	$-L_1$	0	0	0	L_1	L_2	L_3

Table 8.1. These nonlinearities greatly help to improve the decoding performance, especially in the error-floor region. There is a very large number of LUTs that are symmetric and monotonic, and hence can be valid LUTs for variable node processing. Making use of the trapping sets that occur in various LDPC codes, a cost function can be used to decide whether each valid LUT is a good candidate to achieve low error floor. Moreover, different LUTs correct the errors associated with different trapping sets. Multiple FAIDs with different LUTs enable even better performance at the cost of higher hardware complexity. For example, the 7 LUTs [115] listed in Table 8.3 can be used one after another in the variable node processing of the FAID to achieve an even lower error floor. Due to the symmetry, the entries in the lower triangle of each table are omitted. Note that the LUT$_0$ listed in Table 8.3 is the same as Table 8.1.

Details of the cost function and LUT selection process are available in [114]. The FAID design so far has been focused on codes with $d_v = 3$, which are most prone to the error-floor problem. The design of the FAID for codes with higher column weight and/or over other types of channels is not available in the literature at the time this book is prepared.

8.3 LDPC decoder architectures

There are three types of decoder architectures: fully-parallel, serial, and partial-parallel. In a fully-parallel architecture, one hardware unit is implemented for each column and each row of the H matrix. The processing of all rows and columns is carried out simultaneously, and hence fully-parallel decoders can achieve the highest throughput. Since there is one dedicated unit for each check node and each variable node, the check node units (CNUs) and variable node units (VNUs) are connected by hard wires to pass the messages. Hence, no penalty is brought to fully-parallel decoders by random-like

TABLE 8.3
List of the LUTs for 7-level FAIDs

LUT$_0$

$-L_3$	$-L_3$	$-L_3$	$-L_3$	$-L_3$	$-L_3$	0
	$-L_3$	$-L_3$	$-L_2$	$-L_2$	$-L_1$	L_1
		$-L_2$	$-L_2$	$-L_1$	0	L_2
			$-L_1$	0	L_1	L_2
				0	L_1	L_2
					L_2	L_3
						L_3

LUT$_1$

$-L_3$	$-L_3$	$-L_3$	$-L_3$	$-L_3$	$-L_3$	0
	$-L_3$	$-L_3$	$-L_3$	$-L_2$	$-L_1$	L_1
		$-L_2$	$-L_1$	$-L_1$	0	L_2
			$-L_1$	0	L_1	L_2
				L_1	L_1	L_2
					L_1	L_2
						L_2

LUT$_2$

$-L_3$	$-L_3$	$-L_3$	$-L_3$	$-L_3$	$-L_3$	$-L_1$
	$-L_3$	$-L_3$	$-L_3$	$-L_3$	$-L_2$	L_1
		$-L_2$	$-L_1$	$-L_1$	$-L_1$	L_1
			$-L_1$	$-L_1$	L_0	L_2
				0	L_1	L_2
					L_1	L_2
						L_3

LUT$_3$

$-L_3$	$-L_3$	$-L_3$	$-L_2$	$-L_2$	$-L_2$	$-L_1$
	$-L_3$	$-L_3$	$-L_2$	$-L_2$	$-L_2$	L_1
		$-L_3$	$-L_2$	$-L_1$	0	L_1
			$-L_1$	0	0	L_3
				0	L_1	L_3
					L_1	L_3
						L_3

LUT$_4$

$-L_3$	$-L_3$	$-L_3$	$-L_3$	$-L_3$	$-L_3$	$-L_1$
	$-L_3$	$-L_3$	$-L_2$	$-L_2$	$-L_1$	L_1
		$-L_2$	$-L_2$	$-L_1$	0	L_1
			$-L_1$	0	L_1	L_1
				L_1	L_1	L_1
					L_1	L_3
						L_3

LUT$_5$

$-L_3$	$-L_3$	$-L_3$	$-L_3$	$-L_3$	$-L_3$	0
	$-L_3$	$-L_3$	$-L_2$	$-L_2$	$-L_2$	L_1
		$-L_2$	$-L_2$	$-L_1$	0	L_2
			$-L_1$	0	0	L_2
				L_1	L_1	L_2
					L_1	L_2
						L_2

LUT$_6$

$-L_3$	$-L_3$	$-L_3$	$-L_3$	$-L_3$	$-L_3$	$-L_1$
	$-L_3$	$-L_3$	$-L_3$	$-L_2$	$-L_2$	L_1
		$-L_2$	$-L_1$	$-L_1$	0	L_1
			$-L_1$	0	0	L_2
				0	L_1	L_2
					L_1	L_2
						L_3

LDPC codes. Nevertheless, the large number of hardware units and complex routing network make it very expensive to implement fully-parallel decoders, especially when the code is long and/or the code rate is low.

The opposite case is the serial architecture, which has only one CNU and one VNU. They implement the processing of rows and columns of H in a time-multiplexed way, and the c2v and v2c messages are stored and read from memories. The routing of the messages in a serial decoder is also simple. However, the location of each nonzero entry of H needs to be recorded to access the c2v and v2c message memories. Storing these locations leads to larger memory requirement if the code is random-like. Despite the low complexity, the achievable throughput of a serial decoder is very limited.

Partial-parallel decoders achieve tradeoffs between throughput and area requirement. In a partial-parallel architecture, multiple CNUs and VNUs are implemented to process groups of rows and columns of H in a time-multiplexed manner. The c2v and v2c messages are also stored in memories. To avoid memory access conflict, the messages for one group of rows and columns are stored in one address location. Since the locations of the nonzero entries in H change from group to group, when a group of messages is read out, they need to be switched before sending to the CNUs and VNUs. Each CNU and VNU should be able to simultaneously handle the maximum possible number of nonzero H entries in a group. The more messages to process at a time, the more complicated the CNUs and VNUs are. For random-like codes, the nonzero entries appear at irregular locations. Hence the number of nonzero entries in a group varies. In this case, the CNUs and VNUs designed to handle the worst case are under-utilized when they process those groups with fewer nonzero entries. As a result, the hardware efficiency is low. Fortunately, QC-LDPC codes [100] have very good error-correcting performance and are adopted in many applications. As it is illustrated in Fig. 8.2, the H matrix of these codes consists of square submatrices that are either zero or cyclically shifted identities. If H is divided into groups consisting of block rows or columns of sub-matrixes, then the CNUs and VNUs in a partial-parallel decoder need to handle the same number of messages each time. Due to the high hardware efficiency and throughout-area tradeoff, most of the LDPC decoders currently adopted in various systems are partial-parallel designs for QC codes. Therefore, the discussions in the remainder of this section will focus on the architectures of these decoders.

Even if the numbers of CNUs and VNUs employed in a decoder are given, the scheduling of the check and variable node processing affects the latency, memory requirement, convergence speed and/or the error-correcting performance of the decoder. Several popular decoding scheduling schemes are presented next before the details of the CNUs and VNUs are introduced.

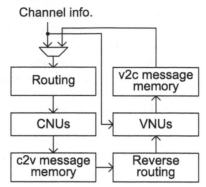

FIGURE 8.4
General partial-parallel decoder architecture adopting the flooding scheme

8.3.1   Scheduling schemes

8.3.1.1   Flooding scheme

In the flooding scheme, the check node processing for all rows of H is completed before variable node processing starts, and the variable node processing is done for all columns of H before the check node processing for the next iteration is carried out. Each decoding iteration is done in two phases.

The general architecture of a partial-parallel flooding decoder is shown in Fig. 8.4. Since the nonzero entries in each block row or column of H appear at different locations, a partial-parallel design needs routing networks to send the v2c messages to the right CNUs. Similarly, the c2v messages are reversely routed before they are sent to VNUs. In Min-sum decoders, instead of the c2v messages, the min1, min2, index for the min1, sign product, and the sign of each v2c message can be stored as compressed c2v messages. Assume that the row weight is d_c and the word length of the messages is w. Storing the c2v messages for a row of H needs a memory of $d_c w$ bits. However, recording the compressed messages only requires $2w + \lceil \log_2 d_c \rceil + 1 + d_c$ bits of memory, which is substantially smaller. The saving further increases when d_c is larger. The real c2v messages are recovered easily from the compressed messages when they are needed in the VNUs. The updated v2c messages are also stored in memory. However, they do not have any compressed format available, and hence the v2c message memory is much larger than the compressed c2v message memory in Min-sum decoders. Besides the c2v and v2c messages, the channel information also needs to be stored since it is added up in the variable node processing for each iteration.

Fig. 8.5 shows the scheduling of the computations in decoders adopting the flooding scheme. It should be noted that, depending on the numbers of CNUs and VNUs adopted, the number of clock cycles needed for the check and variable node processing for the entire H may be unbalanced. Since the

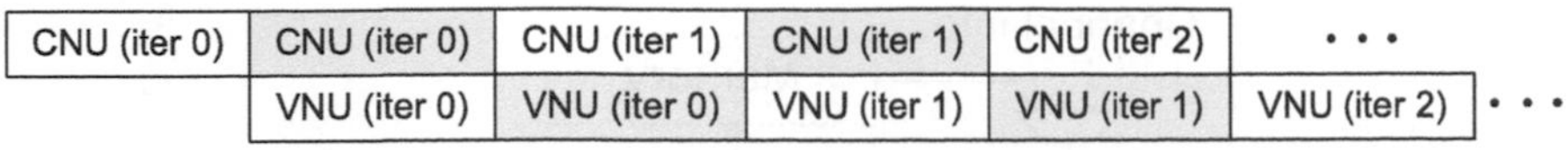

CNU (iter 0)	CNU (iter 0)	CNU (iter 1)	CNU (iter 1)	CNU (iter 2)	• • •

	VNU (iter 0)	VNU (iter 0)	VNU (iter 1)	VNU (iter 1)	VNU (iter 2)	• • •

FIGURE 8.5
Computation scheduling in interleaved decoding adopting the flooding scheme

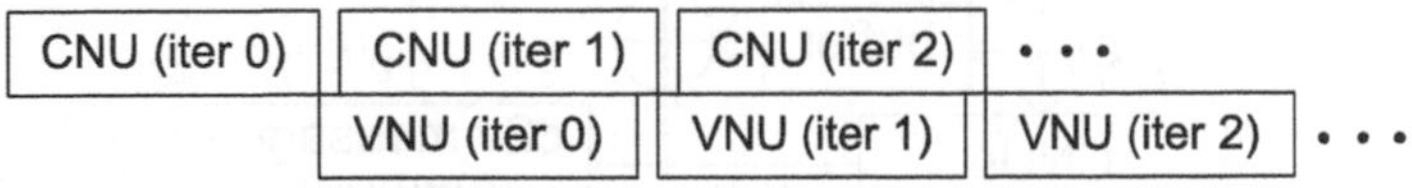

CNU (iter 0)	CNU (iter 1)	CNU (iter 2)	• • •

	VNU (iter 0)	VNU (iter 1)	VNU (iter 2)	• • •

FIGURE 8.6
Computation scheduling in sliced message-passing decoding

check and variable node processing do not overlap, the CNUs and VNUs are idle half of the time. Two blocks of data can be interleaved to more efficiently utilize the CNUs and VNUs as shown in Fig. 8.5. The clear and shaded bars in this figure differentiate the two data blocks. Although the memory needs to be doubled to store the messages for both data blocks, the throughput is substantially increased. In the case that the latencies of the check and variable node processing are the same, as it is shown in Fig. 8.5, the throughput is doubled by interleaving.

8.3.1.2 Sliced message-passing scheme

In the sliced message-passing scheduling scheme [116], the check node processing for all rows of H is carried out in parallel, and the variable node processing is done for one or more block columns of H at a time. Due to the large number of CNUs, this scheme is more appropriate for high-rate codes. Unlike the flooding method, the sliced message-passing scheme carries out the variable node processing simultaneously as the check node processing for the next decoding iteration. After the min1, min2 *etc.* for all rows are computed by the CNUs for decoding iteration i, the variable node processing for the same iteration starts. The v2c messages for one or more block columns of H are updated by using the compressed c2v messages, and the updated v2c messages are sent to the CNUs immediately to compute the min1 and min2 for decoding iteration $i + 1$. Since the v2c messages are consumed right after they are generated, they do not need to be stored. Moreover, as shown in Fig. 8.6, the check node processing for decoding iteration $i + 1$ starts right after the first group of updated v2c messages is available. Therefore, the variable node processing for iteration i can be almost completely overlapped with the check node processing for iteration $i + 1$.

Assume that the QC H matrix consists of $r \times t$ cyclic permutation matrixes of dimension $e \times e$. Then the sliced message-passing decoder has er CNUs. If

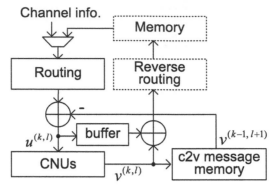

FIGURE 8.7
General single-block-row layered decoder architecture

one column of sub-matrixes are processed simultaneously, then e VNUs are needed. In this case, each decoding iteration takes t clock cycles using the sliced message-passing scheme. The achievable throughput is almost the same as that of the interleaved flooding decoder with er CNUs and e VNUs. However, unlike the interleaved flooding decoder, the sliced message-passing decoder does not store the v2c messages or require duplicated message memories.

8.3.1.3   Row-layered decoding scheme

There are both row-layered [117] and column-layered [118] decoding schemes. The column-layered scheme is also referred to as the shuffled scheme and will be discussed later in this section. By layered decoding, we are referring to the row-layered scheme in this book. In layered decoding, the H matrix is divided into blocks of rows, also called layers. Instead of updating the v2c messages once in each iteration, the c2v messages derived from the decoding of a layer are utilized to update the v2c messages and are used right away in the decoding of the next layer. Since the v2c messages get updated more often, the decoding converges faster and hence takes fewer iterations. It has been shown that layered decoding reduces the number of iterations by around 50%.

Traditionally, in the layered decoding of QC-LDPC codes, each layer consists of one row of sub-matrixes of H. In this case, there is at most one c2v message for each variable node during the decoding of a layer. Denote the v2c and c2v messages for block row l in decoding iteration k by $u^{(k,l)}$ and $v^{(k,l)}$, respectively. For brevity, the variable and check node indexes are omitted from these notations. Let γ be the channel information, and the total number of layers is r. Since the v2c messages are generated from the most updated c2v messages,

$$u^{(k,l+1)} = \gamma + \sum_{j=0}^{l} v^{(k,j)} + \sum_{j=l+2}^{r-1} v^{(k-1,j)}. \qquad (8.9)$$

From the above equation, it can be derived that

$$u^{(k,l+1)} = (u^{(k,l)} + v^{(k,l)}) - v^{(k-1,l+1)}. \qquad (8.10)$$

The channel information is used as the v2c messages for the first layer in the first decoding iteration. After that, the v2c messages are computed using (8.10). As a result, the layered decoder can be implemented according to the architecture shown in Fig. 8.7.

The CNUs process one or more block columns of H in a layer at a time. The inputs of the CNUs are buffered until they are added up with the corresponding CNU outputs according to (8.10). Assume that the messages output from the adder on the right in Fig. 8.7 are for block column n at a certain clock cycle. If the submatrix in block column n of the next layer is nonzero, then the adder outputs are consumed in the computations of the next layer right after, and they do not need to be recorded in layered decoding. Otherwise, the adder outputs need to be stored until a later layer that has a nonzero submatrix in block column n. Therefore, if the H matrix does not have any nonzero submatrix, then the memory in the top right part of Fig. 8.7 can be eliminated. Layered decoding also has the advantage that the channel information is incorporated in the updated v2c messages and does not need to be stored. This not only reduces the memory requirement, but also enables the combination of the reverse routing and routing networks. Moreover, a v2c message only takes two additions to derive, as shown in Fig. 8.7. In previous schemes, a v2c message is generated by adding up $d_v - 1$ c2v messages and the channel information when the column weight is d_v, although intermediate results can be shared in the computations of multiple v2c messages. In layered decoding, the c2v messages for the previous decoding iteration, $v^{(k-1,l+1)}$, need to be stored. However, the size of the memory required for this is the same as that in sliced message-passing decoders.

The number of clock cycles needed for each decoding iteration in a layered decoder is decided by the number of layers and the number of clock cycles spent on each layer. Although fewer iterations are needed for the decoding to converge, the layers are processed one after another in each decoding iteration. This serial processing of layers imposes throughput limitation on the layered decoding. To improve the throughput, multiple block columns can be processed simultaneously within each layer. However, the H matrix needs to be designed so that reading the messages for multiple block columns at a time does not cause memory access conflicts. When p block columns are processed at a time, each CNU for a Min-sum decoder would need to compare p inputs with the intermediate min1 and min2 values to find the updated min1 and min2. The complexities of these comparisons increase with p. The detailed ar-

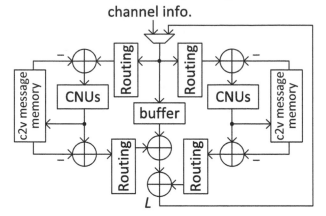

FIGURE 8.8
A 2-block-row layered decoder architecture (modified from [120])

chitectures for the CNUs are presented later in this chapter. Further speedup can be achieved by putting multiple block rows in a layer [119].

When a layer consists of multiple block rows, the v2c messages should still be derived based on the most updated c2v messages. Assume that each layer consists of p block rows of H, and the total number of layers is $r' = r/p$. Then for $l = 0, 1 \dots, r' - 2$ and $i = 0, 1, \dots, p - 1$,

$$u^{(k,lp+i)} = \gamma + \sum_{j=0}^{l-1}\sum_{i'=0}^{p-1} v^{(k,jp+i')} + \sum_{i'=0,i'\neq i}^{p-1} v^{(k-1,lp+i')} + \sum_{j=l+1}^{r'-2}\sum_{i'=0}^{p-1} v^{(k-1,jp+i')}.$$

Compared to (8.9), the difference in the above equation is that the c2v messages for block rows $lp + i'$ ($i' = 0, 1, \dots, p - 1$) are still those from iteration $k - 1$, since p block rows are updated at a time. As a result, the v2c message updating rule should be changed accordingly when each layer consists of multiple block rows. In [119], the v2c message updating is done with the assistance of an intrinsic message L, which is initialized as the channel information, and is updated during the decoding for layer l in iteration k as

$$L \Leftarrow L + \sum_{i=0}^{p-1} \left(v^{(k,lp+i)} - v^{(k-1,lp+i)} \right). \tag{8.11}$$

Then the v2c messages for layer $l + 1$ of the same decoding iteration and $i = 0, 1, \dots, p - 1$ are computed as

$$u^{(k,(l+1)p+i)} = L - v^{(k-1,(l+1)p+i)}. \tag{8.12}$$

As an example, a $p = 2$-block-row layered decoder can be implemented as shown in Fig. 8.8 according to (8.11) and (8.12). In this architecture, it is assumed that the p submatrices in each block column of every layer are not all zero. Otherwise, memories need to be added to store the updated L until it is consumed in the computations of a later layer. The two groups of CNUs take care of the check node processing for even and odd block rows. Since the nonzero entries appear at different locations in the two block rows, a separate routing network is needed for each group of CNUs. Similarly, if the Min-sum algorithm is employed, the min1 and min2 *etc.* computed in the CNUs are stored in memories as compressed c2v messages to be used in updating the v2c messages for later layers. Each pair of messages involved in the subtraction in (8.11) are c2v messages of the same row but different iterations. Hence, no reverse routing is needed before the subtraction is carried out. Also no permutation is necessary on $v^{(k-1,(l+1)p+i)}$ to compute $u^{(k,(l+1)p+i)}$ in (8.12) as long as the intrinsic messages L are routed according to block row $(l+1)p+i$. Nevertheless, because the v2c messages have been permuted differently for different block rows, reverse routing is required to align the c2v messages for the same variable node before the differences for $i = 0, 1, \ldots, p-1$ in (8.11) are added up.

As mentioned previously, in the single-block-row layered decoder shown in Fig. 8.7, the reverse routing can be combined with the routing network. Hence, compared to the single-block-row layered decoder, the 2-block-row layered decoder illustrated in Fig. 8.8 needs twice the CNUs and memories, three times the adder units, and four times the complexity for the routing networks. Nevertheless, the size of the buffer, which can account for a significant proportion of the overall decoder area, does not increase. Although two block rows are processed at a time, the achievable throughout of the 2-block-row layered decoder is not doubled compared to that of single-block-row layered decoder. First of all, it can be observed from Fig. 8.8 that the critical path of the 2-block-row layered design is much longer than that in Fig. 8.7. Hence, the achievable clock frequency is lower. Alternatively, if pipelining is applied to achieve the same clock frequency, the number of clock cycles needed for the decoding of each layer in the 2-block-row design is larger because there are more pipelining stages. In addition, it has been shown in [119] that the more block rows in a layer, the slower the convergence of the decoding. Apparently, this is because the v2c messages get updated less frequently if more block rows are put into a layer. As a result, as the number of block rows in a layer increases, the multi-block-row layered decoding becomes a less effective way to increase the throughput. If the H matrix can be designed to avoid memory access conflicts and the throughput requirement can be met, it is more hardware-efficient to have one block row in each layer and process multiple block columns simultaneously within a layer rather than putting multiple block rows in one layer and processing one clock column in each layer at a time.

8.3.1.4   Shuffled decoding scheme

Shuffled decoding [118] is also called column-layered decoding. In this scheme, the columns of the H matrix are divided into groups or layers, and the decoding is done on one column layer at a time. For QC-LDPC codes, each layer usually consists of one block column. In the decoding of a column layer, the v2c messages for all check nodes are computed. They are then used to update all c2v messages, and the most updated c2v messages are used to compute the v2c messages for the next column layer. Similar to row-layered decoding, the number of iterations needed is reduced in column-layered decoding since the messages are updated more often. Nevertheless, the differences between the row- and column-layered decoding are not limited to whether the H matrix is divided row- or column-wise. In log-domain LDPC decoding algorithms, such as the Min-sum algorithm, the v2c messages are computed as the sum of c2v messages. When a new c2v message is available, the v2c message can be updated by subtracting the old value and adding up the new value as is done in row-layered decoding. However, the c2v messages are derived from v2c messages using a more complex function. When a v2c message is changed, it is very difficult to take out the effect of the old value and incorporate the new value without resorting to the information provided by other variable nodes. A perfect column-layered decoder that generates exact c2v messages often needs to run the entire check node processing again every time a v2c message is changed. On the other hand, approximations can be done to simplify the c2v message update with small performance loss as will be explained in the next paragraph. Despite the higher hardware complexity and performance degradation brought by inexact c2v message update, the column-layered decoding enables faster speed to be achieved compared to row-layered decoding since all rows can be processed simultaneously.

A method for reducing the complexity of the c2v message update in column-layered decoding for the Min-sum algorithm was proposed in [121]. The min1 and min2 messages are not re-computed every time a new v2c message is generated. Instead, they are updated using the newly available v2c message. At the beginning of the processing of a column layer, if the min1 or min2 message of a check node is from these columns, it is deleted. Then the newly computed v2c message for this column layer is compared with the remaining min1 and/or min2 messages to find the minimum and second minimum magnitudes. The signs of the messages are also updated in a similar way. Since only the min1 and min2 messages are recorded, the min2 computed in this way may not necessarily be the real second minimum magnitude as in the original column-layered decoder in [118]. As a result, the approach in [121] leads to noticeable error-correcting performance loss despite the complexity reduction. To make up for the coding gain loss, more messages can be sorted out during the check node processing and stored. However, this unavoidably brings larger memory requirement and overheads to the CNUs.

TABLE 8.4

Comparisons of decoding schemes with word length w for a QC-LDPC code whose H matrix consists of $d_v \times d_c$ cyclic permutation matrixes of dimension $e \times e$

	flooding	sliced message-passing	row-layered	shuffled [121]
logic units				
# of CNUs	ed_v	ed_v	e	$ed_v^{\ddagger}$
# of VNUs	e	e	e^*	e
# of routing networks	$2d_v$	$2d_v$	1	$2d_v$
storage requirement				
channel information(bits)	ed_cw	ed_cw	none	ed_cw
v2c messages or signs(bits)	ed_cd_vw	ed_cd_v	ed_cd_v	ed_cd_v
# of compressed c2v messages†	ed_v	$2ed_v$	$e(d_v+1)$	ed_v
CNU input buffer(bits)	none	none	ed_cw	none
latencies				
# of clks/iteration (without pipelining latency)	$2d_c$	$d_c^{\diamond}$	d_vd_c	d_c
normalized # of iterations	1	1	$\sim50\%$	$\sim50\%$

‡ The CNUs in the shuffled decoders are more complicated.

* The pair of adder and subtractor for v2c message update is counted as a VNU for the layered decoder.

† Each compressed c2v message has $2w + \log_2 d_c + 1$ bits for min1, min2, the index for min1, and the sign product.

⋄ The shuffled decoder can not be pipelined when there are adjacent nonzero submatrices in the same block row of H.

8.3.1.5 Scheduling scheme comparisons

Assume that the H matrix of a QC-LDPC code consists of $d_v \times d_c$ cyclic permutation matrixes of dimension $e \times e$, and the word length used for the messages is w. The complexities of the Min-sum decoders employing the four decoding schemes discussed above are compared in Table 8.4. It is assumed that the simplified check node processing introduced in [121] is adopted in the shuffled decoder.

The achievable reduction on the average number of decoding iterations becomes less significant when multiple block rows or columns are included in a layer in the row-layered or shuffled decoders. Hence, the row-layered and shuffled decoders that have one block row and column, respectively, in each layer are considered in the comparisons in Table 8.4. Although a sliced message-passing decoder has ed_v CNUs, multiple block columns can be processed at a time, and hence the number of VNUs can vary. The complexity of the CNUs increases with the number of block columns processed simultaneously. Similarly, in a row-layered decoder, multiple block columns can be taken care of in parallel within each layer at the cost of more complicated CNUs. Of course, extra constraints may need to be added to the code construction to enable efficient simultaneous processing of multiple block columns. For the purpose of comparison, it is assumed that the sliced message-passing and row-layered decoders also process one block column at a time as in the shuffled decoder. To eliminate the need of storing the v2c messages, all rows in a block column are processed in parallel in the shuffled decoder. The numbers of CNUs and VNUs in a decoder adopting the flooding scheme are flexible. However, in order to balance the speed of the variable and check node processing and compare with the other architectures, e VNUs and ed_v CNUs are employed to process all rows and one block column at a time. In this case, the CNUs in the flooding, sliced message-passing, and layered decoders are the same. Nevertheless, in the shuffled decoder of [121], if the min1 or min2 message is from a column that is currently being processed, it needs to be eliminated. This elimination leads to slightly higher complexity in the CNUs for the shuffled decoder in [121] because of extra control and switching. Unlike the other decoders, the row-layered decoder does not have explicit VNUs. Each v2c message is generated by a pair of adder and subtractor. On the other hand, as will be shown later in this section, each VNU in the other decoders computes d_v v2c messages in parallel using $2d_v$ adders and subtractors.

The routing networks are responsible for switching the v2c messages to the correct CNUs. Similarly, reverse routing networks are needed to send the c2v messages to the right VNUs. Since these two types of networks have the same architecture, they are listed as one item in Table 8.4. The number of routing networks needed for each decoder is decided by the number of CNUs and VNUs employed. Each network listed in Table 8.4 is capable of taking care of the routing corresponding to one $e \times e$ sub-matrix of H. Each of them consists of an e-input barrel shifter. In a row-layered decoder, the channel information only serves as the initial v2c messages, and does not participate in the v2c message computation for later layers and iterations. Accordingly, the reverse routing and routing networks are combined.

Except in the flooding scheme, the v2c messages generated from the min1 and min2 are consumed right away in the calculation of the min1 and min2 for the next decoding iteration or layer, and hence do not need to be stored. Instead of the actual c2v messages, the min1, min2, index of the min1 and the sign product are stored as compressed c2v messages. Each CNU requires storage units for one set of such compressed messages. Moreover, in the sliced message-passing, the compressed messages are used to calculate the v2c messages, while the CNUs are computing the min1 and min2 *etc* for the new v2c messages. Hence, storage is required for $2ed_v$ copies of the compressed messages. There are e CNUs in the row-layered decoder. However, the c2v messages for layer l from the previous iteration need to be stored until they are consumed in updating the v2c messages for layer l in the current iteration. Therefore, a total of $e(d_v+1)$ compressed messages need to be recorded. Using the simplified check node processing proposed in [121], the shuffled decoder updates the min1 and min2 *etc.* each time, and the old values are not needed afterwards. Hence, unlike the sliced message-passing decoder, the shuffled decoder records ed_v compressed c2v messages.

The signs of the v2c messages need to be stored to recover the actual c2v messages from the compressed version. The sliced message-passing decoder stores twice compressed c2v messages compared to the flooding decoder because of concurrent check and variable node processing. However, the signs of the updated v2c messages can replace the signs of the old v2c messages in the memory. Hence, the size of v2c sign memory in the sliced message-passing decoder is ed_cd_v. Similarly, the new v2c signs take the places of the old v2c signs in the row-layered and shuffled decoders, and their v2c sign memories are of the same size. Although the channel information does not need to be stored, the row-layered decoder needs a buffer to hold the inputs of the CNUs until they are added up with the outputs as shown in Fig. 8.7. Since the c2v messages for a layer are only generated after all columns in a layer are processed, this buffer needs to hold the v2c messages for a layer.

In the flooding, sliced message-passing and shuffled decoders, all rows of one block column in the H matrix are processed at a time. Therefore, it takes d_c clock cycles to go over once an H matrix of d_c block columns. Since the check and variable node processing is carried out in two steps in the flooding scheme, the number of clock cycles needed in each decoding iteration is almost doubled in this decoder. The row-layered decoder processes one block column in one block row at a time. Therefore, it needs around d_vd_c clock cycles for each iteration. Nevertheless, it has been shown that the number of iterations needed for convergence is reduced to around 50% by row-layered and shuffled decoding. In the shuffled decoder, the most updated c2v messages are needed to carry out the v2c message computation of the next block column. If the submatrices in block columns j and $j+1$ of the same row in H are both nonzero, then the updated c2v messages derived from the processing of block column j are needed to start the computation for block column $j+1$. In this case, the shuffled decoder can not be pipelined to achieve higher throughput. If there are at least ξ zero submatrices between any nonzero submatrices in the same row, then the shuffled decoder can be pipelined into $\xi+1$ stages. On the other hand, the flooding, sliced message-passing and row-layered decoders can be always pipelined. Pipelining the flooding and sliced message-passing decoders into ξ stages only adds ξ clock cycles to the decoding of each iteration, while ξ pipelining stages causes ξ extra clock cycles to the decoding of each layer in the layered decoder. The achievable clock frequency or the number of pipelining stages needed to achieve a given clock frequency is dependent on the critical path. In each of the VNUs in the flooding, sliced message-passing and shuffled decoders, d_v c2v messages and the channel information need to be added up. On the contrary, the v2c messages are computed by pairs of adders and subtractors in the layered decoder. Adding up more messages leads to a longer data path. Also there is only one combined routing network in the data path of the layered decoder. Considering these, the shuffled decoder has the lowest achievable clock frequency and whether it can be pipelined is subject to the adjacencies of the nonzero submatrices in H. The layered decoder has the shorted critical path, and hence can achieve the highest clock frequency or requires the smallest number of pipelining stages to reach a given clock frequency.

8.3.2   VLSI architectures for CNUs and VNUs

Since the architectures for the OSMLGD and bit-flipping decoders are straightforward, and the Min-sum algorithm is currently adopted in many LDPC decoders used in practical systems, we focus on the implementation architectures of the Min-sum decoding algorithm. Compared to the flooding and shuffled schemes, the sliced message-passing and row-layered methods have lower hardware complexity, and will be the emphasis of the discussion next. The overall architecture of row-layered decoders is shown in Fig. 8.7. The overall architecture of the sliced message-passing decoder is similar to that shown in Fig. 8.4 for the flooding scheme, except that the v2c message memory is not needed and the computations in the CNUs and VNUs overlap. For QC-LDPC codes, the routing networks are implemented as barrel shifters. The same CNU designs can be used for row-layered and sliced message-passing decoders. Row-layered decoders do not have formal VNUs. Each v2c message is derived by a pair of adder and subtractor. The v2c messages in sliced message-passing decoders are also computed by additions. However, intermediate results can be shared if all v2c messages from the same variable node need to be computed in parallel. In the following, the CNU and VNU designs for sliced message-passing and row-layered Min-sum decoding are presented. The FAID is different from the Min-sum algorithm in the variable node processing. The implementation of the VNUs for the FAID is also briefly introduced.

8.3.2.1   CNU architectures

According to Algorithm 23, a CNU for the Min-sum algorithm first finds the min1 and min2 among the magnitudes of all d_c input v2c messages, records the index of the min1 message as idx, and computes the product of the signs of the v2c messages. Then the magnitude of each c2v message is selected by a multiplexor from the min1 and min2 values based on whether the index of the variable node equals idx. Also the signs of the c2v messages are derived by XORing the sign product with the signs of the v2c messages. The implementation of the second part of the CNU is straightforward. However, the sorter for finding the min1 and min2 values can be implemented by various architectures that have speed-area tradeoffs.

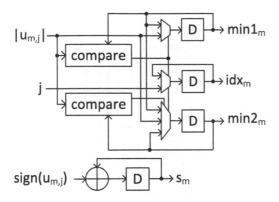

FIGURE 8.9
Min1 and min2 sorter architecture that processes one v2c message at a time

Single and double-input sorter architectures If the CNU processes one v2c message at a time, which is the case when one block column of a QC H matrix is taken care of in each clock cycle, min1, min2 and idx can be found by the architecture illustrated in Fig. 8.9. The intermediate min1, min2, and idx are held in the registers. Every time a v2c message is input, its magnitude is compared with the intermediate min1 and min2, and these intermediate values are updated based on the comparison results. Also if the input v2c magnitude is smaller than min1, then idx is replaced by the index of the current input message. The sign product is computed iteratively by a simple feedback loop with one XOR gate.

Higher speed can be achieved by processing more block columns of H at a time. If two v2c messages are input at a time, the min1 and min2 sorter can be implemented by the architecture shown in Fig. 8.10. The small blocks at the outputs of the comparators consist of simple logic to generate the control signals of the multiplexors. Comparing this architecture with that in Fig. 8.9 processing one v2c message at a time, the number of comparators is more than doubled. If all necessary comparisons are done in parallel, the area requirement of the min1 and min2 sorter increases fast with the number of v2c messages input each time.

Multi-input tree sorter architectures To achieve a given throughput requirement, multiple block columns of H may need to be processed simultaneously in LDPC decoding. This requires each CNU to process multiple v2c messages in each clock cycle. A multi-input sorter can be broken down into smaller sorters that compare fewer inputs at a time, and the smaller sorters are connected in a tree structure. The min1 is found by routing the minimum value from each sorter to the sorter in the next level of the tree. Taking a sorter with 8 inputs, $x_0, x_1, \ldots, x_7$, as an example. The min1 can be calculated by the tree structure in Fig. 8.11(a). For brevity, the logic for computing the sign

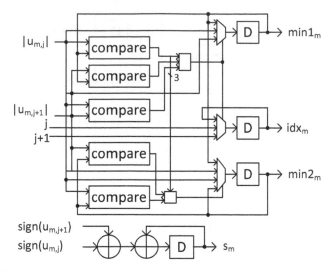

FIGURE 8.10
Min1 and min2 sorter architecture that processes two v2c messages at a time

product and deriving *idx* are omitted from this figure. Since each compare-and-select (CSel) unit takes care of the comparison of two inputs, it is referred to as a radix-2 architecture. The details of the CSel unit are shown in the bottom right corner of Fig. 8.11(a). It routes the smaller input to the output using the comparison result.

Different approaches are available to derive the min2 value. The number of comparators needed to find the min2 can be reduced by making use of the comparison results from the min1 computation according to the algorithm in [122]. For example, if min1 equals x_4, whose path to the min1 output in the tree is highlighted by the dashed lines in Fig. 8.11(a), then the min2 is the minimum among $\min_{10}$, $\min_{03}$, and x_5. These three values are the larger inputs of the CSel units in the path that generates the final min1. Each of them can be generated from the 8 inputs by a 8-to-1 multiplexor whose control signals are derived using the 'c' outputs of the CSel units. From the three values, only two additional Csel units are needed to find the min2 as shown in Fig. 8.11(b). In this figure, each of the select signals s_0, s_1 and s_2 has three bits for an 8-input architecture. Assume that the 'c' signal from the CSel unit is asserted if the upper input is smaller than the lower input. Let the first multiplexor in Fig. 8.11(b) output the larger input of the Csel unit in the last level of the tree. Then one of x_0, x_1, x_2 and x_3 should be output of this multiplexor if $c_{20} =$'0'. Otherwise, one of x_4, x_5, x_6 and x_7 is the output. Therefore, the most significant bit (MSB) of s_0 is c_{20}. As another example, when $c_{20} =$'0', either x_2 or x_3 should be the multiplexor output if $c_{10} =$'0'. In addition, either x_6 or x_7 should be the multiplexor output when $c_{20} =$'1' and

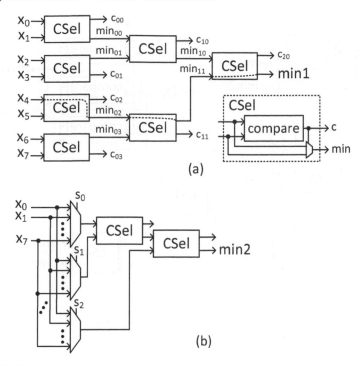

FIGURE 8.11
8-input radix-2 min1 and min2 tree sorter architecture. (a) min1 computation architecture; (b) min2 computation architecture

$c_{11} = {}^\prime 0^\prime$. Accordingly, the second bit of s_0 should be $c_{20}^\prime c_{10}^\prime + c_{20} c_{11}^\prime$. The last bit of s_0, as well as s_1 and s_2 can be derived similarly.

Although the approach in [122] requires fewer comparators, a large number of multiplexors is needed for the min2 computation as shown in Fig. 8.11(b). As the number of inputs increases, these multiplexors cause higher overhead. It has been shown in [123] that the overall complexity and critical path of the tree-based min1 and min2 sorters can be reduced by employing units to find the min1 and min2 from a pair of min1 and min2 each time. Such an architecture is shown in Fig. 8.12. Each of the 2-input compare-and-switch (CS2) units in the first level of the tree compares two inputs, and switches the smaller input to the top based on the comparison result. The other levels of the tree use 4-input compare-and-switch (CS4) units, each of which takes 2 pairs of min1 and min2 generated from the units in the previous level, and finds the min1 and min2 among the four inputs. Since min1 is less than min2 in each input pair, each min1 input only needs to be compared with the min2 input in the other pair to find the candidates of the min2 output. Let the two pairs of inputs be {min1a, min2a} and {min1b, min2b}. If min1a <min1b, then

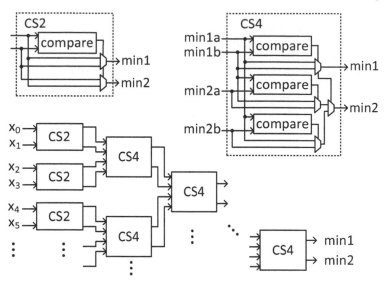

FIGURE 8.12
Parallel radix-2 min1 and min2 tree sorter architecture

min2 must be the smaller of min1*b* and min2*a*. Otherwise, min2 equals the smaller of min1*a* and min2*b*. As a result, only three comparators are needed in a CS4 unit.

Both the designs in [122] and [123] are radix-2. In other words, each unit in these designs only takes care of two inputs or two pairs of inputs at a time. Higher-radix and mixed-radix approaches were exploited in [124]. It was assumed that, in each unit, all necessary comparisons needed to find the min1 and min2 are carried out simultaneously. For the first level of the tree, a radix-r unit finds the min1 and min2 among r inputs. Every input needs to be compared with each other. Hence, such a unit needs $(r-1)+(r-2)+\cdots+1 = r(r-1)/2$ comparators. For the second and other levels of the tree, a radix-r units find the min1 and min2 among r pairs of min1 and min2 computed in the previous level. The min1 output comes from the r min1 inputs. Hence, finding the min1 needs $r(r-1)/2$ comparators. To find the min2 output, each min1 input needs to be compared with the min2 in each of the other input pairs. $r(r-1)$ comparators are needed for this purpose. Based on this analysis, the total number of comparators needed in a sorter tree can be calculated. Different radix may be adopted for different levels in the tree. It was found in [124] that allowing radix higher than 2 and mixed radix in the tree levels leads to further improvements on the area requirement and latency compared to the radix-2 architectures.

When the number of inputs processed at a time is large, the tree-based sorter architectures in Figs. 8.11 and 8.12, as well as those according to the

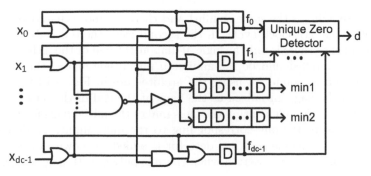

FIGURE 8.13
Bit-serial multi-input min1 and min2 sorter architecture (modified from [115])

higher or mixed-radix designs in [124], have very high complexity. In addition, to achieve high clock frequency, multiple stages of pipelining need to be applied. Pipelining not only incurs extra latency, but leads to area overhead caused by registers.

Bit-serial multi-input sorter architectures In the case that a large number of v2c messages need to be processed at a time, the min1 and min2 values can be found more efficiently using a bit-serial architecture shown in Fig. 8.13 [115]. Except the unique zero detector, this architecture is almost the same as the bit-serial CNU design in [125] that only finds the min1 value. This architecture takes in d_c v2c message magnitudes simultaneously, and each magnitude is input bit by bit, starting from the MSB. Each input is associated with a flag f_i. The flags are initialized as zero. At the beginning, if the bits in all inputs are '0', then the outputs of the three columns of OR, AND, and OR gates in Fig. 8.13 are all '0', and hence the flags remain at '0'. In the case that the bits in all input are '1', the output of the NAND gate in the middle is '0'. Through the columns of AND and OR gates, it makes the flags stay at zero. When there are both '0's and '1's at the inputs, the NAND gate outputs '1'. At this time, if an input bit is '0', the corresponding flag will still be '0'. However, the flags corresponding to those '1' input bits are flipped to '1', and they will be kept at '1' in the following clock cycles by the feedback loops around the registers. Since the magnitudes are sent in serially starting from the MSBs, the inputs with '1' are larger than those with '0' in the same bit position. Hence, a '1' flag means that the corresponding input magnitude is no longer a candidate of min1. In addition, the NOT gate output follows the input bits with zero flags if all of them are the same, and is '0' otherwise. Therefore, assuming that w bits are used to represent the v2c magnitudes, the min1 value is output by the NOT gate bit by bit in w clock cycles.

In [125], min2 was approximated as min1+1, and this approximation leads to noticeable performance loss. To compute min2, two cases need to be considered. In the first case, more than one input has the minimum magnitude.

In this case, min2 is the same as min1 and there are more than one '0' flag at the end of the wth clock cycle. In the second case, there is only one unique min1 and hence there is only a single '0' flag at the end of the wth clock cycle. In this case, min2 should be the minimum magnitude among the rest $d_c - 1$ inputs. It can be found by using the architecture in Fig. 8.13 for a second round with the flags initialized as the flags at the end of the wth clock cycle complemented. Complementing the flags makes the initial flags for the second round '0', except the one corresponding to the min1 input. To differentiate the two cases, a unique zero flag detector is added as shown in Fig. 8.13. Its output signal, which is asserted when there is only one '0' flag, is derived by

$$d = \sum_{i=1}^{d_c} f_i' \prod_{j \neq i} f_j.$$

The add and multiply in the above equation denote logic OR and AND, respectively. The derivation for min2 also takes w clock cycles if it is carried out. Hence, in the worst case, it takes $2w$ clock cycles to compute both min1 and min2 using the architecture in Fig 8.13.

The bit-serial min1 and min2 sorter in [126] has shorter latency than that in [115] at the cost of including a second set of flags and more complicated logic to assemble min1 and min2 values. The overall architecture is similar to that in Fig. 8.13 and is not drawn separately. Besides f_i, another set of flags, pf_i, are kept. pf_i equals the f_i in the previous clock cycle. f_i are still used to mask the inputs through the first column of OR gates in Fig. 8.13. The unique zero detector continuously monitor all f_i. If the number of f_i that are '0' is never reduced to one, then the first case as aforementioned happened. There are multiple equal minimum magnitudes, and min2 equals the min1 found at the end of the wth clock cycle. If the number of f_i equaling to '0' is reduced to one, which corresponds to the second aforementioned case, then a unique min1 is found in that clock cycle. Similarly, min2 should be the smallest among the rest magnitudes. Instead of re-starting the process, the flags are updated as $f_i \Leftarrow (f_i \oplus pf_i)'$. The updated flags mask out the input of min1 and those whose flags have been flipped to '1' previously. In addition, the unique zero detector is disabled after its output becomes asserted so that the flags will not get updated again when a unique min2 value is located. As a result, if the comparisons are continued for the rest clock cycles, min2 will be also available by the end of the wth clock cycle.

Example 44 *Consider that the sorter from [126] has 4 inputs, and they are '01001', '01101', '01010' and '01011'. Relevant values of the signals in the sorter are listed for each clock cycle in Table 8.5.*

In the bit-serial CNU architecture of [115], the values of the min1 and min2 are available at the output of the NOT gate in Fig. 8.13 from the two rounds of comparisons. Extra effort is needed to assemble the min1 and min2 values in the design of [126]. Min1 and min2 share the same bits up to the clock cycle

TABLE 8.5
Example signal values for the bit-serial sorter

clock cycle	1	2	3	4	5
x_0	0	1	0	0	1
x_1	0	1	1	0	1
x_2	0	1	0	1	0
x_3	0	1	0	1	1
f_0, pf_0	0,-	0,0	0,0	0,0→ 1,-	1,1
f_1, pf_1	0,-	0,0	1,0	1,1→ 1,-	1,1
f_2, pf_2	0,-	0,0	0,0	1,0→ 0,-	0,0
f_3, pf_3	0,-	0,0	0,0	1,0→ 0,-	1,0
NAND output	1	0	1	1	1
NOT output	0	1	0	0	0
d	0	0	0	1	-

prior to the one where the unique zero detector output, d, becomes asserted. These bits are available from the NOT gate output. The min1 should have '0' in the next bit position, and min2 should have '1' as the next bit. The remaining bits of min2 are derived from the NOT gate. However, the rest of the bits of min1 should be loaded from the corresponding input directly.

Example 45 *From Table 8.5, d becomes asserted in clock cycle 4. Hence, min1 and min2 are the same in the three MSBs, and they are '010' from the NOT gate output. The fourth bits for min1 and min2 are '0' and '1', respectively. The remaining bits for min2 are from the output of the NOT gate in the remaining clock cycle, which is '0' in the 5th clock cycle. However, the remaining bits for min1 are loaded from the input x_0, whose 5th bit is '1'. As a result, min1='01001' and min2='01010'.*

8.3.2.2 VNU architectures

VNU architecture for Min-sum decoder In decoding schemes other than the row-layered scheme, a v2c message is generated as the sum of the channel LLR and $d_v - 1$ c2v messages as shown in Algorithm 23. This would take $d_v - 1$ adders for each v2c message. In addition, the *a posteriori* message needs to be derived in order to make hard decisions. If all the v2c messages from a variable node need to be computed in parallel, the complexity of the VNU can be reduced by first computing the *a posteriori* message, and then subtracting individual c2v messages as shown in Fig. 8.14. This architecture requires $2d_v$ 2-input adders in total, and is simpler than that of computing each v2c message separately. The CNU architectures for the Min-sum algorithm work on the signs and magnitudes of the v2c messages. On the other hand, the additions

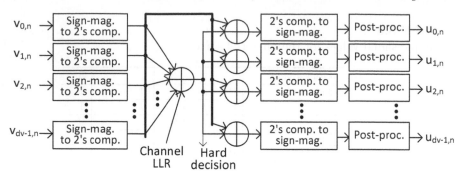

FIGURE 8.14
Parallel VNU architecture

required for variable processing are done more efficiently in 2's complement representation. Hence, converters between these two formats are added before and after the computations of the v2c messages.

In the offset Min-sum algorithm, the v2c messages need to be subtracted by a pre-determined value according to (8.7). If the difference is negative, zero is sent as the v2c message instead. In the normalized Min-sum algorithm, the v2c messages are multiplied by a scaler. The complexity of a scaler multiplication is determined by the number of nonzero bits in the binary representation of the scaler. Although the optimal scaler is derived from simulations, for the purpose of reducing the hardware complexity, it can usually be approximated by a value whose binary representation has fewer nonzero bits without any noticeable performance degradation. Furthermore, if a v2c message becomes larger than the maximum value that can be represented by a given word length, it needs to be saturated to the maximum value. The logics needed for the offset subtraction, scaler multiplication, and/or saturation are done in the post-processing blocks in Fig. 8.14.

For layered decoders, there are no explicit VNUs, and the v2c messages are updated through pairs of adders and subtractors. The post-processing can be done at the output of the adder on the left in Fig. 8.7, and the conversions between the sign magnitude and 2's complement representations are carried out at the inputs and outputs of the CNUs.

VNU architecture for FAID The FAID has the same check node processing as the Min-sum decoder, although the shorter word length required in the FAID makes the bit-serial CNUs more attractive for this decoder. However, the variable node processing in the FAID is different. Instead of adding up c2v messages, it is done by LUTs. Fig. 8.15 shows a VNU architecture for the FAID. Since the LUTs are small for limited alphabet size, they can be mapped to Boolean functions and efficiently implemented using combinational logic. The symmetry of the LUTs helps to simplify the logic expression of each

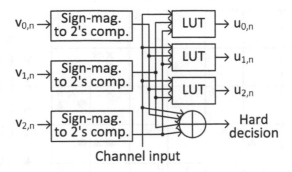

FIGURE 8.15
VNU architecture for FAIDs

output bit through making use of Karnaugh-map reduction. Besides the LUTs, there are other differences between the VNUs for the FAID and those for the Min-sum decoder. The VNUs in the FAID also compute *a posteriori* information, which is needed for making hard decisions, by adding up all c2v messages and channel information. Hence, the messages in sign-magnitude format from the CNUs do need to be converted to 2's complement representation for this purpose like those in the Min-sum VNUs. However, the alphabet levels in the LUTs for the FAID VNUs can be encoded to sign-magnitude representations. Accordingly, unlike the VNUs shown in Fig. 8.14 for the Min-sum decoder, no converter is included at the outputs of the VNUs for the FAID. In addition, any post-processing such as offset subtraction, scaler multiplication, and/or saturation has been already incorporated in the LUTs and their outputs are pre-defined limited-size alphabets. Hence post-processing is not necessary in the VNUs for the FAID. Due to these reasons, the VNUs of the FAID have a lower gate count and a shorter critical path compared to those for the Min-sum decoders.

Various LUTs correct different error patterns. When a FAID fails, FAIDs with other different LUTs can be tried one after another to improve the error-correcting performance. Since only the LUTs are different, the other parts of the decoder hardware remain unchanged to implement multiple FAIDs one after another. In addition, the LUTs do not need to be implemented separately. It can be seen from the example LUTs given in Table 8.3 that most of the entries in the seven LUTs are the same. Hence, the implementations of different LUTs share many logic gates. The entries that are different in the seven LUTs are highlighted in Fig. 8.16. Each square block in this figure corresponds to an entry in the LUTs. Using LUT_0 as the base LUT, if an LUT has an entry different from that in LUT_0, then a symbol is put into the corresponding block. Different types of symbols correspond to different LUTs in Fig. 8.16.

Although the symbols in Fig. 8.16 indicate the locations of different entries

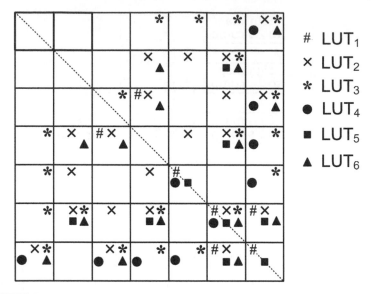

FIGURE 8.16
Differences among the FAID LUTs (modified from [115])

between $LUT_{1\sim6}$ and LUT_0, they also help to discover the common entries shared by $LUT_{1\sim6}$. For example, the same four symbols are found in the 7th blocks of row 1 and 3, as well as the first and third blocks in row 7. It means that LUT_2, LUT_3, LUT_4 and LUT_6 are different from LUT_0 in these entries. Moreover, from Table 8.3, these entries are the same in the four LUTs. Accordingly, logic is shared in the implementations of these entries for LUT_2, LUT_3, LUT_4 and LUT_6. Similarly, other sharable terms are found by identifying the other blocks in Fig. 8.16 that have common symbols and examining the corresponding entries in Table 8.3. Under the same timing constraint, synthesis results showed that implementing the seven LUTs requires less than twice the silicon area for implementing LUT_0 [115].

8.3.2.3   Message compression

Adders, comparators, and multiplexors are the major computation units in LDPC decoders. The longer the word length used to represent the messages, the more logic gates are in these computation units and the longer their critical paths. In addition, longer word length also leads to larger memory for storing the messages. On the other hand, the word length of the messages also affects the error-correcting performance of the decoder. Usually the Min-sum decoder needs 5-bit word length to achieve good performance.

The word length of the messages can not be further reduced when they are involved in additions or subtractions without bringing any performance loss.

However, the minimum and second minimum of the v2c magnitudes are what need to be found in the CNUs. Using less precision to represent those larger magnitudes in the CNUs may not hurt the min1 and min2 results. Taking this into consideration, nonuniform compression schemes can be applied to the messages in LDPC decoders [127]. At the inputs of the CNUs, the v2c messages are in compressed format with lower precision allocated to those larger magnitudes. Expansion is applied to the c2v messages to restore the original word length before any addition or subtraction is carried out. In the decoder architecture shown in Fig. 8.4, which is applicable to the flooding, sliced message-passing and shuffled decoding schemes, the compression and expansion can be added to the outputs and inputs of the VNUs, respectively. In this case, every block except the VNUs are simplified because of the shorter word length, and the memories for storing the messages are also smaller. Compression and expansion units are implemented by simple logic. As a result, the overall decoder complexity is reduced by employing nonuniform message compression schemes.

For layered decoders, whose architecture is depicted in Fig. 8.7, the compression and expansion units can be added at the input of the CNUs and on the min1 and min2 values, respectively. In this case, only the min1 and min2 sorters in the CNUs are simplified despite the overhead brought by the compression and expansion circuitry. Alternatively, expansion can be done after the c2v messages are generated from the min1 and min2 in order to take advantage of the shorter word length in the c2v message memory. However, two sets of the expansion units are needed for this case to restore the two c2v message inputs to the adder and subtractor in Fig. 8.7. Therefore, the complexity reduction achievable by employing message compression is less significant in layered decoders.

Table 8.6 shows an example of the compression and expansion of 4-bit message magnitudes in the Min-sum decoder [127]. Smaller magnitudes are allocated with higher precision, and larger magnitudes are represented with lower precision in this table. The optimal boundaries for breaking up the 4-bit magnitudes into groups to be represented by each possible 3-bit compressed magnitude can be determined from simulations.

8.4 Low-power LDPC decoder design

For power-constraint devices such as wireless handsets and sensor nodes, it is very important to design LDPC decoders with low power consumption. The dynamic power consumed during the decoding of a received word can be estimated as

$$P = C_{total} V_{dd}^2 f_{clk} T.$$

TABLE 8.6

4-bit magnitude nonuniform compression and expansion table

4-bit uniform magnitudes	3-bit non-uniformly compressed magnitudes	4-bit expanded magnitudes
0000	000	0000
0001	001	0001
0010	010	0010
0011	011	0011
0100		
0101	100	0101
0110		
0111	101	0111
1000		
1001		
1010		
1011	110	1011
1100		
1101		
1110		
1111	111	1111

C_{total} is the total switching capacitance of the circuit, V_{dd} is the power supply voltage, f_{clk} is the clock frequency, and T is the period of time that the circuit is active. Designs with smaller silicon area usually have smaller C_{total}, and hence consume less power. Therefore, the decoder architecture, number of CNUs, and number of VNUs are usually chosen so that the system throughput requirement is met with the smallest area.

Besides circuit and architectural level techniques, various algorithmic methods can be used to lower the power consumption of LDPC decoders through reducing C_{total} and/or T. A maximum iteration number, I_{max}, is set to limit the worst-case latency of the LDPC decoding. However, the decoder does not have to run until the last iteration. The entire decoder can be turned off using clock gating once the decoding converges. Further power saving is achieved by deactivating those units whose outputs are unlikely to change in the remaining iterations, even before the entire decoding converges. An example is that a variable node can be deactivated after its reliability becomes higher than a threshold [128]. However, such schemes often leads to performance degradation. It is also helpful to identify those decodings that will never converge and terminate them in earlier iterations [129].

Since the power consumption is proportional to V_{dd}^2, reducing V_{dd} leads to more effective power reduction compared to disabling or simplifying the circuit, which translates to reducing C_{total} or T. The propagation delay of a combinational circuit can be estimated as [21]

$$T_{pd} = \frac{C_{critical}V_{dd}}{k(V_{dd} - V_{th})^2}.$$ (8.13)

Here $C_{critical}$ is the capacitance in the critical path to be charged/discharged in each clock cycle. V_{th} is the threshold voltage of the transistors and k is a technology-dependent parameter. If the setup and hold times of the registers are ignored, the minimum achievable clock period, T_{clk}, is approximately equal to T_{pd}. Pipelining and parallel processing are techniques that may help reduce V_{dd} [21]. Pipelining leads to shorter critical path and hence smaller $C_{critical}$. Accordingly V_{dd} can be reduced while keeping T_{pd} the same. However, the computations in LDPC decoders are iterative. Adding more pipelining stages would increase the number of clock cycles needed for the decoding. If J-parallel processing is applied, T_{pd} can be increased by J times without lowering the overall throughput. If $C_{critical}$ is increased by less than J times, which happens unless there are feedback loops in the architecture and $C_{critical}$ equals the iteration bound, then lower V_{dd} can be adopted in the parallel architecture. Nevertheless, parallel processing needs duplicated copies of the hardware and significantly increases the decoder area.

The LDPC decoder receives codewords at fixed time intervals in many communication systems. In these systems, the worst-case latency, T_{max}, is allocated for the decoding to run I_{max} iterations for each word. However, the decoding of most words takes fewer iterations. In this case, the clock frequency can be relaxed so that V_{dd} is reduced while the decoding is still finished in T_{max}. This voltage scaling technique for LDPC decoder power reduction was investigated in [130]. To make the implementation of the voltage scaling easier, only a few scaled voltages are made available. Denote these voltages by $\{V_i\}$, $i = 0, 1, \ldots, d$, where V_0 equals the original V_{dd}, and $V_i > V_{i+1}$. For each voltage, the clock signal with the corresponding frequency is also available. Assume that the decoding of a received word needs $I \leq I_{max}$ iterations, and each iteration requires n clock cycles. In addition, the voltage supply and clock frequency remain unchanged during each decoding iteration. Let the power supply voltage and clock period adopted in iteration k be $V(k)$ and $T_{clk}(k)$, respectively. The decoding with scaled voltages is subject to the restriction that $\sum_{k=1}^{I} nT_{clk}(k) \leq T_{max}$, and the total energy consumption is

$$P = nC_{total} \sum_{k=1}^{I} V^2(k).$$

Alternatively, the energy consumption would be $nC_{total}IV_{dd}^2$ if V_{dd} is used for the entire decoding process and the decoder is shut down after convergence. Since $V(k) \leq V_{dd}$, adopting lower supply voltage and making full use of the available time for decoding leads to significant power reduction. The saving is more significant when I is smaller, which leads to lower $V(k)$. It only takes a few clock cycles and negligible power to switch from one supply voltage to

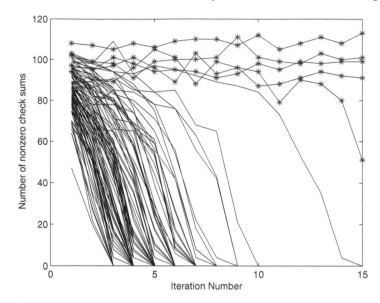

FIGURE 8.17
Number of nonzero check sums in each decoding iteration for an example LDPC code

another. Since each LDPC decoding iteration typically takes dozens or even hundreds of clock cycles, the latency penalty resulting from switching between supply voltages is very small.

Now the question is how to tell the number of iterations needed before the decoding starts in order to decide which scaled voltage supply to use. It turns out that the number of unsatisfied check equations, or the number of nonzero check sums (NNCS), can be used to predict the number of remaining iterations in LDPC decoding [130]. According to the predictions, the minimum scaled voltage supply and hence clock frequency are chosen so that the decoding is still finished in time T_{max}.

Fig. 8.17 shows the NNCS in each iteration of the Min-sum decoding for an example QC-LDPC code under the additive white Gaussian noise (AWGN) channel with $I_{max} = 15$. When all the check sums become zero, the decoding converges. If the SNR is higher, there will be fewer undecodable cases. From Fig. 8.17, the NNCS fluctuates and never goes down to zero for those undecodable cases. This observation was also exploited to shut off the decoder in early iterations for the undecodable cases in [129]. If the received word is decodable, the NNCS may oscillate in the beginning, but will then follow piecewise constant slopes to descend to zero. To take out the effect of oscillation, a large threshold, D, is applied to the drop of NNCS in each iteration.

Only when the decrease of the NNCS is more than D for two consecutive iterations, the slope of the NNCS is computed and used to predict the number of remaining decoding iterations. The slope may change. Hence, it is recalculated in each iteration to adjust the prediction. The predictions may be too optimistic, according to which the voltages would be over-scaled. As a result, the error-correcting performance is degraded because previously decodable words may become undecodable in T_{max}. In order to prevent this from happening, a factor $\beta > 1$ is multiplied to the number of remaining iterations predicted from the NNCS to allow some margin. The predictions are more conservative and the achievable power reduction is less if larger β is used. On the other hand, β that is too small causes performance loss. The optimal β can be found from simulations.

Given the predicted number of remaining iterations, the following scheme is used to decide which V_i the power supply voltage should be scaled to. The T_{pd} corresponding to each V_i, which is denoted by $T_{pd,i}$, is pre-computed using (8.13). Assume that the clock period is kept the same during each decoding iteration, and it is predicted in iteration k that there are $I(k)$ iterations left. Then the longest clock period should not exceed $(T_{max} - n\sum_{j=1}^{k} T_{clk}(j))/(nI(k))$ and the corresponding V_i is chosen as the power supply voltage to be used in the next decoding iteration. Since this prediction is made and adjusted during the decoding, it is called an *in situ* prediction. Let the NNCS in iteration k be $s(k)$, and $\Delta s(k) = s(k) - s(k-1)$. Denote the hard-decision vector by z. Algorithm 26 summarizes the *in situ* prediction and voltage scaling procedure. In this algorithm, a larger D means a more stringent condition to make predictions, and accordingly the achievable power reduction is less. On the other hand, if D is too small, predictions may be made when the NNCS is still oscillating. The value of D is optimized from simulations.

Algorithm 26 In Situ Prediction
initialization: $V(1) = V_0$
for $k = 1 : I_{max}$
    Compute zH^T; *Stop if* $zH^T = 0$
1: *if* $\Delta s(k) > D$ *&* $\Delta s(k-1) > D$
2:    $I(k) = \beta s(k)/((\Delta s(k) + \Delta s(k-1))/2)$
3:    $V(k) = \min\{V_i | T_{pd,i} \leq (T_{max} - n\sum_{j=1}^{k} T_{clk}(j))/(nI(k))\}$
    else
     $V(k) = V(1)$

As shown in Fig. 8.17, the slopes of the NNCS can be steep. Also, predictions are only made when there are two consecutive iterations with a large drop in the NNCS. As a result, when the prediction is made, only a few iterations may be left in the decoding process. The achievable power reduction is limited if the supply voltage is only reduced in the last few iterations.

To further reduce the power consumption, the supply voltage needs to be scaled down from the beginning of the decoding. From Fig. 8.17, it can be also observed that $s(0)$, the NNCS at the beginning of the decoding, is usually smaller for the decoding that converges in fewer iterations. Taking this into consideration, an initial prediction is made at the beginning of the decoding by utilizing $s(0)$. A set of $I_{max}-1$ threshold values sth_i $(i = 1, 2, \ldots, I_{max}-1)$ are decided such that $s(0) \geq sth_i$ for all decoding converges after iteration i. For a given LDPC code, decoding algorithm and channel model, this set can be pre-determined from simulations. At the beginning of the decoding, $s(0)$ is compared with these thresholds. If $s(0)$ is in the range of $[sth_i, sth_{i+1})$, then the preliminary prediction is that the decoding will take $i + 1$ iterations. Accordingly, the scaled voltage is chosen using a method similar to that for the *in situ* prediction. However, to make such an initial prediction, at least $\lceil \log_2(I_{max} - 1) \rceil$ comparisons are needed.

Alternatively, the scaled voltage can be decided directly from $s(0)$ instead of through the initially predicted iteration number. The maximum number of iterations that can be finished in time T_{max} under supply voltage V_i $(i = 1, 2, \ldots, d-1)$, denoted by I_i, is first calculated according to (8.13). Then from simulation results, $d-1$ thresholds $sVth_i$ are determined so that $s(0) \geq sVth_i$ for all decoding that converges after iteration I_i. Since $V_i > V_{i+1}$, $sVth_i > sVth_{i+1}$. Then V_i is selected if $sVth_i > s(0) > sVth_{i+1}$. Because only a few scaled voltages are used, deciding the initial scaled voltage from $s(0)$ this way requires a very small number of comparisons.

The initial prediction can be combined with the *in situ* prediction. The combined prediction scheme is implemented by replacing the V_0 in the initialization of Algorithm 26 with the scaled voltage chosen according to the initial prediction. Since the thresholds used for the initial prediction are pre-computed, implementing this prediction only needs registers to store the $d - 1$ thresholds, $d - 1$ comparators, and simple combinational logic to decide the initial voltage based on the comparison results. $d = 3$ or 4 scaled voltages can be usually adopted. Hence, the implementation of the initial prediction causes very small overhead. For the *in situ* prediction, the major complexity lies in step 3 of Algorithm 26. Since $T_{max} = nI_{max}T_{pd,0}$, steps 2 and 3 can be combined as finding the largest $T_{pd,i}$ so that $T_{pd,i}\beta s(k) < (I_{max}T_{pd,0} - \sum_{j=1}^{k} T_{clk}(j))(\Delta s(k) + \Delta s(k-1))/2$. $T_{clk}(j)$ can be accumulated iteratively in each iteration to compute $\sum_{j=1}^{k} T_{clk}(j)$. Hence a few subtractions, multiplications, and comparisons are all that are needed to find the largest $T_{pd,i}$. These hardware units only account for a very small proportion of the overall LDPC decoder area.

If β and D are chosen properly, these iteration number prediction schemes do not cause any noticeable error-correcting performance loss. As mentioned previously, if the NNCS follows steep slopes to go down to zero, the power reduction achievable by the *in situ* prediction alone is limited, since only a few decoding iterations are left when the prediction is made. The slopes do not change much with the SNR. However, they may be dependent on the code construction and decoding algorithm. Moreover, the average NNCS is smaller when the SNR is higher. This would lead to even fewer remaining iterations when the *in situ* prediction is made. On the contrary, the power reduction contributed by the initial prediction accounts for a larger proportion when the SNR is higher. This is because more decoding converges fast at higher SNR, and hence the voltage can be scaled down from the beginning more often using the initial prediction.

The achievable power reduction and error-correcting performance are also affected by the set of scaled voltages employed. V_0 equals V_{dd}, and the lowest scaled voltage should be reasonably higher than the threshold voltage. Moreover, the inverse square of the supply voltage appears at the formula of the propagation delay in (8.13). Therefore, in order to make the difference between each pair of $T_{pd,i}$ and $T_{pd,i+1}$ similar, $V_i - V_{i+1}$ should be smaller than $V_{i+1} - V_{i+2}$. Lower scaled voltages lead to more power saving. However, if they are too low, some decodable cases may become undecodable. The optimal scaled supply voltages and the value of β need to be jointly decided from simulations.

9

Non-binary LDPC decoder architectures

CONTENTS

Although binary low-density parity-check (LDPC) codes approach the channel capacity theoretically, they need to have very long codeword length in order to do so. Longer codeword length translates to larger buffer for holding the messages, larger silicon area to implement the encoders and decoders, and/or longer encoding and decoding latency. For moderate codeword length, non-binary (NB)-LDPC codes constructed over $GF(q)$ ($q > 2$) [131] have better error-correcting performance. In addition, they are less prone to the error-floor problem and are good at correcting bursts of errors. It was also shown in [132] that a type of regular but random-like NB-LDPC codes, called DaVinci codes, outperform the duo-binary Turbo codes by 0.5dB.

Despite the error-correcting performance advantages, NB-LDPC codes suffer from high decoding complexity. For a code over $GF(q)$, each received symbol can be any of the q elements in $GF(q)$. Hence, vectors of q messages representing the probability that the received symbol equals each element need to be computed, stored, and passed between the check and variable nodes during the decoding. Since the number of sequences of symbols that satisfy a non-binary check equation is much larger than that for a binary check equation, the check node processing in NB-LDPC decoding is much more complicated. Therefore, the two major bottlenecks of NB-LDPC decoding are the complex check node processing and large memory requirement.

The overall decoding schemes for binary LDPC codes, such as layered decoding and sliced message-passing, can be applied to NB-LDPC codes. Moreover, low-power design techniques developed for binary LDPC decoders, including power supply voltage scaling and turning off the hardware units when the decoding converges, can be directly applied to NB-LDPC decoders. However, the check node units (CNUs) in NB-LDPC decoders are fundamentally different from those in binary decoders because of the non-binary check equations. Many decoder architecture designs have been focusing on simplifying the CNUs, whose implementation also greatly affects the messages need to be stored and hence the memory requirement of the decoder. This chapter reviews NB-LDPC decoder architectures, with emphasis given to algorithmic and architectural modification techniques for the CNUs.

9.1 Non-binary LDPC codes and decoding algorithms

In the parity check matrix, H, of a NB-LDPC code over $GF(q)$, the nonzero entries are elements of $GF(q)$. Also each information and codeword symbol is an element of $GF(q)$. For communication and data storage systems, $GF(2^p)$ ($p \in Z^+$) is usually adopted. In this case, each symbol can be represented by a p-bit binary tuple. Quasi-cyclic (QC) codes enable more efficient hardware implementations compared to other NB-LDPC codes. The H matrix of a QCNB-LDPC code constructed by the methods in [133] consists of submatrices that are either zero or α-multiplied cyclic permutation matrixes (CPMs). Let α be a primitive element of $GF(q)$. An α-multiplied CPM is a cyclic permutation matrix whose nonzero entries are replaced by the elements of $GF(q)$, and the nonzero entry in a row equals that in the previous row multiplied by α.

The belief propagation (BP) for binary LDPC decoding [104] can be extended to decode NB-LDPC codes. In non-binary BP, the check node processing also computes sums of products of probabilities. The computations in non-binary check node processing can be considered as convolutions of the incoming messages [134]. In frequency domain, convolutions are converted to term-by-term multiplications. Nevertheless, multiplications are still expensive to implement in hardware. Compared to probability-domain decoders, decoders in log domain are less sensitive to quantization noise. More importantly, multiplications in probability domain are implemented as additions in log domain. A log-domain NB-LDPC decoding algorithm was proposed in [135]. However, it is difficult to compute the Fourier transform in log domain. A mixed-domain decoder [136] carries out the Fourier transform in probability domain and performs check node processing in log domain to take advantage of both domains. The drawback of this design is that conversions are needed between the two domains, and large look-up tables (LUTs) are required for this purpose. To reduce the complexity of NB-LDPC decoding, approximations of the BP have been developed. The most prominent ones are the extended Min-sum algorithm (EMS) [137] and the Min-max algorithm [138]. Both of them are log-domain algorithms. Although the approximations lead to slight performance loss compared to the BP, the hardware complexity has been greatly reduced in these algorithms, and only additions on the reliability messages are involved in their check node processing. More details on these NB-LDPC decoding algorithms are given below.

9.1.1 Belief propagation decoding algorithms

For a NB-LDPC code over $GF(q)$, given the observation of the nth received symbol, y_n, the nth transmitted symbols, x_n, can be any of the q elements in $GF(q)$. Therefore, the messages passed between the check and variables nodes in the decoding process are vectors of q messages. Let α be a primitive element of $GF(q)$. Then all the elements of $GF(q)$ are expressed as $0, 1, \alpha, \alpha^2, \ldots, \alpha^{q-2}$. Denote the probability information from the channel for the nth received symbol by $\gamma_n = [\gamma_n(0), \gamma_n(1), \gamma_n(\alpha), \ldots, \gamma_n(\alpha^{q-2})]$, where $\gamma_n(\beta) = P(x_n = \beta|y_n)$ for $\beta = 0, 1, \alpha, \alpha^2, \ldots, \alpha^{q-2}$. In this chapter, (β) is added to the notation of a vector to denote the message corresponding to finite field element β in that vector when necessary. Let the message vector from variable node n to check node m be $u_{m,n}$, and that from check node m to variable node n by $v_{m,n}$. Also let $S_c(n)$ $(S_v(m))$ be the set of check (variable) nodes connected to variable (check) node n (m). Let $\mathcal{L}(m|a_n = \beta)$ be the set of sequences of finite field elements (a_j) $(j \in S_v(m) \backslash n)$ such that $\sum_{j \in S_v(m) \backslash n} h_{m,j} a_j = h_{m,n} \beta$. Here $h_{i,j}$

is the entry of the H matrix in the *ith* row and *jth* column. Such a set is also called a configuration set.

Algorithm 27 Belief Propagation for NB-LDPC Decoding
input: γ_n
initialization: $u_{m,n}(\beta) = \gamma_n(\beta)$; $z_n = \arg\max_\beta(\gamma_n(\beta))$
for $k = 1 : I_{max}$
    Compute zH^T; *Stop if* $zH^T = 0$
    check node processing
    for each check node m and each $n \in S_v(m)$

$$v_{m,n}(\beta) = \sum_{(a_j) \in \mathcal{L}(m|a_n=\beta)} \prod_{j \in S_v(m) \backslash n} u_{m,j}(a_j) \qquad (9.1)$$

    variable node processing
    for each variable node node n and each $m \in S_c(n)$

$$u_{m,n}(\beta) = \gamma_n(\beta) \prod_{i \in S_c(n) \backslash m} v_{i,n}(\beta)$$

    a posteriori information computation & tentative decision
    for each variable node node n

$$\tilde{\gamma}_n(\beta) = \gamma_n(\beta) \prod_{i \in S_c(n)} v_{i,n}(\beta)$$

$$z_n = \arg\max_\beta(\tilde{\gamma}_n(\beta))$$

Assuming that z_n is the hard-decision of the *nth* received symbol, the probability-domain BP for NB-LDPC decoding can be carried out according to Algorithm 27. In this algorithm, each variable-to-check (v2c) message $u_{m,n}(\beta)$ and *a posteriori* message $\tilde{\gamma}_n(\beta)$ are computed as multiplying up the check-to-variable (c2v) messages of the same finite field element. Hence, these two steps are direct extensions of those in the BP for binary LDPC decoding, and the complexity is q times for a code over $GF(q)$.

The check node processing in non-binary BP still computes sums of products. However, when the check equations are non-binary, the cardinality of the configuration set is much larger, and hence more products need to be added up. For example, consider a binary check equation $x_1 + x_3 + x_4 = 0$. To derive the probability that this check equation is satisfied given that $x_1 = 1$, there are only two cases need to be considered for binary LDPC codes: $x_3 = 1, x_4 = 0$ and $x_3 = 0, x_4 = 1$. On the other hand, for a code over $GF(q)$, there are q pairs of values for x_3 and x_4 that satisfy this check equation given that $x_1 = 1$. The cardinality of a configuration set is actually $\mathcal{O}(q^{d_c-2})$ for a code over $GF(q)$ with row weight d_c. As a result, substantially more products need to be summed up to compute each c2v message and the check node processing in NB-LDPC decoding is much more complicated .

It is difficult to enumerate or record all the sequences in each configuration set. Instead, we can consider all possible pairs of finite field elements from two message vectors at a time, and iteratively find out which β the sum $\sum_{j \in S_v(m) \backslash n} h_{m,j} a_j$ corresponds to. Based on this idea, a forward-backward approach [131] was proposed to simplify the check node processing. For a code of row weight d_c, the forward-backward check node processing for the BP in Algorithm 27 is carried out according to Algorithm 28. To make the notations simpler, the variable nodes connected to check node m are indexed by $0, 1, \ldots, d_c - 1$.

Algorithm 28 Forward-backward Check Node Processing for Probability-domain BP

input: $u_{m,n}$ $(n = 0, 1, \ldots, d_c - 1)$

initialization: $f_0 = u_{m,0}$; $b_{d_c-1} = u_{m,d_c-1}$

forward process

 for $j = 1$ to $d_c - 2$

$$f_j(\beta) = \sum_{h_{m,j}\beta = h_{m,j-1}\beta_1 + h_{m,j}\beta_2} (f_{j-1}(\beta_1) \times u_{m,j}(\beta_2))$$

backward process

 for $j = d_c - 2$ down to 1

$$b_j(\beta) = \sum_{h_{m,j}\beta = h_{m,j+1}\beta_1 + h_{m,j}\beta_2} (b_{j+1}(\beta_1) \times u_{m,j}(\beta_2))$$

merging process

$$v_{m,0}(\beta) = b_1(\beta)$$
$$v_{m,d_c-1}(\beta) = f_{d_c-2}(\beta)$$
 for each $0 < j < d_c - 1$
$$v_{m,j}(\beta) = \sum_{h_{m,j}\beta = h_{m,j-1}\beta_1 + h_{m,j+1}\beta_2} (f_{j-1}(\beta_1), b_{j+1}(\beta_2))$$

It should be noted that the backward and forward processes can be switched, since they are independent. When hardware resources and duplicated memories are available, they can be also carried out in parallel. The computations involved in all three processes are identical. This enables implementing the check node processing with very regular architectures. However,

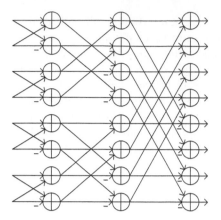

FIGURE 9.1
Butterfly architecture for the Fourier transform of message vectors over $GF(2^3)$

the intermediate results from the forward and backward processes need to be stored and require memory.

9.1.1.1   Frequency-domain decoding algorithm

The sum-of-product computations in the check node processing in (9.1) is equivalent to a convolution. This convolution can be implemented more efficiently as term-by-term multiplications in frequency domain. Denote the Fourier and inverse Fourier transform over a vector by $\mathcal{F}$ and $\mathcal{F}^{-1}$, respectively. The check node processing in (9.1) can be done instead as

$$v_{m,n} = \mathcal{F}^{-1} \left(\dot{\prod}_{j \in S_v(m) \backslash n} \mathcal{F}(u_{m,j}) \right),$$

where $\dot{\prod}$ represents term-by-term multiplications.

When the messages are probability mass functions defined over $GF(2^p)$, the Fourier transform of a message vector is a p-dimension two-point Fourier transform. An element of $GF(2^p)$ can be represented by a p-bit binary tuple $b_0 b_1 \ldots b_{p-1}$. Let the inputs to the lth-dimension Fourier transform be $I_{[b_0 b_1 \ldots b_{p-1}]}$. Then the outputs of the Fourier transform of this dimension are generated as

$$O_{[b_0 \ldots b_{l-1} 0 \, b_{l+1} \ldots b_{p-1}]} = 1/\sqrt{2} \left(I_{[b_0 \ldots b_{l-1} 0 \, b_{l+1} \ldots b_{p-1}]} + I_{[b_0 \ldots b_{l-1} 1 \, b_{l+1} \ldots b_{p-1}]} \right)$$

$$O_{[b_0 \ldots b_{l-1} 1 \, b_{l+1} \ldots b_{p-1}]} = 1/\sqrt{2} \left(I_{[b_0 \ldots b_{l-1} 0 \, b_{l+1} \ldots b_{p-1}]} - I_{[b_0 \ldots b_{l-1} 1 \, b_{l+1} \ldots b_{p-1}]} \right)$$

Such a Fourier transform is implemented by a butterfly structure similar to that for a fast Fourier transform over vectors of real numbers. Fig. 9.1 shows an example architecture that implements the Fourier transform of a message vector over $GF(2^3)$. For clarity, the multiplications with the scaler $1/\sqrt{2}$ are not shown. The inverse Fourier transform can be implemented in a similar way.

9.1.1.2 Log-domain decoding algorithm

The number of multiplications in the check node processing is reduced in the frequency-domain approach. However, as in the original BP, probabilities are used as messages. Probabilities may require a longer word length to represent, and the quantization scheme needs to be carefully designed to prevent performance loss. These issues are addressed by log-domain decoders. They are less sensitive to quantization noise, and shorter word length is possible. More importantly, multiplications are converted to additions in log domain, and adders take much less hardware to implement than multipliers.

A log-domain BP was developed in [135]. In this decoder, the messages are represented as log-likelihood ratios (LLRs). Let $L(\beta) = \ln(P(\beta)/P(0))$. Then the multiplications in the variable node processing and *a posteriori* information computation steps of the probability-domain BP are converted to additions of LLRs. Moreover, the backward-forward scheme for the probability-domain check node processing can also be applied to log-domain check node processing [135].

In log domain, each elementary step in the forward-backward approach is to compute an LLR vector $L_o(\beta)$ from two LLR vectors $L_1(\beta_1)$ and $L_2(\beta_2)$ subject to the constraint that $h_{m,o}\beta = h_{m,1}\beta_1 + h_{m,2}\beta_2$. It has been derived in [135] that

$$
\begin{aligned}
&L_o(h_{m,1}\beta_1 + h_{m,2}\beta_2 = \delta) \\
&= \ln\left(\frac{P(h_{m,1}\beta_1 + h_{m,2}\beta_2 = \delta)}{P(h_{m,1}\beta_1 + h_{m,2}\beta_2 = 0)}\right) \\
&= \ln\left(\frac{\sum_{\omega \in GF(q)} P(\beta_1 = \omega)P(\beta_2 = h_{m,2}^{-1}(\delta + h_{m,1}\omega))}{\sum_{\omega \in GF(q)} P(\beta_1 = \omega)P(\beta_2 = h_{m,2}^{-1}h_{m,1}\omega)}\right) \\
&= \ln\left(\frac{\sum_{\omega \in GF(q)} \frac{P(\beta_1=\omega)P(\beta_2=h_{m,2}^{-1}(\delta+h_{m,1}\omega))}{P(\beta_1=0)P(\beta_2=0)}}{1 + \sum_{\omega \in GF(q)\backslash 0} \frac{P(\beta_1=\omega)P(\beta_2=h_{m,2}^{-1}h_{m,1}\omega)}{P(\beta_1=0)P(\beta_2=0)}}\right)
\end{aligned}
$$

Accordingly, the output LLR vector, L_o, is rewritten in terms of the input

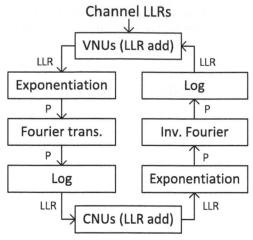

FIGURE 9.2

Block diagram of mixed-domain NB-LDPC decoder

vectors, L_1 and L_2, as

$$L_o(h_{m,1}\beta_1 + h_{m,2}\beta_2 = \delta)$$

$$= \ln\left(e^{L_1(h_{m,1}^{-1}\delta)} + e^{L_2(h_{m,2}^{-1}\delta)} + \sum_{\omega \in GF(q)\setminus\{0, h_{m,1}^{-1}\delta\}} (e^{L_1(\omega)} + e^{L_2(h_{m,2}^{-1}(\delta + h_{m,1}\omega))})\right)$$

$$- \ln\left(1 + \sum_{\omega \in GF(q)\setminus 0} e^{L_1(\omega) + L_2(h_{m,2}^{-1}h_{m,1}\omega)}\right)$$

From the above equation, the computations needed to derive L_o can be broken down as iterative calculations of the Jacobian logarithm

$$\overset{*}{\max}(x_1, x_2) \doteq \ln(e^{x_1} + e^{x_2}) = \max(x_1, x_2) + \ln(1 + e^{-|x_1 - x_2|}).$$

Besides one comparison, one subtraction and one addition, the max* operation needs a LUT to implement the $\ln(\cdot)$ function. In addition, $2(q - 1)$ max* computations are required to derive one message in L_o. Hence, the check node processing in log domain still has very high complexity.

9.1.1.3   Mixed-domain decoding algorithm

The check node processing complexity can be further reduced if the frequency and log-domain approaches are combined. When the Fourier transform is applied, the check node processing is done as term-by-term multiplications. If logarithm is applied to the Fourier-transformed v2c messages, then only additions are required in the check node processing. However, since the Fourier transform involves additions, carrying out the Fourier transform in log domain would need the Jacobian logarithm too. This issue is solved in a mixed-domain decoder [136], whose block diagram is shown in Fig. 9.2.

In Fig. 9.2, the 'LLR's and 'P's denote that the corresponding signals are LLRs and probabilities, respectively. Represent the channel information by LLRs. The variable node units (VNUs) add channel LLRs and c2v LLRs, and generate v2c messages in LLRs. Then the v2c LLRs are converted to probabilities by taking exponentiations, and the Fourier transform is carried out in probability domain. After the Fourier transform, logarithm is taken to represent the transformed v2c messages as LLRs, over which additions are done for the check node processing. The outputs of the CNUs are LLRs. They are converted to probabilities in order to have the inverse Fourier transform done to derive the c2v messages. Then, logarithm is taken to convert those c2v probabilities into LLRs before they are used in the variable node processing for the next decoding iteration.

Both the variable node and check node processings are implemented as simple additions in the mixed-domain decoder. However, LUTs are need to convert back and forth between probabilities and LLRs. Partial-parallel or fully-parallel LDPC decoder designs are necessary to achieve practical speed. A great number of LUTs are needed in these designs, and would cause large area overhead to the decoder.

9.1.2 Extended Min-sum and Min-max algorithms

It is apparent that the NB-LDPC decoders in probability domain have much higher complexity than frequency-domain and log-domain decoders. In addition, log-domain decoders win over frequency-domain decoders in that they are less sensitive to quantization noise, and adders are much easier to implement than multipliers. The only drawback of log-domain decoders is that their check node processing needs to compute Jacobian logarithm. This bottleneck is eliminated by adopting approximations in the check node processing. Examples are the EMS algorithm [137] and the Min-max algorithm [138]. Both of them are log-domain algorithms, and have substantially lower complexity than the BP with only small degradation in the error-correcting performance.

In the EMS and Min-max algorithms, the LLRs are defined with respect to the most likely finite field element for each received symbol, $\ln(P(x_n = \hat{\beta}_n)/P(x_n = \beta))$, where $\hat{\beta}_n$ is the most likely element for the nth symbol. Using this definition, all LLRs are non-negative, and the LLR for the most likely field element is always zero in each vector. Furthermore, the smaller an LLR, the more likely that the received symbol equals the corresponding field element. The pseudo codes of the EMS algorithm are shown in Algorithm 29.

Algorithm 29 Extended Min-sum Decoding Algorithm
input: γ_n
initialization: $u_{m,n}(\beta) = \gamma_n(\beta)$; $z_n = \arg\min_\beta(\gamma_n(\beta))$
for $k = 1 : I_{max}$
      Compute zH^T; Stop if $zH^T = 0$
      check node processing
      for each check node m and each $n \in S_v(m)$

$$v_{m,n}(\beta) = \min_{(a_j) \in \mathcal{L}(m|a_n=\beta)} \sum_{j \in S_v(m) \backslash n} u_{m,j}(a_j) \qquad (9.2)$$

      variable node processing
      for each variable node n and each $m \in S_c(n)$

$$u'_{m,n}(\beta) = \gamma_n(\beta) + \sum_{i \in S_c(n) \backslash m} v_{i,n}(\beta)$$

$$u_{m,n}(\beta) = u'_{m,n}(\beta) - \min_{\omega \in GF(q)} (u'_{m,n}(\omega))$$

      a posteriori information computation & tentative decision
      for each variable node n

$$\tilde{\gamma}_n(\beta) = \gamma_n(\beta) + \sum_{i \in S_c(n)} v_{i,n}(\beta)$$

$$z_n = \arg\min_\beta(\tilde{\gamma}_n(\beta))$$

     In the EMS algorithm, the normalization in the variable node processing is required to force the smallest LLR in each vector back to zero. The approximations adopted to derive the check node processing in (9.2) actually lead to underestimation of the LLRs. To compensate for the underestimation, offsetting or scaling can be applied to the sums of the c2v messages before they are added up with the channel LLRs to compute the v2c messages and *a posteriori* information. This is analogous to the offset and normalized Min-sum decoding for binary LDPC codes. With proper scaling or offsetting, the EMS algorithm only leads to less than 0.1dB coding gain loss compared to the original BP.

     In the Min-max algorithm [138], the computations in each step are the same as those in Algorithm 29, except that the 'sum' computation in the check node processing is replaced by 'max' comparisons. A 'max' comparator requires fewer logic gates to implement than an adder. Moreover, the 'max' of multiple messages equals one of the messages. This enables additional simplifications to be made compared to the min-sum computations. As a result, the

Mix-max decoder has even lower hardware complexity than the EMS decoder. Simulation results showed that the achievable coding gain of the Min-max algorithm is slightly less than that of the EMS algorithm.

9.1.3   Iterative reliability-based majority-logic decoding

Iterative reliability-based majority-logic decoding (MLGD) algorithms are also available for NB-LDPC codes [139]. Analogous to those for binary LDPC codes, these algorithms have lower complexity but inferior error-correcting capability compared to the EMS and Min-max algorithms. The basic ideas in the MLGD algorithms for binary and non-binary codes are similar, and the non-binary algorithms also perform well only when the column weight of the code is not small.

The superscript (k) is added to the messages when needed to denote the values in iteration k. Define LLRs with respect to the zero field element as $\ln(P(\beta)/P(0))$ Let γ_j be the LLR vector computed from the channel information for the jth received symbol. An iterative soft-reliability based (ISRB)-MLGD algorithm for NB-LDPC codes is shown in Algorithm 30 [139]. In this algorithm, λ is a scaling factor used for optimizing the performance. Since $\phi_{i,j}$ does not include any contribution from variable node j, it is an extrinsic reliability measure from the channel for the jth symbol. $\psi_j(\beta)$ is the extrinsic reliability that the jth received symbol equals β accumulated in a decoding iteration. ψ_j is analogous to the E_j in Algorithm 25 in Chapter 8.

Algorithm 30 ISRB-MLGD Algorithm for NB-LDPC Codes
input: γ_j
initialization: $z_j^{(0)} = \arg\max_\beta(\gamma_j(\beta))$
$\qquad\qquad R_j^{(0)}(\beta) = \lambda\gamma_j(\beta)$
$\qquad\qquad \phi_{i,j} = \min\limits_{j' \in S_v(i)\backslash j} \max\limits_{\beta} \gamma_{j'}(\beta)$
for $k = 1 : I_{max}$
$\qquad$ *stop if $z^{(k-1)}H^T = 0$*
$\qquad$ *for each variable node j*
$\qquad\qquad$ *for each $i \in S_c(j)$*
$\qquad\qquad\qquad \sigma_{i,j} = h_{i,j}^{-1} \sum_{u \in S_v(i)\backslash j} z_u^{(k-1)} h_{i,u}$
$\qquad\qquad \psi_j(\beta) = \sum_{\sigma_{i,j}=\beta, i \in S_c(j)} \phi_{i,j}$
$\qquad\qquad R_j^{(k)}(\beta) = R_j^{(k-1)}(\beta) + \psi_j(\beta)$
$\qquad\qquad z_j^{(k)} = \arg\max_\beta(R_j^{(k)}(\beta))$

Algorithm 31 IHRB-MLGD Algorithm for NB-LDPC Codes
input: z_j
initialization: $z_j^{(0)} = z_j$

$$R_j^{(0)}(\beta) = \gamma \text{ if } \beta = z_j^{(0)}$$
$$R_j^{(0)}(\beta) = 0 \text{ if } \beta \neq z_j^{(0)}$$

for $k = 1 : I_{max}$
     stop if $z^{(k-1)} H^T = 0$
     for each variable node j
         for each $i \in S_c(j)$

$$\sigma_{i,j} = h_{i,j}^{-1} \sum_{u \in S_v(i) \backslash j} z_u^{(k-1)} h_{i,u}$$
$$\xi_j(\beta) = |\{\sigma_{i,j} | \sigma_{i,j} = \beta, i \in S_c(j)\}|$$
$$R_j^{(k)}(\beta) = R_j^{(k-1)}(\beta) + \xi_j(\beta)$$
$$z_j^{(k)} = \arg\max_\beta R_j^{(k)}(\beta)$$

When only hard decisions, z_j, are available from the channel, an iterative hard-reliability based (IHRB)-MLGD algorithm [139] can be used for NB-LDPC decoding. This algorithm is listed in Algorithm 31. Compared to the ISRB algorithm, the major differences in the IHRB algorithm lie in the initial messages and the accumulated extrinsic reliabilities. In the IHRB algorithm, the LLR in a vector for the most likely finite field element is initialized as a pre-set positive integer γ, and all the other LLRs in the same vector are set to zero. If $\phi_{i,j}$ as in the ISRB algorithm were calculated for the IHRB algorithm, all of them would be the same and equal to γ. Therefore, the accumulated extrinsic reliabilities are simply counts of $\sigma_{i,j}$ that equals each β in the IHRB algorithm.

As expected, the IHRB algorithm for NB-LDPC codes does not perform as well as the ISRB algorithm because it does not utilize soft information. On the other hand, its complexity is lower since the accumulated extrinsic reliabilities are easier to compute and shorter word length is needed to represent the LLRs.

Since the complexity of the original BP in either probability, frequency, log or mixed domain is prohibitive for practical applications, the remainder of this chapter will focus on the VLSI architectures for the EMS, Min-max and MLGD algorithms.

9.2 Min-max decoder architectures

As discussed previously, there is only one difference between the EMS and Min-max algorithms. The check node processing in the EMS algorithm, which is described by (9.2), computes sums of the v2c messages, while that in the Min-max algorithm calculates the maximum of the v2c messages. As a result, every simplification scheme developed for the EMS algorithm can be extended to the Min-max algorithm by replacing the adders with comparators. On the other hand, the maximum of the v2c messages equals one of the v2c messages. Making use of this property, additional optimizations for the check node processing in the Min-max algorithm have been proposed. Since this property does not apply to the sum computation, those architectures originally designed for the Min-max algorithm may not be directly extended to the EMS algorithm. Also comparators cost less hardware to implement than adders, and hence Min-max decoders require smaller silicon area to implement than EMS decoders. The error-correcting performance loss of the Min-max algorithm compared to the EMS algorithm is very small. Therefore, this section first focuses on the architectures for the Min-max algorithm. Those non-trivial extensions for implementing the EMS algorithm will be discussed in the next section.

Scheduling schemes similar to those for binary LDPC decoders discussed in Chapter 8, such as flooding, sliced message-passing, layered and shuffled schemes, can be applied to NB-LDPC decoders as well. Hence, they will not be repeated in this chapter. The major bottlenecks of NB-LDPC decoder implementations are the complicated check node processing and large memory for storing the message vectors. From Chapter 8, the layered scheme has advantages over the other schemes because the decoding converges faster and the channel information does not need to be stored. Although shuffled decoders also converge fast, they make the complex check node processing in NB-LDPC decoding even more complicated. Next, algorithmic and architectural simplification schemes for the NB CNUs and VNUs are detailed before a memory reduction scheme is introduced by using layered decoding as an example. The CNUs and VNUs presented in this section can be also adopted in other scheduling schemes.

9.2.1 Forward-backward Min-max check node processing

In the Min-max algorithm, the check node processing computes

$$v_{m,n}(\beta) = \min_{(a_j)\in\mathcal{L}(m|a_n=\beta)} \max_{j\in S_v(m)\backslash n} u_{m,j}(a_j). \qquad (9.3)$$

As aforementioned, the configuration set $\mathcal{L}(m|a_n = \beta)$, which is a set of sequences of finite field elements (a_j) such that $\sum_{j\in S_v(m)\backslash n} h_{m,j}a_j = h_{m,n}\beta$, is difficult to record or enumerate, especially when the row weight and/or the order of the finite field is high. Instead, the backward-forward approach [135] can be employed to compute the c2v messages iteratively without explicitly deriving the configuration set. In addition, the multiplication of $h_{m,j}$ can be taken care of separately before the check node processing in order to reduce the overall number of finite field multiplications required. Of course, $h_{m,n}$ needs to be divided after the check node processing to recover the actual c2v message. In this case, the configuration set is modified to $\mathcal{L}'(m|a_n = \beta)$, which is the set of sequences of finite field elements (a_j) such that $\sum_{j \in S_v(m)\setminus n} a_j = \beta$. As a result, the log-domain forward-backward check node processing for the Min-max algorithm is carried out as in Algorithm 32

Algorithm 32 Forward-backward Min-max Check Node Processing
input: $u_{m,n}$ $(n = 0, 1, \ldots, d_c - 1)$
initialization: $f_0 = u_{m,0}$; $b_{d_c-1} = u_{m,d_c-1}$
forward process
    for $j = 1$ to $d_c - 2$

$$f_j(\beta) = \min_{\beta = \beta_1 + \beta_2} \max(f_{j-1}(\beta_1), u_{m,j}(\beta_2))$$

backward process
    for $j = d_c - 2$ down to 1

$$b_j(\beta) = \min_{\beta = \beta_1 + \beta_2} \max(b_{j+1}(\beta_1), u_{m,j}(\beta_2))$$

merging process
    $v_{m,0}(\beta) = b_1(\beta)$
    $v_{m,d_c-1}(\beta) = f_{d_c-2}(\beta)$
    for each $0 < j < d_c - 1$

$$v_{m,j}(\beta) = \min_{\beta = \beta_1 + \beta_2} \max(f_{j-1}(\beta_1), b_{j+1}(\beta_2))$$

From Algorithm 32, the forward-backward scheme for the Min-max check node processing is broken down into elementary steps, each of which computes an output vector, L_o, from two input vectors, L_1 and L_2, according to

$$L_o(\beta) = \min_{\beta = \beta_1 + \beta_2} \max(L_1(\beta_1), L_2(\beta_2)). \tag{9.4}$$

When the order of the finite field, q, is not large, all the q messages in a vector can be computed and stored without causing overwhelming complexity. In addition, keeping all q messages leads to very regular VLSI architectures with simpler control logic for the CNUs. However, the complexity of NB-LDPC decoders, especially that of the CNUs, increases very fast with q. When q is

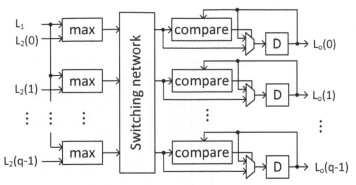

FIGURE 9.3
An L_1-serial Min-max CNU elementary step architecture using basis representation

large, it is impractical to compute and store all q messages. Instead, only the $q' < q$ most reliable messages in each vector can be kept without causing noticeable performance loss if proper compensation schemes are adopted. When all q messages are kept, the messages in the inputs and/or output vectors of an elementary step can be processed simultaneously, and a higher level of parallelism can be achieved at the cost of larger area. In the case that $q' < q$ messages are kept, sorters are involved to find the q' most reliable messages in each vector. Since parallel sorters are very expensive, it is more suitable to process the messages in a vector serially in this case. The CNU design is largely affected by whether all q messages are kept in each vector. Both types of CNU architectures will be discussed in this chapter.

9.2.1.1 Elementary step architecture with all q messages and polynomial representation

First consider the case that all q messages are kept in each vector. In this case, each elementary step for the Min-max algorithm needs to compare q^2 pairs of messages from the two input vectors to find the ones with larger LLRs. Then $L_o(\beta)$ is the minimum among the larger LLRs that correspond to the same finite field element β. Taking care of all q^2 comparisons and generating all q messages in the output vector in parallel is hardware-demanding. Alternatively, one of the input vectors or the output vector can be processed or generated serially to achieve speed-area tradeoff.

In the CNU design of [140], the messages from the L_1 vector are input serially, and all messages in L_2 are available at the same time. This architecture is shown in Fig. 9.3. In each clock cycle, one message from L_1 is compared with each of the q messages in L_2 simultaneously. Then the outputs of the 'max' units are sent to the feedback loops to be compared and used to update the values stored in the registers, which are intermediate minimum messages.

TABLE 9.1

Sums of elements in $GF(8)$ in polynomial representation.

		β_1							
		0	1	2	3	4	5	6	7
	0	0	1	2	3	4	5	6	7
	1	1	0	3	2	5	4	7	6
	2	2	3	0	1	6	7	4	5
	3	3	2	1	0	7	6	5	4
β_2	4	4	5	6	7	0	1	2	3
	5	5	4	7	6	1	0	3	2
	6	6	7	4	5	2	3	0	1
	7	7	6	5	4	3	2	1	0

No valid output message is generated in the first $q - 1$ clock cycles. After the qth clock cycle, all the messages in the output vector, L_o, are derived. Each minimum message stored in a register corresponds to a fixed finite field element, so does each message from the L_2 input vector. However, the message from the L_1 vector input in each clock cycle corresponds to a different finite field element. From (9.4), field element additions are taken for the two input messages included in the 'max' comparisons. Hence, the field element corresponding to the output of each 'max' unit is different in each clock cycle. Therefore, a switching network is needed between the 'max' units and the feedback loops so that the output of each 'max' unit is compared with the intermediate minimum message of the same field element in each clock cycle.

Taking $GF(8)$ as an example, the polynomial representation of each element can be also written as a 3-bit binary tuple. The eight elements are '000', '001', '010', '011', '100', '101', '110' and '111'. If these binary tuples are converted to decimal numbers, they are $0, 1, 2, \ldots, 7$. In [140], the messages in each vector are stored according to their field elements in polynomial representation. The ith message in a vector is for the finite field element whose corresponding decimal value is i. The min-max computations in the elementary steps are subject to the constraint $\beta = \beta_1 + \beta_2$. Table 9.1 lists the sums of all possible pairs of elements in $GF(8)$ if the additions are done by interpreting the 3-bit tuples as polynomial representations. When a message corresponding to β_1 in the L_1 vector is input, β_1 is added up with the field element corresponding to each message in L_2. The corresponding sums, β, are those listed in the column for β_1 in Table 9.1. As can be observed from this table, the order of the field elements are different in each column. From (9.4), the 'min' in the elementary steps of the forward-backward computations are done over the 'max' values corresponding to the same field element. Therefore, the switching network needs to route the outputs of the 'max' units according to

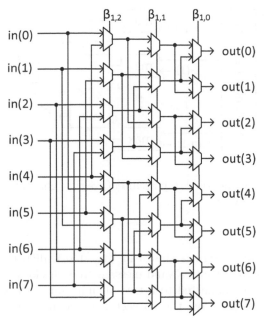

FIGURE 9.4
$GF(8)$ switching network for the elementary step architecture

Table 9.1 in order to compare the 'max' values and intermediate minimum messages with the same field elements.

Due to the symmetry of the switching table, the switching network is implementable by regular architectures. Take Table 9.1 for $GF(8)$ as an example. This table can be divided into four 4×4 squares, the one in the top left is the same as the one in the bottom right, and the top right square is the same as the bottom left square. Each square can be further divided into four sub-squares with the same symmetry. Taking advantage of this property, the switching network corresponding to Table 9.1 is shown in Fig. 9.4. Denote the three bits in the polynomial representation of β_1 by $\beta_{1,2}\beta_{1,1}\beta_{1,0}$, where $\beta_{1,2}$ is the most significant bit. Assume that the output of a multiplexor equals its upper input when the select signal is '0'. If the multiplexors are connected as shown in Fig. 9.4, then the select signals of the multiplexors in the three columns are simply the three bits of β_1. For finite fields larger than $GF(8)$, the tables listing the sums of every possible pair of finite field elements satisfy the same symmetry property as Table 9.1. For every layer of decomposition into four squares, one column of multiplexors are needed. Therefore, the switching network for codes over $GF(q)$ can be implemented by $\log_2 q$ columns of multiplexors, and each column has q 2-to-1 multiplexors.

Alternatively, all the q pairs of messages from L_1 and L_2 whose corresponding β_1 and β_2 add up to be the same β can be compared at a time

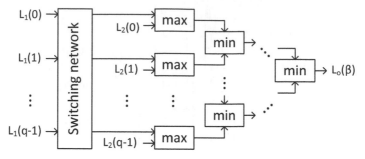

FIGURE 9.5

A Min-max CNU elementary step architecture that generates one output message in each clock cycle

to find $L_o(\beta)$. This design is shown in Fig. 9.5. Unlike the architecture in Fig. 9.3, this design outputs a valid message in each clock cycle. In addition, the 'max' units are connected to a tree of 'min' units through fixed wiring. Although the messages from the L_2 vector are connected to the 'max' units directly, the messages from L_1 need to pass a switching network so that the pairs of messages compared at all 'max' units have the same sum of finite field elements in each clock cycle.

Assuming that the messages in each vector are ordered according to their polynomial representation, Table 9.1 can also be used to decide the switching network in Fig. 9.5 for codes over $GF(8)$. For example, to output $L_o(0)$, the pairs of β_1 and β_2 whose sums are zero need to be sent to the 'max' units. Since $L_2(0), L_2(1), \ldots, L_2(7)$ are connected to the 'max' units, the L_1 messages at the inputs of the 'max' units should be $L_1(0), L_1(1), \ldots, L_1(7)$ in this clock cycle. In the clock cycle for computing $L_o(1)$, the '1' sums need to be identified from Table 9.1. If the '1' in row β_2 is located at column β_1, then $L_1(\beta_1)$ needs to be sent to the $\beta_2 th$ 'max' unit. Hence, the max units should take $L_1(1), L_1(0), L_1(3), L_1(2), L_1(5), L_1(4), L_1(7), L_1(6)$ in the clock cycle for computing $L_o(1)$. The outputs of the switching network in other clock cycles is decided in a similar way. Such a switching network is also implemented by columns of multiplexors like that in Fig. 9.4, except that the connections among the multiplexors and the select signals should be changed accordingly.

9.2.1.2   Elementary step architecture with all q messages and power representation

Both of the architectures in Figs. 9.3 and 9.5 for the elementary step of the Min-max check node processing need a switching network, which adds significant hardware overhead. This switching network can be eliminated from both designs by making use of the power representation of finite field elements [142]. Let α be a primitive element of $GF(q)$. Then each nonzero element of

$GF(q)$ can be expressed as a power of α. In the design of [142], the messages in a vector are arranged according to their finite field elements in the order of $1, \alpha, \ldots, \alpha^{q-2}, 0$. If α^i is multiplied to the vector $[1, \alpha, \ldots, \alpha^{q-2}]$, which consists of all nonzero field elements of $GF(q)$, then the product vector is the multiplicand vector cyclically shifted by i positions. As mentioned previously, the min-max computations in the elementary step is subject to the constraint $\beta = \beta_1 + \beta_2$, which is satisfied if and only if $\alpha\beta = \alpha\beta_1 + \alpha\beta_2$. Therefore, an equivalent implementation of the elementary step is to compute

$$L_o(\alpha\beta) = \min_{\beta = \beta_1 + \beta_2} (\max(L_1(\alpha\beta_1), L_2(\alpha\beta_2))).$$

From the above equation, if the two input LLR vectors are both cyclically shifted by one position, then the constraint remains the same if the output vector is also cyclically shifted by one position. Accordingly, if the input and output vectors are cyclically shifted by the same number of positions, the connections between the 'max' and 'min' units in Fig. 9.3 and those between the L_1 and 'max' units in Fig. 9.5 can remain unchanged in each clock cycle. As a result, the switching network is eliminated. Note that the message whose field element is zero should be excluded from the vector when cyclical shifting is applied.

When the messages in a vector are ordered according to their finite field elements in polynomial representation, including the multiplications with the H matrix entries in the constraint would make the elementary step architecture much more complicated. Hence, these finite field multiplications are taken care of by separate multipliers when the elementary step architectures in Figs. 9.3 and 9.5 are used. When the messages in a vector are arranged according to the power representation of finite field elements, multiplying a vector by a field element leads to a vector shifting. As a result, the multiplication by an entry of H can be done in one clock cycle by a barrel shifter, which has $\log_2 q$ stages of q multiplexors. Alternatively, the shifting can be done through registers. If registers are already employed to store the message vector, implementing the shifting only requires one multiplexor to be added for each of the q registers to link them together. This area overhead is much smaller than that of the barrel shifter. Although i clock cycles are needed to multiply α^i through register shifting, this latency penalty can be hidden in the forward-backward process for check node processing.

In (9.4), it is assumed that the multiplications of the message vectors with the entries of H have been taken care of by separate units. β, β_1 and β_2 are used to index the messages in the input and output vectors in this equation. Consider the processing of row m in H. Let α^{h_j} be the jth nonzero entry in this row. Define $\beta = \alpha^{h_j}\beta'$ if β is used as the message index for vectors related to the jth nonzero entry, such as $u_{m,j}$, f_j and b_j in Algorithm 32. Then β' is the index of the messages associated with the jth original v2c vector. If the original v2c vector $u_{m,j}$ is directly sent to the CNU, then the jth step of the forward process should compute

$$f_j(\beta) = \min_{\beta = \beta_1 + \alpha^{h_j} \beta_2'} (\max(f_{j-1}(\beta_1), u_{m,j}(\beta_2'))). \tag{9.5}$$

The constraint in the above equation can be rewritten as $\alpha^{h_j}\beta' = \alpha^{h_{j-1}}\beta_1' + \alpha^{h_j}\beta_2'$. Accordingly, it is equivalent to carry out the forward process as

$$f_j(\beta') = \min_{\beta' = \alpha^{(h_{j-1} - h_j)} \beta_1' + \beta_2'} (\max(f_{j-1}(\beta_1'), u_{m,j}(\beta_2'))). \tag{9.6}$$

Similarly, the elementary step of the backward process is done instead by

$$b_j(\beta') = \min_{\beta' = \alpha^{(h_{j+1} - h_j)} \beta_1' + \beta_2'} (\max(b_{j+1}(\beta_1'), u_{m,j}(\beta_2'))). \tag{9.7}$$

In (9.5), β_2' is multiplied with α^{h_j} in the constraint. Hence, the original v2c message vector needs to be cyclically shifted to implement this multiplication. On the other hand, powers of α are multiplied to β_1' in (9.6) and (9.7), and hence the intermediate message vectors need to be shifted. Therefore, by carrying out the elementary steps as (9.6) and (9.7), the shifting has been migrated from the original v2c vector to the intermediate message vectors.

Shifting a different vector does not make (9.6) achieve any hardware saving over (9.5) for the forward process. Nevertheless, (9.6) enables the shifting to be eliminated from the merging process. If the forward-backward process is carried out with the multiplications of the H entries done separately, then the c2v vector is computed as

$$v_{m,j}(\beta') = \min_{\alpha^{h_j} \beta' = \beta_1 + \beta_2} (\max(f_{j-1}(\beta_1), b_{j+1}(\beta_2))). \tag{9.8}$$

In other words, after merging the intermediate vectors derived from the forward and backward processes, the result vector needs to be cyclically shifted by $-h_j$ positions to become the actual c2v vector. Alternatively, (9.8) is rewritten as

$$v_{m,j}(\beta') = \min_{\beta' = \alpha^{(h_{j-1} - h_j)} \beta_1' + \alpha^{(h_{j+1} - h_j)} \beta_2'} (\max(f_{j-1}(\beta_1'), b_{j+1}(\beta_2'))).$$

From the above equation, if the intermediate vectors from the forward and backward processes according to (9.6) and (9.7) are cyclically shifted by $(h_{j-1} - h_j)$ and $(h_{j+1} - h_j)$ positions, respectively, before they are used in the merging step, then the results of the merging are actual c2v vectors and no more shifting is needed. The shifted intermediate vectors, $f_{j-1}(\beta'') = f_{j-1}(\alpha^{(h_{j-1} - h_j)}\beta')$ and $b_{j+1}(\beta'') = b_{j+1}(\alpha^{(h_{j+1} - h_j)}\beta')$, are already being used as the inputs to the elementary steps in the forward and backward processes in (9.6) and (9.7). They can be stored when available without causing any overhead. As a result, the elementary steps of the forward-backward process

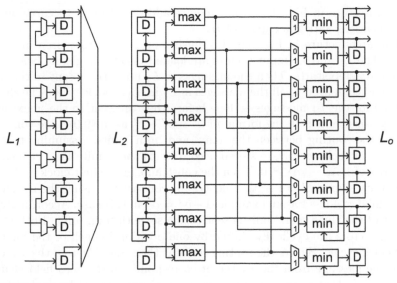

FIGURE 9.6
An L_1-serial Min-max CNU elementary step architecture using power representation (modified from [141])

with multiplications and divisions of the H matrix entries incorporated are carried out as

$$f_j(\beta') = \min_{\beta'=\beta_1''+\beta_2'} (\max(f_{j-1}(\beta_1''), u_{m,j}(\beta_2')))$$

$$b_j(\beta') = \min_{\beta'=\beta_1''+\beta_2'} (\max(b_{j+1}(\beta_1''), u_{m,j}(\beta_2'))) \qquad (9.9)$$

$$v_{m,j}(\beta') = \min_{\beta'=\beta_1''+\beta_2''} (\max(f_{j-1}(\beta_1''), b_{j+1}(\beta_2''))).$$

These elementary steps have the same format as those for the original forward-backward process in Algorithm 32, and the results of the merging process do not need to be shifted to get the actual c2v messages. Next, architectures employing power representation and these modified elementary steps are introduced. It will be shown that, except in the first and last elementary steps, the shifting of the intermediate result vectors does not cause extra latency.

Fig. 9.6 shows a Min-max CNU elementary step architecture for codes over $GF(8)$ resulting from applying power representation and the modified elementary steps to the architecture in Fig. 9.3. As it can be observed, the switching network consisting of $q \log_2 q$ 2-to-1 multiplexors in Fig. 9.3 is replaced by one single column of q multiplexors in Fig. 9.6. Although a q-to-1 multiplexor has been added to select a message from the L_1 vector in Fig. 9.6, its overhead is smaller than the saving achieved by eliminating the switching network. Moreover, unlike the decoders using the design in Fig. 9.3, the decoders adopting the architecture in Fig. 9.6 do not require additional units to carry out the multiplications with the entries of H.

The architecture in Fig. 9.6 also finishes one elementary step in q clock cycles. The L_1 and L_2 input vectors are loaded into the registers before each elementary step starts. Then the messages in L_1 are serially sent to the 'max' units, and compared to all the messages in L_2 in each clock cycle. The shifting required for the L_1 messages is done before they are sent to the 'max' units, and the constraint is in the format of $\beta' = \beta_1'' + \beta_2'$ as in (9.9). The constraint decides the routing between the 'max' and 'min' units. Assuming that the messages in each vector are stored according to their finite field elements in the order of $1, \alpha, \ldots, \alpha^{q-2}, 0$. Then the constraint in the first clock cycle of an elementary step is $\beta' = 1 + \beta_2'$. Take $GF(8)$ constructed using the irreducible polynomial $x^3 + x + 1$ as an example. $1 + \alpha^2 = \alpha^6$ in this case. Accordingly, the output of the third 'max' unit is sent to the seventh 'min' unit. Both the L_2 and L_o vectors are cyclically shifted by one position in each of the following $q - 2$ clock cycles. Hence, the routing remains the same. Then in the qth clock cycle, the message from the L_1 vector corresponds to the zero field element. $0 + \beta_2' = \beta_2'$. Hence, the output of each 'max' unit should be routed to the 'min' unit of the same row in this clock cycle. This different routing is taken care of by the column of multiplexors before the 'min' units shown in Fig. 9.6. The select signals of these multiplexors are '1' in the first $q - 1$ clock cycles and '0' in the last clock cycle. The messages with zero field element in L_2 and L_o are located in the last registers of those columns, and they are not involved in the shifting.

In the beginning of each forward and backward elementary step, the L_o computed in the previous step is loaded into the register column on the left in Fig. 9.6 to become the L_1 vector for the current step. The L_1 vector needs to be cyclically shifted by $(h_{j-1} - h_j)$ or $(h_{j+1} - h_j)$ positions according to (9.6) and (9.7). Cyclically shifting L_1 by x positions means that the first message of L_1 sent to the 'max' units should be its xth entry. Instead of waiting for the shifting, the xth entry is selected by the q-to-1 multiplexor to avoid extra latency. The messages with nonzero field elements in L_1 are cyclically shifted by one position in each clock cycle. Therefore, by keeping the select signal of this multiplexor unchanged, the other messages of L_1 with nonzero field elements are sent to the 'max' units in the next $q - 1$ clock cycles. In addition, after L_1 has been shifted by $\alpha^{(h_{j-1} - h_j)}$ positions in the jth forward elementary step, or $\alpha^{(h_{j+1} - h_j)}$ positions in the jth backward step, the shifted vector is stored into memory to be used in the merging process. In this design, the shifting needed on the intermediate vector does not cause extra latency, and the elementary steps in the merging process do not need to have their input or output vectors shifted to derive the actual c2v message vectors. Nevertheless, the vectors output from the last elementary steps of the forward and backward processes need to be shifted by $h_{d_c-2} - h_{d_c-1}$ and $h_1 - h_0$ positions, respectively, before they become the last and first c2v message vectors. If these shifting is not overlapped with other computations, they bring

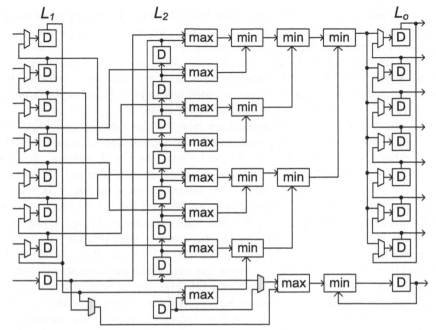

FIGURE 9.7

An L_o-serial Min-max CNU elementary step architecture using power representation (modified from [141])

an extra latency of $(h_1 - h_0) \mod (q-1) + (h_{d_c-2} - h_{d_c-1}) \mod (q-1)$ clock cycles.

Power representation and the modified elementary steps can also be adopted to develop an architecture that takes in all messages in the L_1 and L_2 vectors and generates one message in the L_o vector at a time. Fig. 9.7 shows such an architecture for $GF(8)$ codes. The input and output vectors are stored in registers. There are exactly q pairs of elements in $GF(q)$ whose sums equal a given field element. In the first clock cycle of each elementary step, the q pairs of messages from the L_1 and L_2 input vectors whose corresponding field elements add up to be $\alpha^0 = 1$ are sent to the q 'max' units. Assuming that the irreducible polynomial $x^3 + x + 1$ is used to construct $GF(8)$. The pairs of elements from $GF(8)$ whose sums equal to 1 are $0+1$, $\alpha^3+\alpha$, $\alpha^6+\alpha^2$, $\alpha+\alpha^3$, $\alpha^5+\alpha^4$, $\alpha^4+\alpha^5$, $\alpha^2+\alpha^6$, and $1+0$. The routing between the L_1 vector and the 'max' units in Fig. 9.7 are decided by these pairs, and $L_o(1)$ is computed in the first clock cycle. The input and output vectors are cyclically shifted by one position in each clock cycle. Therefore, the routing does not need to be changed to generate $L_o(\alpha), L_o(\alpha^2), \ldots, L_o(\alpha^6)$ serially in the next six clock cycles. $L_o(0)$ is calculated iteratively using the units on the bottom of Fig. 9.7. In the first $q-1$ clock cycles, the messages with the same nonzero field element from L_1 and L_2 are passed through the two multiplexors in gray color to the 'max' unit on the bottom right corner. Then, these two multiplexors select the pair of messages with zero field elements in the last clock cycle. Therefore, the architecture in Fig. 9.7 also takes q clock cycles to complete an elementary step. Additionally, it has the same number of 'max' and 'min' units as that in Fig. 9.5. However, this architecture does not require the complex routing network between the L_1 vector and the 'max' units.

In the beginning of each elementary step for the forward and backward processes, the L_o vector computed from the previous step is loaded into the first column of registers in Fig. 9.7 to be used as the L_1 vector. L_1 also needs to be shifted by $\alpha^{(h_{j-1}-h_j)}$ positions in the jth forward elementary step, or $\alpha^{(h_{j+1}-h_j)}$ positions in the jth backward step, before it is sent to the 'max' units or stored in the memory to be used for the merging process. Different from the design in Fig. 9.6, this shifting is done through controlling the write location of the messages in the L_o vector in the previous step instead of adding multiplexors to the outputs of the L_1 vector. If L_1 needs to be shifted by x positions, then each output message in the previous step is written into the xth register of the L_o vector while the messages in this vector are cyclically shifted by one position in each clock cycle. The vector shifting does not incur extra latency except in the first elementary steps of the forward and backward processes. If the shifting is not overlapped with other computations, the latency overhead is also $(h_1 - h_0) \bmod (q-1) + (h_{d_c-2} - h_{d_c-1}) \bmod (q-1)$ clock cycles, which is the same as that of the architecture in Fig. 9.6.

The latency overhead brought by the vector shifting in the designs of Fig. 9.6 and 9.7 can be reduced by manipulating the entries of the H matrix. The constraints to be satisfied in the check node processing are not changed if every entry in a row of H is multiplied with the same nonzero scaler. Therefore, a constant α^δ may be multiplied to a row of H to minimize the latency overhead resulting from the message vector shifting. Such scaler multiplications do not affect the error-correcting performance of the LDPC code. Let $\Delta h_0 = (h_1 - h_0) \bmod (q - 1)$ and $\Delta h_{d_c-1} = (h_{d_c-2} - h_{d_c-1}) \bmod (q - 1)$. Both Δh_0 and Δh_{d_c-1} are nonnegative. For finite field $GF(q)$, modulo $q-1$ is taken over the exponent of every computation result in power representation. Since the goal is to minimize $\Delta h_0 + \Delta h_{d_c-1}$, the α^δ to be multiplied is dependent on the relative distance of Δh_0 and Δh_{d_c-1} modulo $q - 1$. The following two cases are considered.

Case 1: $|\Delta h_0 - \Delta h_{d_c-1}| < (q - 1)/2$. The latency overhead is minimized to $|\Delta h_0 - \Delta h_{d_c-1}|$ by letting $\delta = -\min\{\Delta h_0, \Delta h_{d_c-1}\}$.
Case 2: $|\Delta h_0 - \Delta h_{d_c-1}| > (q - 1)/2$. The latency overhead is minimized to $(q - 1) - |\Delta h_0 - \Delta h_{d_c-1}|$ when $\delta = (q - 1) - \max\{\Delta h_0, \Delta h_{d_c-1}\}$.
From these two cases, the latency overhead can be reduced to at most $\lfloor (q - 1)/2 \rfloor$ if the rows of H are multiplied with proper scalers.

For a NB-LDPC code over $GF(32)$, Table 9.2 shows the δ selected for some rows of the H matrix and the shifting latency overhead reduction achieved by multiplying α^δ to each entry in the row. Note that the computations and

TABLE 9.2
Latency overhead reduction examples for a NB-LDPC code over $GF(32)$

	$h_1 - h_0,$ $h_{d_c-2} - h_{d_c-1}$	$\Delta h_0,$ Δh_{d_c-1}	Original latency	Case #	δ	Reduced latency
example row 1	-8, -13	23, 18	23+18	1	-18	5
example row 2	-18, 24	13, 24	13+24	1	-13	11
example row 3	25, -20	25, 11	25+11	1	-11	14
example row 4	2, 29	2, 29	2+29	2	2	4

multiplications of the scalers are done off line, and hence do not incur hardware complexity. Although adopting power representation and the modified elementary stepa leads to extra latency for the vector shifting, it is only a very small part of the number of clock cycles needed for the entire decoding process.

9.2.1.3   Elementary step architecture with $< q$ messages

When q is not small, computing and storing all the q messages in each vector have very high hardware complexity. Instead, only the $q' < q$ most reliable messages in each vector can be kept [137]. For example, for codes over $GF(32)$, the number of messages kept for each vector can be reduced to 16 without causing any noticeable performance loss if proper compensation schemes are applied to those unstored messages. In the case that not all q messages are kept, it would be quite wasteful to assign dummy values to those unstored messages and use the architectures designed for vectors with all q messages. In addition, the finite field elements corresponding to the most reliable messages in each vector are different and change with the computations. Hence, it is more efficient to process the messages in a vector serially when not all q messages are kept. After all, since the number of involved messages is smaller, the latency of a serial design becomes more manageable.

When not all q messages are kept, the finite field element for each message needs to be stored explicitly. Denote the the two LLR input vectors to the check node processing elementary step by $L_1 = [L_1(0), L_1(1), \ldots, L_1(q' - 1)]$ and $L_2 = [L_2(0), L_2(1), \ldots, L_2(q' - 1)]$. The corresponding finite field element vectors are represented by $\alpha_{L_1} = [\alpha_{L_1}(0), \alpha_{L_1}(1), \ldots, \alpha_{L_1}(q' - 1)]$ and $\alpha_{L_2} = [\alpha_{L_2}(0), \alpha_{L_2}(1), \ldots, \alpha_{L_2}(q' - 1)]$. The LLR and field element vectors of the output are denoted by $L_o = [L_o(0), L_o(1), \ldots, L_o(q' - 1)]$ and $\alpha_{L_o} = [\alpha_{L_o}(0), \alpha_{L_o}(1), \ldots, \alpha_{L_o}(q' - 1)]$, respectively. For the Min-max decoding algorithm, the entries in L_o are the q' minimum values of $\max(L_1(i), L_2(j))$ with different $\alpha_{L_1}(i) + \alpha_{L_2}(j)$ for any combination of i and j less than q'. Let $\{L_o, \alpha_{L_o}\} \leftarrow \{L, \alpha_L\}$ denote storing an LLR L and its associated field element α_L into the output vector. The elementary step of the check node processing in the Min-max algorithm can be carried out according to Algorithm 33 [143] when $q' < q$ messages are kept for each vector. In this algorithm, it is assumed that the messages in a vector are stored in the order of increasing LLR.

Algorithm 33 Elementary step of forward-backward Min-max check node processing when $q' < q$ messages are kept
input: $L_1, \alpha_{L_1}, L_2, \alpha_{L_2}$
initialization: $i = 0$, $j = 1$, $num = 0$, $l = 0$
while $(l < l_{max})$
{   *if* $(L_1(i) < L_2(j))$
                    for $k = 0$ *to* $j - 1$
                    { $l = l + 1$
                            if $((\alpha_{L_1}(i) + \alpha_{L_2}(k)) \notin \alpha_{L_o})$
                                $\{L_o, \alpha_{L_o}\} \leftarrow \{L_1(i), \alpha_{L_1}(i) + \alpha_{L_2}(k)\}$
                                $num = num + 1$
                                if $(num = q')$, *go to stop*
                    }
                    $i = i + 1$
        else
                    for $k = 0$ *to* $i - 1$
                    { $l = l + 1$
                            if $((\alpha_{L_1}(k) + \alpha_{L_2}(j)) \notin \alpha_{L_o})$
                                $\{L_o, \alpha_{L_o}\} \leftarrow \{L_2(j), \alpha_{L_1}(k) + \alpha_{L_2}(j)\}$
                                $num = num + 1$
                                if $(num = q')$, *go to stop*
                    }
                    $j = j + 1$
}
stop:

In Algorithm 33, num tracks the number of messages inserted into the output vector and l_{max} is used to control the maximum number of clock cycles allowed for each elementary step. The operation of this algorithm is explained via the assistance of the matrix shown in Fig. 9.8. Two example input vectors for a code over $GF(32)$ are considered in this figure. The finite field elements are denoted by the exponents in their power representations, and 31 represents the zero field element. The entries in the matrix are the maximum values of the corresponding LLRs in the L_1 and L_2 vectors. Starting from the top left corner of the matrix, a boundary is drawn to follow the comparison results of the LLRs. It goes down when the LLR in the L_1 vector is less than that in the L_2 vector, and goes to the right otherwise. It can be observed that, in the

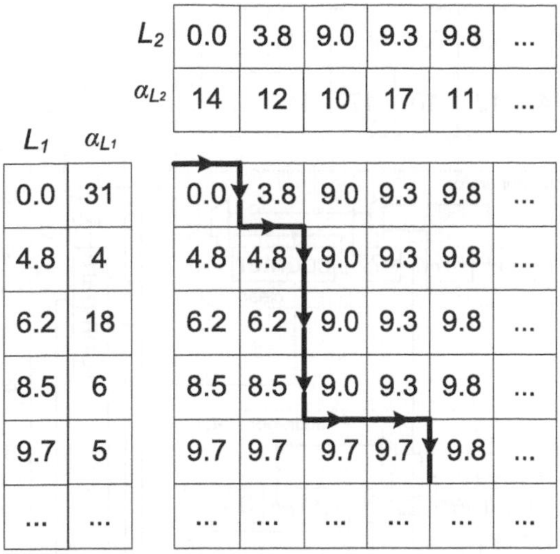

L_2	0.0	3.8	9.0	9.3	9.8	...
α_{L_2}	14	12	10	17	11	...

L_1	α_{L_1}						
0.0	31	0.0	3.8	9.0	9.3	9.8	...
4.8	4	4.8	4.8	9.0	9.3	9.8	...
6.2	18	6.2	6.2	9.0	9.3	9.8	...
8.5	6	8.5	8.5	9.0	9.3	9.8	...
9.7	5	9.7	9.7	9.7	9.7	9.8	...
...	...	...	...	...	...	...	...

FIGURE 9.8
Example of maximum values of pairs of LLRs from two vectors

same column, the LLRs above the horizontal segment of the boundary are the same, and they are smaller than those below the segment. Similarly, in the same row, the LLRs to the left of the vertical segment are the same, and they are smaller than those to the right of the segment. Therefore, starting from the top left corner, the smallest entries of the matrix can be distilled by taking the entries to the left of the vertical segments and those above the horizonal segments of the boundary. The testing result of whether $L_1(i) < L_2(j)$ in Algorithm 33 actually tells the direction of the segments in the boundary. Although all cells in the same row of the matrix to the left of the boundary are the same, they correspond to different finite field elements. Similarly, the cells in the same column also correspond to different field elements. If the finite field element corresponding to an entry in the matrix is the same as that of an entry previously inserted into the output vector, the new candidate entry will not be inserted, since the previous entry has smaller LLR.

Algorithm 33 can be implemented by the architecture in Fig. 9.9. In algorithm 33, i and j are the addresses of the entries in the L_1 and L_2 vectors, respectively. They either do not change or increase by one in each loop. To shorten the data path of the address generation, the additions by one are precomputed as shown in Fig. 9.9. When $L_1(i) < L_2(j)$, the counter in the address generator for the L_2 vector is cleared. As the counter increases, $\alpha_{L_2}(k)$ for $k = 0, 1, 2 \ldots$ are read out serially and added up with $\alpha_{L_1}(i)$. This continues until the counter reaches j, whose value is available at the output of the multiplexor in the middle of the address generator. In the case of $L_1(i) \geq L_2(j)$, the counter in the address generator for the L_1 vector is cleared and then in-

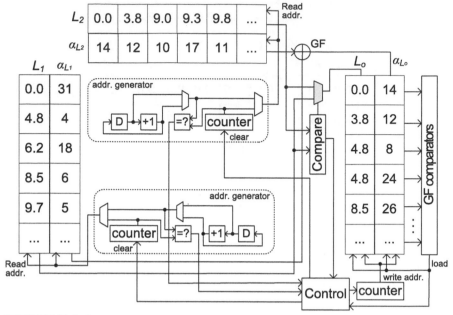

FIGURE 9.9
Min-max CNU elementary step architecture with $< q$ messages (modified from [143])

creased by one each time to read out $\alpha_{L_1}(k)$ for $k = 0, 1, \ldots, i-1$. The write address of the L_o vector is increased by one every time a new entry needs to be inserted. It is generated by a counter whose enable signal comes from the GF comparator block in Fig. 9.9. This block compares the finite field element of the candidate entry with those already in L_o, and its output is asserted if the candidate element does not equal any of the existing elements. This output also serves as the 'load' signal of the L_o vector. To reduce the latency, the comparisons with all the field elements in L_o are done in parallel. If the vectors are stored in RAMs, it is necessary to feed the field element of the candidate entry also to a register vector when it is inserted into L_o in order to enable simultaneous comparisons.

The values of some key signals in the architecture depicted in Fig. 9.9 are listed in Table 9.3 for each clock cycle. The Mux in this table refers to the one in grey in the figure. The finite field additions are done based on the irreducible polynomial $x^5 + x^2 + 1$ being used to construct $GF(32)$.

The maximum number of iterations in Algorithm 33, and hence the maximum number of clock cycles that can be spent on each elementary step is limited by the parameter l_{max}. At the end of iteration l_{max}, the output vector may have less than q' valid entries. In this case, the rest of the entries in L_o

TABLE 9.3
Example signal values for the Min-max CNU elementary step architecture with $< q$ messages

l	Compare inputs		Mux output	GF adder inputs		GF adder output	i	j	k	load	num
1	0.0	3.8	0.0	31	14	14	0	1	0	1	1
2	4.8	3.8	3.8	31	12	12	1	1	0	1	2
3	4.8	9.0	4.8	4	14	8	1	2	0	1	3
4	4.8	9.0	4.8	4	12	24	1	2	1	1	4
5	6.2	9.0	6.2	18	14	24	2	2	0	0	4
6	6.2	9.0	6.2	18	12	8	2	2	1	0	4
7	8.5	9.0	8.5	6	14	26	3	2	0	1	5
⋮	⋮	⋮	⋮	⋮	⋮	⋮	⋮	⋮	⋮	⋮	

are set to the last LLR inserted into this vector. Filling the rest of the entries requires extra clock cycles. Alternatively, the address of the last inserted entry is stored. It is available in the counter for generating the output vector write address at the end of the l_{max} clock cycle. When the rest of the entries are needed, the LLR is read from this stored address. To decide the finite field elements should be used for the rest of the entries, three schemes have been tested in [143]. In the first scheme, the limitation on the maximum number of iterations can be spent on an elementary step is removed, and Algorithm 33 is carried out until q' entries are inserted into the output vector. The second scheme sets l_{max} to $2q'$. If the output vector is not filled with q' entries at the end, zero is used as the finite field elements for the rest entries. The last scheme also has $l_{max} = 2q'$. Nevertheless, the finite field elements for the rest entries are set to be the same as that of the last inserted entry. Simulation results showed that the error-correcting performance is almost the same using these three schemes. Actually, the finite field elements of the rest of the entries can be set to any arbitrary elements without affecting the performance. The reason is that the rest of the messages in the output vector are less reliable. In addition, most of the time, they will not participate in later computations if only q' messages are kept for each vector, and hence they have minor effect on the decoding. To reduce the hardware complexity, the second scheme is employed.

9.2.1.4 Scheduling of elementary steps

The forward-backward check node processing involves three processes. The elementary steps in the forward and backward processes are iterative, and the merging process utilizes the intermediate vectors generated from the forward and backward processes. The scheduling of the computations in these three processes affects the number of elementary step units needed, the size of

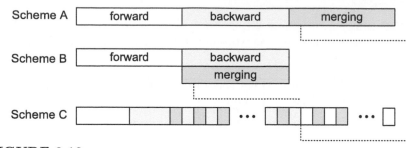

FIGURE 9.10
Scheduling schemes for forward-backward check node processing

the memory for storing the intermediate message vectors, and the achievable throughput of the check node processing. Next, several schemes for scheduling the elementary steps in the three processes are discussed and their hardware complexities are compared. The differences of the scheduling schemes lie in the order of the elementary steps carried out. Any of the previously presented elementary step architectures can be adopted in these schemes.

The straightforward way is to carry out the three processes one after another using one elementary step unit for each check node. Let us call this Scheme A. This scheduling is shown in Fig. 9.10. Assume that the row weight of the check node is d_c, and an elementary step takes T_e clock cycles to complete. Then each of the forward, backward and merging processes takes $(d_c-2)T_e$ clock cycles. Since the merging starts after the forward and backward processes are done, all intermediate vectors from the forward and backward processes need to be stored. Therefore, the memory requirement is $2(d_c - 2)$ vectors.

In Scheme B shown in Fig. 9.10, two elementary step units are employed for each check node. The backward process starts after the forward process is finished. In parallel to the backward process, the merging is carried out. Every time a new intermediate vector is computed by the backward process, it is consumed right away with a forward intermediate vector in the merging process to derive the corresponding c2v vector. Therefore, the check node processing using Scheme B is finished in $2(d_c-2)T_e$ clock cycles. Although two elementary step units are needed, the intermediate vectors from the backward process do not need to be stored. Hence, the memory requirement is reduced to $d_c - 2$ vectors.

The memory requirement can also be reduced without increasing the number of elementary step units as shown by Scheme C in Fig. 9.10. The forward process is only finished for the first half of the d_c v2c input vectors before the backward process is carried out. After the backward process is completed for the second half of the v2c input vectors, the intermediate vectors for the merging to generate the c2v vector in the middle are available. Then the elementary step unit alternates between the backward process for the first half

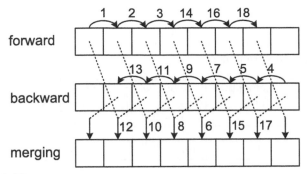

FIGURE 9.11
Order of elementary steps in Scheme C for an example case of $d_c = 8$

of the v2c vectors and the merging for computing the first half of the c2v vectors. After that, the forward process and merging for the second half of the vectors are carried out alternatively in a similar way. Since the bars for Scheme C in Fig. 9.10 are narrow, the forward, backward and merging steps are differentiated by the shades on the bars. Once an intermediate vector is used in the corresponding merging, it is erased if no longer needed for the later iterative forward or backward steps. As a result, at most $d_c - 2$ vectors need to be stored at any time in Scheme C, and only one elementary step unit is employed.

To clarify Scheme C, an example is given in Fig. 9.11 for the case of $d_c = 8$. Each block in the rows for the forward and backward processes in this figure represents a v2c message vector, and those in the merging row denote c2v vectors. The starting and ending points of the arrowed lines indicate the vectors involved in each elementary step, The numbers by the lines tell the order of the elementary steps. For example, the first elementary step takes the first and second input v2c message vectors and generates the first intermediate vector of the forward process. The second step takes the first forward intermediate vector and the third v2c vector, and computes the second forward intermediate vector. In elementary step 6, the third forward intermediate vector and second backward intermediate vector are used to generate the 5th c2v vector. Following the order of the elementary steps, the maximum number of intermediate vectors need to be stored is $6 = d_c - 2$, and it happens at the end of the 7th elementary step.

The three scheduling schemes are compared in Table 9.4. Scheme B has a shorter latency and a smaller memory than Scheme A, although two times the elementary step units are needed. The complexity of the elementary step architecture does not change with d_c, while the memory requirement increases linearly with d_c. In the case that d_c is large, Scheme B may require a smaller area than Scheme A. Scheme C has lower hardware complexity than Scheme A and achieves the same throughput. Although Scheme C is slower than Scheme B, its area is smaller. The drawback of Scheme C is that the c2v vectors are

TABLE 9.4

Comparisons of forward-backward check node processing scheduling schemes for a code with row weight d_c

Scheme	# of elementary units	# of clks	memory (# of intermediate vectors)	in-order output?
A	1	$3(d_c - 2)T_e$	$2(d_c - 2)$	yes
B	2	$2(d_c - 2)T_e$	$d_c - 2$	yes
C	1	$3(d_c - 2)T_e$	$d_c - 2$	no

generated out of order. As shown in Fig. 9.11, the first output vector available is the one in the middle. This poses disadvantages when the computations in the decoding of different iterations or different segments of H need to be overlapped to increase the throughput, such as in the case of sliced message-passing or layered decoders. The forward and backward processes for the same v2c message vectors can not be done in parallel in order to avoid conflicts in accessing the memory storing the v2c vectors. This is reflected in Fig. 9.10. However, the processes for different layers or different decoding iterations can be done simultaneously using additional elementary step units. As shown by the dashed bars for Scheme A and B in Fig. 9.10, once a c2v vector is generated by the merging, it can be consumed in the computation of the v2c vector for the next layer or decoding iteration right away, and the processes for the next layer or iteration starts. Also, the order of the forward and backward processes for the same v2c vectors can be switched. In Scheme A and B, the merging process generates c2v vectors in the order of $d_c - 1, d_c - 2, \ldots, 0$. Therefore, the backward process for the next layer or decoding iteration overlaps with the current merging process in these two schemes. Nevertheless, the forward and backward processes have to start from the first and the last v2c message vectors, respectively. Hence, when Scheme C is adopted, the processes of the next layer needs to wait until these two c2v message vectors are available. This may require extra memory to store the out-of-order c2v message vectors generated depending on the decoder architecture.

9.2.2   Trellis-based path-construction Min-max check node processing

Although the forward-backward check node processing can be implemented by identical elementary steps, it needs to store a large number of intermediate message vectors. Moreover, the iterative nature of the forward and backward processes limits the achievable throughput. Even though the number of 'min' and 'max' comparisons is reduced by keeping $q' < q$ messages for each vector, many of them are still redundant in the forward-backward check node processing. To eliminate the redundancy, the v2c messages can be represented

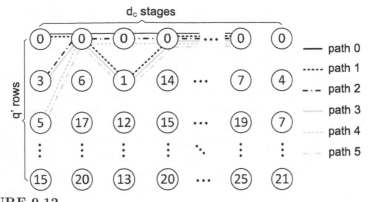

FIGURE 9.12
Example trellis of v2c message vectors (modified from [144])

by the nodes in a trellis. Then the c2v message computations are mapped to constructing paths involving a limited number of nodes in the trellis [144]. Only these nodes need to be stored as intermediate results to carry out the path construction, and the path construction for different c2v vectors can be done in parallel. This trellis-based path construction check node processing method is detailed next.

The configuration set $\mathcal{L}(m|a_n = \beta)$ has multiple sequences of finite field elements. From (9.3), the Min-max check node processing first finds the largest v2c LLR corresponding to the field elements in each sequence. Then among all sequences in the configuration set, the smallest of those largest LLRs is selected to be $v_{m,n}(\beta)$. Therefore, in the Min-max algorithm, $v_{m,n}(\beta)$ equals one of the v2c LLRs input to the check node processing. In addition, since the minimum is taken over the LLRs, the c2v LLRs equal to the smallest of all input v2c LLRs. From this observation, it was proposed in [144] to first sort out the smallest of all input v2c LLRs, which are also the most reliable messages. Then all c2v messages are computed from the sorted values.

It is not a simple task to compute $v_{m,n}$ from the most reliable v2c messages. In addition, when only the $q' < q$ most reliable entries of $v_{m,n}$ are needed, it is desirable not to waste computations on the remaining unreliable entries. This goal is achieved by utilizing a trellis representation of the v2c message vectors and mapping the c2v message computations to constructing paths passing through the nodes in the trellis. An example of such a trellis is shown in Fig. 9.12. For a check node with degree d_c, this trellis has d_c stages, one for each v2c input vector. If q' messages are kept for each vector, then each stage has q' nodes, representing the q' entries in the corresponding vector. The numbers in each node are example values of quantized LLRs. The finite field elements of the nodes are not labeled in Fig. 9.12, and the nodes in the same row do not necessarily have the same field element. In the trellis-based check node processing scheme [144], the messages in each vector are ordered according to increasing LLR. Normalization similar to that listed in Algorithm 29 for the EMS algorithm is also used in the Min-max algorithm to force the smallest LLR in each v2c vector to zero. As a result, the nodes in the first row of the trellis have zero LLR. A sequence of finite field elements (a_j) in the configuration set $\mathcal{L}(m|a_n = \beta)$ can be represented by a path in the trellis. This path passes exactly one node from each stage, except stage n, and the node passed in the jth stage is the one with finite field element a_j. Define the LLR of a path as the maximum LLR of all nodes in the path. Let the finite field element of a path be the sum of the finite field elements of all the nodes in the path. Then the q' entries of $v_{m,n}$ with the smallest LLR can be computed through finding the q' paths passing exactly one node in each stage, except stage n, with the smallest LLRs and different finite field elements.

It is apparent that the path with the smallest LLR is the one passing only the zero-LLR nodes in the first row of the trellis. This path is labeled by path 0 in Fig 9.12. Sort the $d_c \times (q' - 1)$ nonzero LLRs of the input v2c messages. Assume that they are listed in non-decreasing order as $x(1), x(2), \cdots$. Let $\alpha(1), \alpha(2), \cdots$ be their associated field elements, and they belong to stages $e(1), e(2), \cdots$. The other paths for computing $v_{m,n}$ can be iteratively constructed through replacing the nodes in path 0 by the nodes with the smallest nonzero LLR not belonging to stage n. In order to construct the paths with the smallest LLRs first and not to waste computations on those unreliable c2v messages that will not be kept, the nonzero-LLR nodes are considered in the path construction starting from the one with $x(1)$. The node with $x(i+1)$ is only considered after all possible paths with $x(i)$ are constructed. When the node with $x(i)$ and $e(i) \neq n$ is considered, new paths are constructed by using this node to replace the node in the same stage in each previously constructed path. This can ensure that all possible paths are constructed. Nevertheless, a path includes only one node from each stage. As a result, if a previously constructed path has a nonzero-LLR node in stage $e(i)$, then replacing that node by the new node would lead to the same path as replacing the zero-LLR node in stage $e(i)$ of another previous path. Accordingly, if a previous path has a nonzero-LLR node in stage $e(i)$, then the path construction of replacing that node by the new node can be skipped. Since the node with $x(i+1)$ is considered after that of $x(i)$, the paths constructed have non-decreasing LLR. It is possible that the finite field element of the newly constructed path is the same as that of a previous path. In this case, the new path is not stored since it has larger LLR than the previous one. This path construction process is repeated until q' entries are derived for the c2v vector or the number of iterations reaches a preset limit.

Example 46 *Use $o_{i,j}$ $(0 \leq i < q', 0 \leq j < d_c)$ to denote the node in the ith row and jth column of the trellis. In the example trellis shown in Fig. 9.12, the nonzero-LLR nodes can be ordered according to nondecreasing LLR as $o_{1,2}, o_{1,0}, o_{1,d_c-1}, o_{2,0}, \cdots$. Accordingly, $x(1) = 1$, $x(2) = 3$, $x(3) = 4$, $x(4) = 5 \ldots$ and $e(1) = 2$, $e(2) = 0$, $e(3) = d_c - 1$, $e(4) = 0 \ldots$. Now consider the path construction for computing v_{m,d_c-1}. Since $e(1) = 2 \neq d_c - 1$, path 1 is* *formed by replacing the node in stage 2 of path 0 by $o_{1,2}$ as shown in Fig. 9.12. Next consider node $o_{1,0}$, which can be also included in the path construction for v_{m,d_c-1} computation because $e(2) = 0 \neq d_c - 1$. Replacing the node in stage 0 of path 0 and path 1 by this node, path 2 and path 3, respectively, are constructed. The next node to be considered is o_{2,d_c-1}. However, this node is from stage $d_c - 1$, and hence can not be used in constructing paths for v_{m,d_c-1}. $o_{2,0}$ is the next node that can be included in the paths. Four paths have been constructed so far. By using $o_{2,0}$ to replace the zero-LLR node in stage 0 of path 0 and 1, path 4 and 5, respectively, are constructed. Nevertheless, path 2 and 3 have a nonzero-LLR node in the same stage as $o_{2,0}$. Replacing $o_{1,0}$ in path 2 and 3 by $o_{2,0}$ would lead to the same path 4 and 5, respectively, and hence these path constructions are skipped. As a result, by including $o_{2,0}$, only two instead of four new paths are constructed.*

Whether path k has a nonzero-LLR node in a certain stage can be tracked by using a d_c-bit binary vector P_k. The jth bit of P_k, $P_k(j)$, is '1' if path k passes a nonzero-LLR node in the jth stage, and is '0' otherwise. In addition, the field element of a newly constructed path does not have to be computed from the field element of each node in the path. If path k is constructed by replacing a zero-LLR node in stage j of path k', then the field element of path k is that of path k' plus the difference of the field elements of the nodes in stage j. Let the field element of the zero-LLR node in stage j be $z(j)$, and $\alpha_{sum} = \sum_{0 \leq j < d_c} z(j)$. The c2v LLR and field element vectors for variable node n, denoted by L_n and α_{L_n}, respectively, can be computed by Algorithm 34 [144].

Algorithm 34 Path Construction Algorithm
input*: $x(i), \alpha(i), e(i)$ $i = 1, 2, 3, \ldots$*
initialization*: $L_n(0) = 0$, $\alpha_{L_n}(0) = \alpha_{sum} + z(n)$*
$\qquad\qquad\quad i = 1$, $num = 1$, $P_0 = [0, 0, \cdots, 0]$
loop: if $(e(i) \neq n)$
$\qquad\quad \kappa = num$
$\qquad\quad$ for $k = 0$ to $\kappa - 1$
$\qquad\qquad\quad \delta = \alpha_{L_n}(k) \oplus z(e(i)) \oplus \alpha(i)$
$\qquad\qquad\quad$ if $(P_k(e(i)) \neq 1)$ & $(\delta \notin \alpha_{L_n})$
$\qquad\qquad\qquad\quad L_n(num) = x(i)$
$\qquad\qquad\qquad\quad \alpha_{L_n}(num) = \delta$
$\qquad\qquad\qquad\quad P_{num} = P_k$
$\qquad\qquad\qquad\quad P_{num}(e(i)) = 1$
$\qquad\qquad\qquad\quad num = num + 1$
$\qquad\qquad\qquad\quad$ if $(num = q')$, stop
$\qquad i = i + 1$; goto loop

In Algorithm 34, i tracks the last nonzero-LLR node that has been considered in the path construction. The variable num records the number of valid entries that have been derived for the output c2v message vector. Since the paths with the same field elements as those of previously constructed paths are not recorded, they do not contribute to later path constructions. As a result, Algorithm 34 may not construct all possible paths. However, it is very unlikely that an unconstructed path has one of the q' smallest LLRs and distinct field element. A maximum iteration number can be set to limit the worst-case latency of Algorithm 34. If less than q' valid entries have been derived for the c2v output vector at the end, then the LLRs of the rest entries are set to the LLR of the last constructed path, and the field elements of rest entries are set to zero. This compensation is similar to that of the forward-backward check node processing with q' messages in each vector discussed previously. Simulation results showed that if the maximum iteration number is set to $2q'$, then the trellis-based path construction check node processing in Algorithm 34 has the same error-correcting performance as the forward-backward scheme [144].

The complexity of sorting the v2c LLRs is dependent on the number of nonzero-LLR nodes in the trellis that will be involved in the path construction. From Algorithm 34, even if $e(i) = n$, one iteration is spent on finding the testing result. One iteration is also carried out in the case that $(P_k(e(i)) = 1)$ or $(\beta \in \alpha_{L_n})$. Hence, the number of iterations that will be spent on constructing paths by considering a new node with $e(i) \neq n$ is equal to the number of valid paths that have already been constructed. Note that those valid paths have distinct finite field elements. Therefore, if fewer new paths with distinct field elements are constructed when new nodes are considered, then more nodes will be involved in the path construction in a given number of iterations. The worst case happens when the $q' - 1$ nodes with the smallest nonzero LLRs all belong to stage n. In this case, after $q' - 1$ iterations are carried out in Algorithm 34, there is still only one valid path, which is path 0. Then path 1 is constructed when the q'^{th} nonzero-LLR node is considered. This path has a field element different from that of path 0, and hence is kept as a valid path. After that, if each new node to be considered is in a different stage and all the paths that can be constructed have the same field element as that of either path 0 or path 1, then two iterations of Algorithm 34 are spent on each new node. Assume that the maximum number of iterations allowed in Algorithm 34 is set to $maxloop$. Then the maximum number of nonzero-LLR nodes that will be involved in the path construction, $maxnum$, can be computed from $q' - 1 + 1 + 2(maxnum - q') = maxloop$. For example, if at most $2q'$ iterations are allowed to be carried out in Algorithm 34, then at most $1.5q'$ nodes are involved in the path construction, and hence only $1.5q'$ v2c messages with the smallest nonzero LLR need to be sorted out.

The implementation architecture of the trellis-based path-construction CNU has two parts. The first part sorts out $1.5q'$ input v2c messages with the smallest nonzero LLR. The second part is a path constructor. It generates the c2v messages from the sorting results according to Algorithm 34.

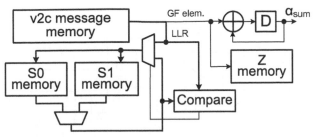

FIGURE 9.13
Serial sorter architecture for trellis-based path-construction check node processing (modified from [144])

The architecture of a sorter that compares the messages in each v2c input vector serially is shown in Fig. 9.13. The pair of S0 and S1 memories are used to store the sorting results in a ping-pong manner. Each of these memory blocks stores $1.5q'$ messages and the indices of the variable nodes from which they are taken. Hence each of them has $1.5q'(w+\log_2 q+\log_2 d_c)$ bits, assuming that each LLR is represented by w bits. The field elements with zero LLR are read from the v2c message memory one at a time to compute α_{sum} using the feedback loop in the top right corner of Fig. 9.13. In addition, they are also stored into the Z memory, whose size is $d_c \times \log_2 q$-bit. The sorting takes d_c rounds, and each round takes care of the comparisons with one v2c vector. In the first round, all the $q'-1$ messages with nonzero LLRs in the first v2c vector are copied into the S0 memory, and zero is written into each entry of S0 as the variable node index of the messages. In the second round, the messages stored in S0 are compared with those in the second v2c vector serially. Since the messages in each v2c vector are stored in the order of nondecreasing LLR, the comparisons start from the first nonzero-LLR messages in the two vectors. The one with smaller LLR is written into S1 with the corresponding variable node index. Then the read address of the vector that contributed the smaller LLR is increased by one in the next clock cycle, so that the next message is read out for comparison. Such a process is repeated until S1, which stores the comparison output vector in this round, has $1.5q'$ entries. Starting from the third round, the S0 and S1 memories store the input and output vectors of the comparisons in a ping-pong manner, and one additional v2c vector is compared in each round. Since $1.5q'$ entries are kept in the comparison output vector, each of the second and later rounds needs $1.5q'$ clock cycles. Moreover, reading the finite field element corresponding to the zero-LLR entry from the v2c message memory takes one clock cycle in each round. Therefore, the entire sorting process needs $q' + (d_c-1)(1.5q'+1)$ clock cycles using the architecture in Fig. 9.13.

The architecture of the path constructor that implements Algorithm 34 is depicted in Fig. 9.14. The $1.5q'$ sorted v2c messages are input to the path constructor one at a time. Note that each sorted message has three parts:

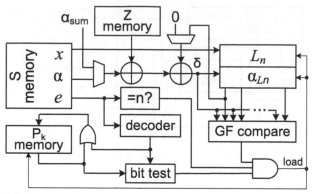

FIGURE 9.14
Path construction architecture (modified from [144])

LLR denoted by $x(i)$, the corresponding finite field element $\alpha(i)$, and $e(i)$, the index of the v2c vector from which the message is taken. During the path construction for computing $v_{m,n}$, whether $e(i) = n$ is first tested. This can be implemented by $\log_2 d_c$ XOR gates whose outputs are sent to an $\log_2 d_c$-input OR gate. Only when $e(i) \neq n$, the path construction is carried out using the sorted message. $e(i)$ is sent to a binary decoder, whose output is a one-hot d_c-bit vector with the $e(i)^{th}$ bit equals to '1' and the others equal to '0'. In the bit test block of Fig. 9.14, bit-wise AND is carried out on the binary decoder output vector and the vector P_k, and the results are sent to a NOR gate. Hence, the output of the bit test block is asserted when $P_k(e(i)) \neq 1$. The field element for the newly constructed path, δ, is derived by two finite field adders. To enable the computation of $\alpha_{L_n}(0)$, two multiplexors have been added to the inputs of the adders as shown in Fig. 9.14. When a new path is constructed, δ needs to be compared to each element in α_{L_n} in parallel to decide whether the new path should be inserted into the output vector. The GF comparator outputs '1' only when δ does not match any of the entries in α_{L_n}. When the outputs of the equality tester, bit test block, and the GF comparator are all '1', the load signal in Fig. 9.14 is asserted, and accordingly δ and $x(i)$ of the sorted messages are loaded into the α_{L_n} and L_n vectors, respectively. In addition, the P_k vector for the new path is generated by the bit-wise OR gates and stored in memory.

Using the trellis-based path construction CNU architecture, only $1.5q'$ sorted v2c messages need to be recorded for each check node. Compared to the intermediate message vectors of the forward and backward processes, these sorted messages require significantly smaller memory to store. In addition, as it can be observed from Figs. 9.13 and 9.14, the logics for implementing the sorter and path constructor are not complex. It has been shown in [144] that, to achieve the same throughput, the path construction CNU requires much smaller area than the forward-backward CNU keeping q' messages in each vec-

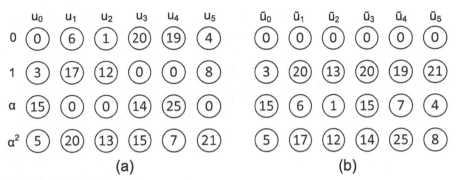

FIGURE 9.15
Example trellises when $d_c = 6$ and all $q = 4$ messages are kept for each vector; (a) original trellis, (b) reordered trellis

tor. Another advantage of the path construction CNU is that once the sorted v2c messages are available, multiple c2v messages can be computed independently in parallel. It does not suffer from the data dependency among the message vectors posed by the forward and backward processes. Nevertheless, despite the lower cost and faster speed compared to forward-backward CNUs, the path construction CNU requires an iterative process to compute the messages in each c2v vector. This iterative process prevents further speedup from being achieved.

9.2.3 Simplified Min-max check node processing

The CNU architectures introduced so far use (9.3) to compute the c2v messages. The configuration set $\mathcal{L}(m|a_n = \beta)$ specifies the constraints to be satisfied in c2v message computations. They are the constraints that were followed in the iterative forward and backward processes and the iterative path constructions. According to the configuration set, exactly one message is taken from each input v2c vector, except for the vector from variable node n, to compute a candidate c2v message. This requirement is broken in the simplified Min-sum Algorithm [145] in order to allow additional hardware saving. The same simplification scheme can be also applied to achieve a simplified Min-max algorithm (SMMA).

The SMMA keeps all q messages in each vector and makes use of a reordered trellis representation of the v2c vectors. To simplify the notations, the check node index m is dropped from the message vector notations when no ambiguity occurs. Let $\hat{\beta}_j$ be the finite field element of the most reliable message in u_j. Define $\tilde{u}_j(\beta) = u_j(\beta + \hat{\beta}_j)$ for each $\beta \in GF(q)$. Since there is a bijective mapping from β to $\hat{\beta}_j + \beta$, the messages in the $\tilde{u}_j$ vector are those in u_j re-ordered. $\tilde{u}_j$ for $j = 0, 1, \ldots, d_c - 1$ can still be represented by a trellis, and such a trellis is referred to as the reordered trellis. Fig. 9.15 shows the original trellis and the corresponding reordered trellis for an example case of $d_c = 6$ and $q = 4$. The numbers in the nodes are quantized LLR values. Since all messages are kept for each vector, the finite field elements for the messages do not need to be explicitly stored if the messages in each vector are recorded according to a pre-determined order. For both of the trellises in Fig. 9.15, the messages are stored according to their field elements in the order of $0, 1, \alpha, \alpha^2$. Take the messages in u_1 as an example. In the original trellis, $u_1(\alpha) = 0$ and $\hat{\beta}_1 = \alpha$. For finite field $GF(4)$, $0 + \alpha = \alpha$, $1 + \alpha = \alpha^2$, $\alpha + \alpha = 0$, and $\alpha^2 + \alpha = 1$. Therefore, $\tilde{u}_1(0) = u_1(\alpha) = 0$, $\tilde{u}_1(1) = u_1(1 + \alpha) = u_1(\alpha^2) = 20$, $\tilde{u}_1(\alpha) = u_1(0) = 6$, and $\tilde{u}_1(\alpha^2) = u_1(\alpha^2 + \alpha) = u_1(1) = 17$. The check node processing can be carried out using the reordered trellis representation provided that the computed c2v messages are reversely reordered. Denote the output vector of the check node processing using the reordered trellis by $\tilde{v}_j$. Assume that the multiplications with the entries of the generator matrix are done separately before and after the check node processing. Then the field element of a c2v message is the sum of the field elements of the involved v2c messages. As a result, $\tilde{v}_j$ should be reversely reordered to derive the actual c2v message vector v_j through $v_j(\beta) = \tilde{v}_j(\beta + d_j)$, where $d_j = \sum_{i \in S_v(m) \setminus j} \hat{\beta}_i$.

As it has been observed in the trellis-based path construction scheme, only a few nodes in the trellis with the smallest nonzero LLR contribute to the c2v messages. Instead of using as many nodes with the smallest nonzero LLR as needed to derive enough number of messages in the output c2v vector or until the maximum iteration number for path construction is reached, the SMMA algorithm puts a limitation on the number of nonzero-LLR nodes involved in the c2v message computation. In addition, as shown in Fig. 9.15 (b), the nodes in the first row of the reordered trellis have zero LLR and zero finite field element. Therefore, if a stage does not have any nonzero-LLR node selected for computing the c2v messages, the finite field element contribution from that stage is also zero. Assume that at most κ nonzero-LLR nodes are involved in the computation of the c2v message vector $\tilde{v}_{m,n}$. The κ nodes can come from any of the d_c stages, except stage n. Denote these stages by $j_1, j_2, \ldots, j_\kappa$, and all possible combinations of the κ stages form a set $S_{n,\kappa}$. Since the messages in each stage are ordered according to their field elements, the relative LLR magnitudes of the messages can not be told from their locations in the stage. Hence, every message in the chosen stages needs to be evaluated in the min-max comparisons to compute the c2v messages. Assume that the field elements of the κ nodes from the κ chosen stages are $\alpha_1, \alpha_2, \ldots, \alpha_\kappa \in GF(q)$. Then (9.3) for the check node processing in the Min-max algorithm is modified as [145]

$$\tilde{v}_{m,n}(\beta) = \min_{\beta = \sum_{l=1}^{\kappa} \alpha_l} \left(\min_{j_1, j_2, \ldots, j_\kappa \in S_{n,\kappa}} \left(\max_{1 \le l \le \kappa} \tilde{u}_{m,j_l}(\alpha_l) \right) \right). \qquad (9.10)$$

The κ nodes are selected from distinct trellis stages in (9.10). Nevertheless, if more than one node is allowed to be chosen from a stage, (9.10) can be approximated as

$$
\begin{aligned}
\tilde{v}_{m,n}(\beta) &\approx \min_{\beta=\sum_{l=1}^{\kappa} \alpha_l} \left(\min_{j_1 \in S_{n,\kappa}} \min_{j_2 \in S_{n,\kappa}} \cdots \min_{j_\kappa \in S_{n,\kappa}} \left(\max_{1 \le l \le \kappa} \tilde{u}_{m,j_l}(\alpha_l) \right) \right) \\
&= \min_{\beta=\sum_{l=1}^{\kappa} \alpha_l} \left(\max_{1 \le l \le \kappa} \left(\min_{j_l \in S_{n,\kappa}} \tilde{u}_{m,j_l}(\alpha_l) \right) \right).
\end{aligned}
\tag{9.11}
$$

Although (9.11) does not follow the definition of the configuration set in the original Min-max check node processing of (9.3), and it does not satisfy the requirement that the v2c messages involved in the computation of a c2v message need to be from distinct stages of the trellis, simulation results show that (9.11) leads to only negligible loss in the error-correcting capability [145]. This can be explained as follows. In the case that two or more nodes from the same stage are used in the computation of a c2v message, the second and other nodes can be considered as nodes with the same field elements from other distinct stages. Since the nodes with the same field elements in other stages may have larger or smaller LLRs, using two or more nodes from the same stage in a c2v message computation results in over- or under-estimation of the c2v LLR. However, the original Min-max algorithm is already an approximation of BP. The over- and under-estimation due to including more than one node from a stage would either compensate or deviate the LLR errors caused by the Min-max approximation. Therefore, including multiple nodes from the same stage in the computation of a c2v message does not lead to noticeable performance loss.

According to (9.11), the v2c messages that contribute to the c2v vector $\tilde{v}_{m,n}$ are $\min_{j_l \in S_{n,\kappa}} \tilde{u}_{m,j_l}(\alpha_l)$. α_l can be any element of $GF(q)$. Therefore, the v2c messages that may contribute to $\tilde{v}_{m,n}$ correspond to the nodes with the smallest LLR in each of the second and later rows of the reordered trellis, excluding those nodes from stage n. A similar idea as that in the Min-sum algorithm for binary LDPC decoding can be adopted to simplify the computations of $\tilde{v}_{m,n}$ with different n. If the node of the smallest LLR in a row of the trellis is in stage n, then the node with the second minimum LLR in that row is used in the computation of $\tilde{v}_{m,n}$. Therefore, to compute $\tilde{v}_{m,n}$ for all $n \in S_v(m)$, only the nodes with the smallest and second smallest LLR need to be found for each nonzero-LLR row of the trellis. These nodes are referred to as the min1 and min2 nodes of the corresponding row. Depending on n, one of them is picked to be $\min_{j_l \in S_{n,\kappa}} \tilde{u}_{m,j_l}(\alpha_l)$. After that, min-max comparisons are carried out as in (9.11) to derive the c2v messages.

Let $\tilde{v}_{m,n}^{(0)}(\alpha_l) = \min_{j_l \in S_{n,\kappa}} \tilde{u}_{m,j_l}(\alpha_l)$. The min and max over $\tilde{v}_{m,n}^{(0)}$ in (9.11) still involve many comparisons. For an $1 < l' < l$, (9.10) can be also approximated as

$$
\tilde{v}_{m,n}(\beta) \approx \min_{\beta=\sum_{l=1}^{\kappa} \alpha_l} \left(\min_{j_1,\dots,j_{l'} \in S_{n,\kappa}} \min_{j_{l'+1},\dots,j_\kappa \in S_{n,\kappa}} \left(\max_{1 \le l \le \kappa} \tilde{u}_{m,j_l}(\alpha_l) \right) \right).
$$

Accordingly, the min-max comparisons are broken up and carried out over two groups. Applying similar approximations iteratively, the min-max comparisons can be carried out in layers, and each layer takes care of the comparisons over a pair of intermediate messages. The first layer computes

$$\tilde{v}_{m,n}^{(1)}(\beta) = \min_{\beta=\alpha_1+\alpha_2} \left(\max(\tilde{v}_{m,n}^{(0)}(\alpha_1), \tilde{v}_{m,n}^{(0)}(\alpha_2)) \right),$$

and the kth layer derives

$$\tilde{v}_{m,n}^{(k)}(\beta) = \min_{\beta=\alpha_1+\alpha_2} \left(\max(\tilde{v}_{m,n}^{(k-1)}(\alpha_1), \tilde{v}_{m,n}^{(k-1)}(\alpha_2)) \right).$$

$\log_2 \kappa$ such layers are needed to derive $\tilde{v}_{m,n}$. The comparisons are made modular by using this iterative process. In addition, fewer comparisons are needed for later layers by using the properties of finite field elements. Take $GF(8)$ as an example, for a given $\beta \neq 0$, there are exactly 8 pairs of $\alpha_1, \alpha_2 \in GF(8)$ such that $\alpha_1 + \alpha_2 = \beta$. However, both input vectors of the first layer comparisons are $\tilde{v}_{m,n}^{(0)}$. The max comparisons do not need to be repeated over the same pair of messages. Hence, the min and max comparisons are done over four pairs of messages for computing each $\tilde{v}_{m,n}^{(1)}(\beta)$ with $\beta \neq 0$ in the first layer. The corresponding pairs of finite field elements, (α_1, α_2), are listed in Table 9.5. In this table, the finite field elements are denoted by the integer values of the corresponding standard basis representation. For example, the element '011' in standard basis representation is denoted by 3. To differentiate the notations, let the constraint for the second layer comparisons be $\beta = \beta_1 + \beta_2$. It seems that four pairs of messages in $\tilde{v}_{m,n}^{(1)}$ with $\beta_1 + \beta_2 = \beta$ need to be compared to derive $\tilde{v}_{m,n}^{(2)}(\beta)$. Nevertheless, $(\alpha_1 + \alpha_2) + (\alpha_3 + \alpha_4) = (\alpha_1 + \alpha_3) + (\alpha_2 + \alpha_4)$. Hence the comparison result for $\beta_1 = \alpha_1 + \alpha_2$ and $\beta_2 = \alpha_3 + \alpha_4$ would be the same as that for $\beta_1 = \alpha_1 + \alpha_3$ and $\beta_2 = \alpha_2 + \alpha_4$. As a result, half of the message pair comparisons can be eliminated in the second layer, and the comparisons only need to be done over the pairs listed in Table 9.6. This table is basically the first two columns of Table 9.5, since any pair in the last two columns of Table 9.5 is covered by those in the first two columns. For example, $\beta_1 = 4$ and $\beta_2 = 5$ can be broken down into $4 = 0 + 4$ and $5 = 2 + 7$. Since $0 + 2 = 2$ and $4 + 7 = 3$, $\beta_1 = 2$ and $\beta_2 = 3$ cover the same constraint. It can be easily derived that the pairs of messages need to be compared are reduced to half in each additional layer.

The CNU for the SMMA can be implemented by the architecture in Fig. 9.16. In this figure, the details of the first min-max units in layer 1 and 2 are shown by using $GF(8)$ as an example. For other finite fields and cases with $\kappa > 4$, tables similar to Table 9.5 and 9.6 can be derived and the inputs of the min-max units are decided according to the tables. The reordering network is implemented by $\log_2 q$ stages of q 2-input multiplexors. The reordered messages $\tilde{u}_{m,j}(\beta)$ for $j = 0, 1, \ldots d_c - 1$ are sent to a min1 & min2 sorter. The same sorter architectures as those discussed in Chapter 8 for binary LDPC decoders can be used here. Denote the minimum and second minimum LLRs among $\tilde{u}_{m,j}(\beta)$ by $min1(\beta)$ and $min2(\beta)$, respectively. In addition, the index

TABLE 9.5

(α_1, α_2) of pairs of messages from $\tilde{v}_{m,n}^{(0)}$ used in computing $\tilde{v}_{m,n}^{(1)}(\beta)$

β	(α_1, α_2)			
1	(0,1)	(2,3)	(4,5)	(6,7)
2	(0,2)	(1,3)	(4,6)	(5,7)
3	(0,3)	(1,2)	(4,7)	(5,6)
4	(0,4)	(1,5)	(2,6)	(3,7)
5	(0,5)	(1,4)	(2,7)	(3,6)
6	(0,6)	(1,7)	(2,4)	(3,5)
7	(0,7)	(1,6)	(2,5)	(3,4)

TABLE 9.6

(β_1, β_2) of pairs of messages from $\tilde{v}_{m,n}^{(1)}$ used in computing $\tilde{v}_{m,n}^{(2)}(\beta)$

β	(β_1, β_2)	
1	(0,1)	(2,3)
2	(0,2)	(1,3)
3	(0,3)	(1,2)
4	(0,4)	(1,5)
5	(0,5)	(1,4)
6	(0,6)	(1,7)
7	(0,7)	(1,6)

of $min1(\beta)$ is recorded as $idx(\beta)$. In the computation of $v_{m,n}$, the min selector in Fig. 9.16 chooses $min1(\beta)$ if $n \neq idx(\beta)$ or $min2(\beta)$ otherwise to be $\tilde{v}_{m,n}^{(0)}(\beta)$. The $\tilde{v}_{m,n}^{(0)}$ vector is sent to each layer 1 min-max unit. Similarly, the outputs of all layer 1 min-max units, which form $\tilde{v}_{m,n}^{(1)}$, are sent to each layer 2 min-max unit. $\tilde{v}_{m,n}$ is available at the outputs of the last-layer min-max units. $d_n = \sum_{j \in S_v(m) \setminus n} \hat{\beta}_j$ decides how $\tilde{v}_{m,n}$ is reversely reordered. To eliminate redundant computations, $\hat{\beta}_{sum} = \sum_{j \in S_v(m)} \hat{\beta}_j$ is derived first. Then d_n is calculated by a single finite field addition as $d_n = \hat{\beta}_{sum} + \hat{\beta}_n$.

9.2.4 Syndrome-based Min-max check node processing

In the check node processing of the SMMA, $min1(\beta)$ and $min2(\beta)$ are the two smallest LLRs in the row for finite field element β in the reordered trellis. Since $min2(\beta)$ is only used in the computations for $\tilde{v}_{m,n}$ when $n = idx(\beta)$, the same computations are involved in calculating the c2v vectors corresponding to the stages that do not have any min1 nodes. Hence, a lot of redundant

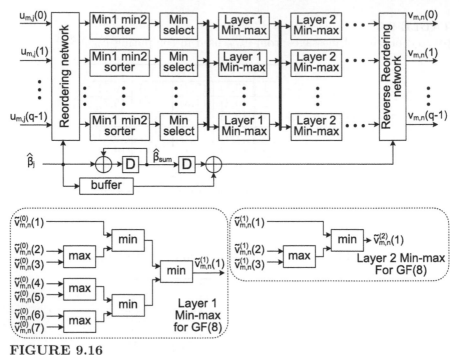

FIGURE 9.16
CNU architecture for the SMMA

computations are carried out in the SMMA. To eliminate these redundancy, a syndrome-based check node processing approach was developed in [146]. An extra column is added to the reordered trellis to represent the syndromes, which include the contributions of the nodes from every stage. Then the contributions of the nodes from stage n, if there are any, are removed from the syndromes to compute $\tilde{v}_{m,n}$. Since the syndromes only need to be computed once for each check node, and all c2v vectors from the same check node are derived by taking care of the differences from the syndromes, the computation complexity is greatly reduced.

The algorithm in [146] is for the EMS algorithm. It can be directly modified for the Min-max algorithm. However, the removal of the contributions of a node from the syndromes is better explained as subtractions of LLRs in the EMS algorithm. Hence the EMS algorithm is used next to explain the syndrome-based check node processing. Let $\mathcal{T}(m|\beta)$ be the configuration set of the sequences of d_c symbols (a_j) $(j \in S_v(m))$ such that $\sum_{j \in S_v(m)} a_j = \beta$. The syndromes for the EMS algorithm are defined as

$$w(\beta) = \min_{(a_j) \in \mathcal{T}(m|\beta)} \left(\sum_{j \in S_v(m)} \tilde{u}_{m,j}(a_j) \right). \tag{9.12}$$

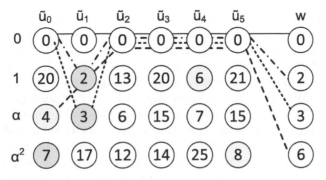

FIGURE 9.17
Example reordered trellis for the syndrome-based check node processing with $n_r = 2$, $n_c = 2$ of a $d_c = 6$ code over $GF(4)$

As analyzed in [144] and [145], only a small number of nonzero-LLR nodes in the trellis contribute to the c2v message vectors. These nodes are called the deviation nodes in [146]. The complexity of the syndrome-based check node processing is adjusted by making use of two parameters n_r and n_c. Only the n_r nodes with the smallest LLR in each nonzero row of the reordered trellis are considered for c2v message computations, and n_c is the maximum number of deviation nodes can be included in a configuration. The configuration set under such constraints is denoted by $\mathcal{T}_{n_r,n_c}(m|\beta)$. n_r and n_c also affect the error-correcting performance of the decoder. Simulation results in [146] showed that taking $n_r = 2$ and small n_c, such as 3 or less, leads to negligible performance degradation compared to the original EMS algorithm.

From the definition of $\mathcal{T}(m|\beta)$, the nodes in a configuration should come from distinct stages of the trellis. Moreover, for codes over finite fields of characteristic two, only those configurations with deviation nodes from distinct rows need to be considered. If two nodes in a configuration (a_j) are from the same row, then the sum of their field elements is zero. Hence, there should be another configuration $(a_j)'$ in $\mathcal{T}(m|\beta)$ that is the same as (a_j) except the two nodes from the same row in (a_j) are replaced by zero-LLR nodes. The deviation nodes have LLRs larger than zero. Since 'min' is taken over all the configurations in $\mathcal{T}(m|\beta)$ in (9.12), a configuration with two deviation nodes in the same row does not lead to smaller min-max result. Accordingly, only the configurations that have deviations nodes from distinct rows need to be considered in the syndrome computation.

Fig. 9.17 illustrates the trellis for an example code over $GF(4)$ with $d_c = 6$. The paths in the trellis show how the syndromes are computed when $n_r = 2$ and $n_c = 2$. The syndrome of $\beta = 0$ should be $w(0) = 0$. Assume that α is a primitive element of $GF(4)$. Then the four elements of $GF(4)$ are $0, 1, \alpha, \alpha^2$ and $1 = \alpha + \alpha^2$. Since at most $n_c = 2$ deviation nodes can be included, the configurations with either one deviation node from row β or two deviation nodes from the other two nonzero rows need to be considered to compute each $w(\beta)$ with $\beta \neq 0$. The minimum is taken over the sum of the LLRs for all configurations according to (9.12). Since the min1 node has smaller LLR than the min2 node of the same row, the only configuration with one deviation node to be considered is the one with min1 node in row β. However, there are multiple possible configurations with two deviation nodes from the other two rows. If the min1 nodes of the two rows are from different stages, then the configuration with those two min1 nodes is valid and it has the smallest LLR among the configurations with two deviations nodes. However, if those two nodes are from the same stage, they can not be both included in a configuration according to (9.12). In this case, the configurations with cross-over min1 and min2 nodes from the two rows need to be considered. Apparently, the configuration with the two min2 nodes has larger LLR, and hence does not contribute to the minimum sum of LLRs. In the example trellis shown in Fig. 9.17, the dark-shaded circles are the min1 nodes, and those light-shaded ones are the min2 nodes. The min1 nodes in row α and α^2 are from different stages. Hence, for the computation of $w(1)$, the configuration with the min1 node from row 1 is compared to the configuration with the two min1 nodes from row α and α^2. Since $2 < 3+7$, $w(1) = 2$. The computation of $w(\alpha)$ is done similarly. However, the min1 nodes in row 1 and α are in the same stage. Therefore, to compute $w(\alpha^2)$, the configurations with cross-over min1 and min2 nodes in these two rows are needed. Since $2 + 4 = 6 < 3 + 6 = 9$ and 6 is smaller than 7, which is the LLR of the configuration with a single min1 node in row α^2, $w(\alpha^2) = 6$. Each path in Fig. 9.17 shows the nodes in the configuration that has the smallest sum of LLRs, and hence leads to the corresponding syndrome.

$\tilde{v}_{m,j}(0)$ for $j = 0, 1, \ldots, d_c - 1$ are zero. The other messages in the $\tilde{v}_{m,j}$ vector can be derived from the syndromes by taking out the contributions of the nodes in stage j. Assume that the configuration corresponding to $w(\beta)$ is $\eta^{(\beta)} = [\eta_0^{(\beta)}, \eta_1^{(\beta)}, \ldots, \eta_{d_c-1}^{(\beta)}]$. The nonzero messages in each $\tilde{v}_{m,j}$ vector are initialized with the maximum possible LLR representable by a given word length. For each syndrome, $w(\beta)$, the following computations are carried out for $j = 0, 1, \ldots, d_c - 1$ to derive the c2v messages [146]

$$\tilde{v}_{m,j}(\beta - \eta_j^{(\beta)}) = \min(\tilde{v}_{m,j}(\beta - \eta_j^{(\beta)}), w(\beta) - \tilde{u}_{m,j}(\eta_j^{(\beta)})). \tag{9.13}$$

Basically, the finite field element and LLR of the node in stage j are subtracted from those of the syndrome to derive a candidate for the c2v message $\tilde{v}_{m,j}(\beta - \eta_j^{(\beta)})$. The candidate LLR, $w(\beta) - \tilde{u}_{m,j}(\eta_j^{(\beta)})$, is compared with the previous LLR, and the smaller one is kept. It should be noted that different β may lead to the same $\beta - \eta_j^{(\beta)}$. Therefore, the computations in (9.13) are repeated iteratively for each syndrome to derive the c2v messages. Additionally, some of the c2v messages may not be covered by the computations in (9.13). If $\eta_j^{(\beta)} \neq 0$, then $\beta - \eta_j^{(\beta)} \neq \beta$. Therefore, $\tilde{v}_{m,j}(\beta)$ for $\beta \neq 0$ is not covered by (9.13) if there is a deviation node in stage j. The solution in [146] is to set

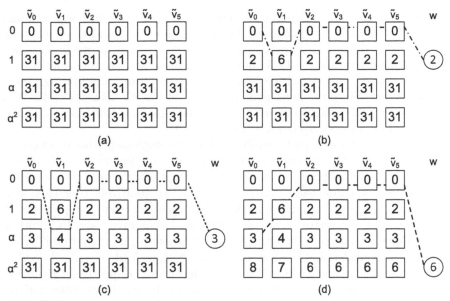

FIGURE 9.18
Example c2v message computations from the syndromes. (a) initial c2v messages; (b) c2v messages after the computations using $w(1)$; (c) c2v messages after the computations using $w(\alpha)$; (d) c2v messages after the computations using $w(\alpha^2)$

$\tilde{v}_{m,j}(\beta)$ as $min1(\beta)$ if this min1 node is not in stage j. Otherwise, $v_{m,j}(\beta)$ is set to $min2(\beta)$.

For the example trellis in Fig. 9.17, the c2v message computations using the syndromes are explained in Fig. 9.18. Part (a) of this figure shows the initial c2v LLR values assuming 5-bit word length is adopted. The syndrome nodes and configurations leading to the syndromes are kept in this figure to better illustrate the updating of the c2v LLRs. The configuration leading to $w(1)$ is $[0, 1, 0, 0, 0, 0]$. Hence, $\tilde{v}_j(1 - 0) = \tilde{v}_j(1)$ are updated as $min(31, w(1) - 0) = 2$ for $j = 0, 2, 3, 4, 5$. For stage $j = 1$, the updating from (9.13) is done on $\tilde{v}_1(1 - 1) = \tilde{v}_1(0)$ instead, since $\tilde{v}_1(0)$ already has the smallest possible LLR, which is zero, it is not changed. Note that $\tilde{v}_1(1)$ has not been assigned any value from (9.13). Since the corresponding node in the v2c message trellis is the min1 node for row 1, $\tilde{v}_1(1)$ is assigned the value of the min2 node in this row, which is 6 according to Fig. 9.17. The values of the c2v messages after these computations are shown in Fig. 9.18(b). The updating of the c2v messages using $w(\alpha)$ is similar, and $\tilde{v}_1(\alpha)$ is assigned the LLR of the min2 node of the v2c messages in row α, which is 4. The updated messages after these steps are depicted in part (c) of the figure. The configuration leading to $w(\alpha^2)$ is $[\alpha, 1, 0, 0, 0, 0]$, and has two deviation nodes. For $j = 2, 3, 4, 5$,

$\tilde{v}_j(\alpha^2 - 0) = \tilde{v}_j(\alpha^2)$ are updated as $\min(31, w(\alpha^2) - 0) = 6$. For $j = 0$ and 1, $w(\alpha^2) = 6$ yields two candidate LLRs, $6 - \tilde{u}_0(\alpha) = 2$ and $6 - \tilde{u}_1(1) = 4$, for updating $\tilde{v}_0(\alpha^2 - \alpha) = \tilde{v}_0(1)$ and $\tilde{v}_1(\alpha^2 - 1) = \tilde{v}_1(\alpha)$, respectively. They are compared with the old LLRs of these two messages, and the smaller ones are kept. $\tilde{v}_0(\alpha^2)$ and $\tilde{v}_1(\alpha^2)$ have not been assigned any value through these computations. Since the min1 node in row α^2 of the v2c message trellis is in stage 0, $\tilde{v}_0(\alpha^2)$ is assigned $min2(\alpha^2) = 8$, and $\tilde{v}_1(\alpha^2) = min1(\alpha^2) = 7$. The final c2v LLR values are shown in Fig. 9.18 (d).

The above process can be easily modified to implement the Min-max algorithm. Although the syndrome-based approach enables intermediate results to be shared among the computations of different c2v vectors, the algorithm in [146] has two drawbacks. Carrying out (9.13) iteratively over each syndrome not only requires a large number of comparators, but also prevents all messages in a c2v vector from being generated simultaneously. In addition, to compute a syndrome, configurations with cross-over min1 and min2 nodes need to be considered. The cross-over computations lead to multiple adders or comparators in the EMS or Min-max algorithms, respectively. By relaxing the constraints on the configurations and analyzing the possible updates involved in c2v message computations, a modified syndrome-based Min-max decoder was developed in [147] to overcome these two issues. It does not require cross-over comparisons, and the iterative operations are eliminated from the c2v message calculations. This modified syndrome-based decoder is presented next.

As it was observed in [145] and discussed earlier in this section, allowing more than one node from the same trellis stage to be included in a configuration only introduces negligible performance loss. This relaxation can be incorporated into the syndrome computation [147]. To carry out the Min-max algorithm, the 'sum' in (9.12) should be replaced by 'max'. In the approach of [146], cross-over maximums of the min1 and min2 LLRs from different rows are computed in case the min1 nodes of the rows belong to the same stage of the trellis. When multiple nodes from a trellis stage are allowed to be in a configuration, all the cross-over computations are eliminated, and the syndromes are derived only from the min1 nodes. In the case of $n_r = 2$ and $n_c = 2$, the number of comparators needed for computing a syndrome is reduced from three to one. The achievable reduction is more significant when n_r is larger.

The simplified method in [147] for computing the c2v messages from the syndromes is explained next using $GF(4)$ as an example. It can be also extended for higher-order finite fields. Since $GF(4)$ has three nonzero elements, n_r is at most 2 for codes over $GF(4)$. Let us start with the case where the configuration corresponding to $w(\beta)$ ($\beta \neq 0$) has only one deviation node in stage j. In other words, $w(\beta) = min1(\beta)$, $\eta_j^{(\beta)} = \beta$, and $\eta_i^{(\beta)} = 0$ for $0 \leq i < d_c$ and $i \neq j$. Initially, $\tilde{v}_{m,i}(\beta)$ for $0 \leq i < d_c$ and $\beta \neq 0$ are set to the largest possible LLR. After the computations in (9.13) are carried out, $\tilde{v}_{m,i}(\beta)$ with $i \neq j$ becomes $w(\beta) = min1(\beta)$. In addition, $\tilde{v}_{m,j}(\beta - \eta_j^{(\beta)}) = \tilde{v}_{m,j}(\beta - \beta) = \tilde{v}_{m,j}(0)$. Since $\tilde{v}_{m,j}(0)$ is zero initially, it will not get updated. No value has been derived for $\tilde{v}_{m,j}(\beta)$ through (9.13), and $\tilde{u}_{m,j}(\beta)$ is the min1 value for row β in the trellis. Using a similar approach as that in [146], $\tilde{v}_{m,j}(\beta)$ is set to $min2(\beta)$ for the Min-max algorithm so that the contribution of the v2c message from variable node j is excluded. From these computations, $\tilde{v}_{m,i}(\beta)$ $(0 \leq i < d_c)$ are derived. They correspond to the same field element as the syndrome $w(\beta)$. For each of the other syndromes that have only one deviation node in the corresponding configuration, the associated computations will yield distinct c2v messages. Therefore, applying (9.13) to the syndromes with one deviation node will not update any c2v message for a second time.

Next consider the case in which there are two deviation nodes in the configuration corresponding to $w(\beta)$ $(\beta \neq 0)$. Assume that they are from stage i and j. Accordingly, $\eta_i^{(\beta)} + \eta_j^{(\beta)} = \beta$. In the Min-max algorithm, $w(\beta) = \max(min1(\eta_i^{(\beta)}), min1(\eta_j^{(\beta)}))$. Excluding the contribution of the v2c messages in stage i from $w(\beta)$, it can be derived that $\tilde{v}_{m,i}(\beta - \eta_i^{(\beta)}) = \tilde{v}_{m,i}(\eta_j^{(\beta)})$ should be updated as $\min(\tilde{v}_{m,i}(\eta_j^{(\beta)}), min1(\eta_j^{(\beta)}))$. Similarly, $\tilde{v}_{m,j}(\eta_i^{(\beta)})$ should be updated as $\min(\tilde{v}_{m,j}(\eta_i^{(\beta)}), min1(\eta_i^{(\beta)}))$. It is possible that $\tilde{v}_{m,i}(\eta_j^{(\beta)})$ and $\tilde{v}_{m,j}(\eta_i^{(\beta)})$ have been computed previously from other syndromes. However, in those previous computations, they were set to $min1(\eta_j^{(\alpha)})$ and $min1(\eta_i^{(\alpha)})$, respectively, as discussed in the prior paragraph. As a result, the updating of $\tilde{v}_{m,i}(\eta_j^{(\beta)})$ and $\tilde{v}_{m,j}(\eta_i^{(\beta)})$ can be skipped because their values will not be changed. For $n \neq i, j$, $\tilde{v}_{m,n}(\beta)$ is updated as $w(\beta)$. Similarly, $\tilde{v}_{m,i}(\beta)$ and $\tilde{v}_{m,j}(\beta)$ are set to $min1(\beta)$ or $min2(\beta)$ based on if $min1(\beta)$ is in stage i or j. Since these c2v messages have the same finite field element as the syndrome, they will not be updated again when other syndromes are utilized to compute c2v messages.

From these analyses, Algorithm 35 [147] can be used to compute c2v messages from the syndromes in the case of $n_r = n_c = 2$. In this algorithm, idx is the stage index of the min1 node. When $n_r = n_c = 2$, excluding the contribution of a v2c message from a syndrome derived according to minimum of sums lead to the same value as that calculated as minimum of max. Hence, Algorithm 35 can be applied to both the Min-max and EMS algorithms. This algorithm generates exactly the same results as (9.13), and does not bring any error-correcting performance loss. Only one of the three values needs to be selected by multiplexors to compute a c2v message. Hence, the hardware complexity of this algorithm is substantially lower than that for implementing (9.13). Moreover, all the messages in a c2v vector can be generated in one clock cycle, and hence much higher throughput is achievable.

Algorithm 35 Simplified Syndrome-based c2v Message Computation

inputs: $w(\beta)$, $min1(\beta)$, $idx(\beta)$, $min2(\beta)$, $\eta^{(\beta)}$

$\quad$ *for each $\beta \neq 0$*

$\quad\quad$ *if there is one deviation node, and it is in stage i:*

$$\tilde{v}_{m,n}(\beta) = \begin{cases} min1(\beta) \text{ if } n \neq i \\ min2(\beta) \text{ if } n = i \end{cases}$$

$\quad\quad$ *if there are two deviation nodes, and they are in stages i and j:*

$$\tilde{v}_{m,n}(\beta) = \begin{cases} w(\beta) \text{ if } n \neq i, j \\ min1(\beta) \text{ if } (n = i \text{ or } j)\&(n \neq idx(\beta)) \\ min2(\beta) \text{ if } (n = i \text{ or } j)\&(n = idx(\beta)) \end{cases}$$

By making use of the property that $GF(4)$ has three nonzero elements, $1, \alpha, \alpha^2$ and $1 + \alpha = \alpha^2$, efficient architectures have been developed in [147] to implement the modified syndrome-based CNU. Since multiple nodes in the same stage of the trellis are allowed in a configuration, only two configurations need to be considered for computing each $w(\beta)$ with $\beta \neq 0$. One of them has a single deviation node $min1(\beta)$, and the other configuration has the other two min1 nodes. Accordingly, the syndromes for the Min-max algorithm are derived as

$$\begin{cases} w(1) = \min(min1(1), \max(min1(\alpha), min1(\alpha^2))) \\ w(\alpha) = \min(min1(\alpha), \max(min1(1), min1(\alpha^2))) \\ w(\alpha^2) = \min(min1(\alpha^2), \max(min1(1), min1(\alpha))) \end{cases}$$

A syndrome equals one of the three min1 values from the above equation. To reduce the hardware complexity, three comparators are employed to compare each pair of min1 values first. Then the comparison results are shared to generate the three syndromes. The architecture in Fig. 9.19 computes $w(1)$. Assume that the output of a 'max' unit is '0' if the upper input is larger than the lower input. Then the select signals of the multiplexors in Fig. 9.19 are $s0 = c$, and $s1 = a + b$. '+' here denotes logic OR. Besides the syndrome, the configuration leading to the syndrome also needs to be recorded. Each syndrome only has two possible configurations. Hence, a 1-bit flag is used to record the configuration for each syndrome. If the configuration with two deviation nodes leads to the syndrome, then the flag is set to '1'. Otherwise, it is '0'. The flag for $w(1)$, denoted by $f(1)$, is generated as $(a + b)'$. The computations of the other two syndromes and corresponding flags share the results of

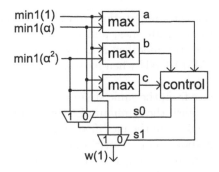

FIGURE 9.19
Syndrome computation architecture for codes over $GF(4)$ (modified from [147])

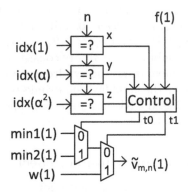

FIGURE 9.20
Simplified syndrome-based c2v message computation for codes over $GF(4)$ (modified from [147])

the comparators in Fig. 9.19 and only require additional pairs of multiplexors and simple logics.

The architecture for computing $\tilde{v}_{m,n}(1)$ from the syndrome is illustrated in Fig. 9.20. n is first tested to see if it equals to the index of any min1 node. A tester outputs '1' if n equals the index. From Algorithm 35, the possible values of $\tilde{v}_{m,n}(1)$ are $min1(1)$, $min2(1)$, and $w(1)$. Which one to choose depends on the configuration of $w(1)$ and if the min1 nodes are in stage n. It can be derived that the control signals for the multiplexors in Fig. 9.20 are $t0 = x$ and $t1 = f(1)(y + x)'$. Similarly, the outputs of the index testers are shared, and another two sets of multiplexors are needed to recover $\tilde{v}_{m,n}(\alpha)$ and $\tilde{v}_{m,n}(\alpha^2)$.

Simulation results in [147] showed that the error-correcting performance of the modified syndrome-based scheme is slightly better than that of the SMMA for $GF(4)$ codes. Comparing the architectures in Figs. 9.19 and 9.20 with that in Fig. 9.16, the modified syndrome-based CNU has lower complexity than the

SMMA CNU. Moreover, the computations of c2v vectors from the same check node can share the same syndromes, and only the parts in Fig. 9.20 need to be duplicated. Hence, the modified syndrome-based CNU has even lower complexity than the SMMA CNU when multiple c2v vectors are computed at a time.

When multiple nodes from the same stage are allowed to be in a configuration, only $n_r = 2$ nodes with the minimum LLRs need to be sorted from each row of the reordered trellis as in the modified syndrome-based check node processing method shown above. This algorithm can be easily extended for codes over higher-order finite fields as long as $n_c = 2$. For $GF(q)$, there are exactly q pairs of finite field elements whose sums equal to a given element. Therefore, when $n_c = 2$, there are q configurations in each $\mathcal{T}_{n_r,n_c}(m|\beta)$. Similarly, the comparisons needed to compute the $q-1$ syndromes can be shared, and hence $\binom{q-1}{2} = (q-1)(q-2)/2$ comparators are needed. If $n_c > 2$, the configuration sets become much larger, and many more comparisons are required. For a code over $GF(q)$, there are $\log_2 q - 1$ layers of min-max operators in the SMMA CNU architecture in Fig. 9.16. Layer i ($i = 1, 2, \ldots, \log_2 q - 1$) has $q(q/2^{i-1} - 2)$ comparators. Hence, the min-max operators consist of $2q(q - 1 - \log_2 q)$ comparators. Besides the comparators, the syndromes and c2v messages are derived by simple logics in the modified syndrome-based CNU. As a result, the modified syndrome-based CNU has a much smaller area than the SMMA CNU, especially when q is larger.

9.2.5   Basis-construction Min-max check node processing

By re-ordering the v2c message trellis and allowing multiple nodes from the same stage of the trellis to be included in a configuration, the SMMA [145] and modified syndrome-based [147] check node processing schemes achieve great complexity reduction. In addition, only the min1 and min2 nodes need to be stored, and hence the memory requirement is also substantially reduced. However, the complexity of both of them is $\mathcal{O}(q^2)$, and it increases fast with q.

Each element in $GF(2^p)$ can be uniquely expressed as a linear combination of p independent field elements. By utilizing the reordered trellis and also allowing multiple nodes from the same stage of the trellis to be used in the computation of a c2v message, an efficient method was developed in [148] to construct a minimum basis that has the p most reliable v2c messages with linearly independent field elements. Then the c2v message vector is derived from the minimum basis through linear combinations. For $q = 2^p$, the complexity of such a process is $\mathcal{O}(q \log_2 q)$. It has been shown in [148] that the CNU architecture based on the basis-construction method has a much lower gate count than the SMMA CNU for a NB-LDPC code over $GF(32)$, and the saving further increases with q. In addition, all the messages in a c2v vector are computed in parallel using a small combinational network in the basis-construction CNU. It overcomes the limitation of the trellis-based path construction CNU [144] that the messages in a c2v vector need to be computed iteratively. With a slightly larger gate count, it achieves a much higher throughput than the path construction CNU.

In the following, the check node index m is dropped to simplify the notations. Denote the minimum basis for computing $\tilde{v}_j$ by B_j. For $GF(2^p)$, B_j is a set consisting of exactly p nodes with the smallest LLRs and independent field elements that are not from the jth stage of the reordered trellis. Note that multiple nodes in B_j can be taken from the same stage of the trellis. Since the field elements of the nodes in B_j form a basis of $GF(2^p)$, any element $\beta \in GF(2^p)$ can be expressed as the sum of these field elements or a subset of them. Also the nodes in B_j have the smallest LLRs. Therefore, according to (9.3) for the Min-max algorithm, $\tilde{v}_j(\beta)$ equals the maximum LLR of the nodes in B_j whose field elements add up to be β.

A 4-step process can be used to construct B_j. Since the min1 and min2 nodes in the $q - 1 = 2^p - 1$ nonzero rows of the reordered trellis have the smallest LLRs, and the min1 and min2 nodes of the same row belong to different stages, all the entries in the minimum basis must be from the min1 and min2 nodes. Therefore, the first step of the minimum basis construction is to find the min1 and min2 nodes of each nonzero row in the reordered trellis. Two vectors $M_{min1} = \{(min1(\beta), idx(\beta))\}$ and $M_{min2} = \{min2(\beta)\}$ $(1 \leq \beta < q)$ are recorded. Here integer values corresponding to the binary strings of polynomial basis representations are used to denote finite field elements. For constructing B_j, the second step takes out the nodes from stage j. A vector $M_j = \{m_j(\beta)\}$ is obtained. $m_j(\beta) = min1(\beta)$ if $j \neq idx(\beta)$ and $m_j(\beta) = min2(\beta)$ otherwise. The third step sorts the entries in M_j and derives a vector $\bar{M}_j = \{(m_j^{(i)}, \alpha_j^{(i)})\}$ $(1 \leq i < q)$. $\alpha_j^{(i)}$ is the field element associated with the sorted LLR $m_j^{(i)}$, and $m_j^{(i)} \leq m_j^{(i')}$ if $i < i'$. In the last step, the minimum basis $B_j = \{(m_j^{'(i)}, \alpha_j^{'(i)})\}$ $(1 \leq i \leq p)$ is extracted from the entries of $\bar{M}_j$ using a function Ψ. Since the entries in $\bar{M}_j$ are sorted, the function Ψ tests one entry of $\bar{M}_j$ at a time starting from the first one. If the field element of the entry is independent of those already in B_j, the entry is added to B_j. This process is repeated until p entries are put in B_j. Using such a function, the nodes in B_j are also ordered according to increasing LLR.

The basis-construction check node processing scheme is summarized in Algorithm 36. Similar notations as those in the SMMA are used. $\hat{\beta}_j$ is the most likely symbol for the v2c message vector u_j, and $\tilde{u}_j$ is the reordered v2c vector. Steps 2-5 derive the minimum basis. Then the c2v messages are recovered from the minimum basis in Step 6. $w_1, w_2, \ldots, w_{p'}$ is a subset of p' indices from $\{1, 2, \ldots, p\}$ and $1 \leq p' \leq p$. Since $\alpha_j^{'(i)}$ $(1 \leq i < p)$ form a basis of $GF(2^p)$, the sums of the $2^p - 1$ possible combinations of the elements from $\alpha_j^{'(i)}$ are the $2^p - 1$ distinct nonzero elements of $GF(2^p)$. The LLR of β in $\tilde{v}_j$ is the maximum of $m_j^{'(i)}$ whose field elements add up to be β. At the end, $\tilde{v}_j$ is reversely reordered in Step 7 to get v_j. Simulation results in

[148] showed that using such a basis-construction check node processing does not result in noticeable performance loss compared to the original Min-max algorithm. Similar to the SMMA and syndrome-based check node processing, only the min1 and min2 nodes need to be stored as intermediate results in the basis-construction check node processing. On the other hand, the simple and parallelizable computations in Step 6 of Algorithm 36 lead to more efficient hardware implementation than the large number of min and max comparators in the SMMA and syndrome-based check node processing when q is not small.

Algorithm 36 Basis-construction Min-max Check Node Processing
input: u_j and $\hat{\beta}_j$ for $0 \leq j < d_c$
$$\hat{\beta}_{sum} = \sum_{0 \leq j < d_c} \hat{\beta}_j$$
start:
1:      $\tilde{u}_j(\beta) = u_j(\beta + \hat{\beta}_j)$ $(0 \leq j < d_c)$
2:      *compute* $M_{min1} = \{(min1(\beta), idx(\beta))\}$ $(1 \leq \beta < q)$
         $M_{min2} = \{min2(\beta)\}$ $(1 \leq \beta < q)$
     for $j=0$ *to* d_c-1

3:      $m_j(\beta) = \begin{cases} min1(\beta), & j \neq idx(\beta) \\ min2(\beta), & j = idx(\beta) \end{cases}$
         $M_j = \{m_j(\beta)\}$ $(1 \leq \beta < q)$
4:      *Sort* M_j *to obtain* $\bar{M}_j = \{(m_j^{(i)}, \alpha_j^{(i)})\}$, $(1 \leq i < q)$
5:      $B_j = \Psi(\bar{M}_j) = \{(m_j'^{(i)}, \alpha_j'^{(i)})\}$ $(1 \leq i \leq p)$
6:      $\tilde{v}_j(0) = 0$
         $\tilde{v}_j(\beta) = max\{m_j'^{(w_1)}, m_j'^{(w_2)}, \ldots m_j'^{(w_{p'})}\}$
         if $\beta = \alpha_j'^{(w_1)} \oplus \alpha_j'^{(w_2)} \oplus \cdots \oplus \alpha_j'^{(w_{p'})}$ $(1 \leq p' \leq p)$
7:      $v_j(\beta) = \tilde{v}_j(\beta + \hat{\beta}_{sum} + \hat{\beta}_j)$

Example 47 *For an example set of min1 and min2 nodes, some key values involved in Steps 3-6 of Algorithm 36 for computing $\tilde{v}_0$ are listed in Table 9.7. This example is for a code over $GF(2^3)$ with $d_c = 6$. Again, a field element is represented by the integer value of the corresponding binary string in polynomial basis. To compute $\tilde{v}_0$, the nodes from stage 0 should be excluded. Hence, for the entry with $\beta = 3$, the LLR from M_{min2} is put into M_0. The entries in M_0 are sorted to get $\bar{M}_0$ in Step 4. Then the $p = 3$ entries of $\bar{M}_0$ with the smallest LLR and independent field elements need to be extracted to form B_0. Since the first and second entries, $(0.1, 2)$ and $(0.2, 6)$, in $\bar{M}_0$ have independent field elements, they are put into B_0. The third entry of $\bar{M}_0$ is examined next. Its field element is 4. Because $4 = 2 \oplus 6$, the third entry of $\bar{M}_0$*

TABLE 9.7
Example values of variables in the basis-construction Min-max check node processing for computing $\tilde{v}_0$

	β	1	2	3	4	5	6	7
	M_{min1}	(0.7,2)	(0.1,3)	(0.4,0)	(0.3,2)	(0.5,1)	(0.2,5)	(1.1,4)
	M_{min2}	1.1	0.8	0.8	1.4	1.4	1.2	1.9
Step 3	M_0	0.7	0.1	0.8	0.3	0.5	0.2	1.1
Step 4	M_0	(0.1,2)	(0.2,6)	(0.3,4)	(0.5,5)	(0.7,1)	(0.8,3)	(1.1,7)
Step 5	B_0	(0.1,2)	(0.2,6)	(0.5,5)				
Step 6	$\tilde{v}_0(0)$	$\tilde{v}_0(1)$	$\tilde{v}_0(2)$	$\tilde{v}_0(3)$	$\tilde{v}_0(4)$	$\tilde{v}_0(5)$	$\tilde{v}_0(6)$	$\tilde{v}_0(7)$
	0	0.5	0.1	0.5	0.2	0.5	0.2	0.5
	-	$2 \oplus 6 \oplus 5$	2	$6 \oplus 5$	$2 \oplus 6$	5	6	$2 \oplus 5$

can not be included in B_0. Here $\oplus$ means XOR, and it is carried out on the binary strings corresponding to the integers. The field element of the fourth entry, which is 5, is independent of 2 and 6. Therefore, (0.5,5) becomes the next entry in B_0. Now B_0 has $p = 3$ entries, and is a valid minimum basis for computing $\tilde{v}_0$. For Step 6 in Table 9.7, the second row shows the value of each $\tilde{v}_0(\beta)$ and the third row lists the field elements of the nodes in B_0 that sum up to be β. Take $\tilde{v}_0(3)$ as an example. $3 = 6 \oplus 5$, the sum of the field elements of the second and third nodes in B_0. Therefore, $\tilde{v}_0(1)$ equals the maximum of the LLRs of these two nodes, which is $max\{0.2, 0.5\} = 0.5$.

The M_j vector varies for different j, since $m_j(\beta)$ takes the value from M_{min2} in the case that the min1 node of row β is in stage j. Even though the M_j vector is only slightly different for each stage, $\bar{M}_j$ and B_j need to be recomputed for each stage and many computations are repeated. Simulation results in [148] showed that using imprecise values for M_{min2} does not hurt the performance much when the finite field is not very small and the codeword length is not very short. Accordingly, M_{min2} can be replaced by compensation values when they are needed. Considering this, a global minimum basis consisting of a few nodes with the minimum LLRs from the entire reordered trellis is derived first. Then each B_j is computed by modifying the global basis [148]. Much calculation is saved by not computing M_j and B_j afresh in this simplified basis construction method. In addition, storing a single global basis instead of d_c minimum basis for each check node further reduces the memory requirement.

Three steps are carried out to compute B_j in the simplified basis construction scheme. The first step finds only the $q-1$ min1 nodes in the $q-1$ nonzero rows of the reordered trellis, and the vector M_{min1} records their LLRs and stage indices. Next, a modified basis construction function Φ takes M_{min1} as input and generates a global basis $B = B_I \bigcup B_{II}$, where B_I is a set of p nodes with the smallest LLRs and linearly independent field elements from M_{min1}, and B_{II} consists of supplementary values used for making up the unrecorded M_{min2}. One possible construction of B_{II} is to include the n_b nodes of M_{min1} with the smallest LLRs that are not included in B_I. The larger n_b, the better the error-correcting performance. In the last step, a function Θ derives each minimum basis B_j through modifying B_I using the extra information kept in B_{II}.

Denote the nodes in B_I and B_{II} by $(m'^{(i)}, \alpha'^{(i)}, I'^{(i)})$ $(1 \leq i \leq p)$ and $(m''^{(i)}, \alpha''^{(i)}, I''^{(i)})$ $(1 \leq i \leq n_b)$, respectively. The function Φ for constructing the global basis $B = B_I \bigcup B_{II}$ from M_{min1} is described by the pseudo codes in Algorithm 37 [148]. Similar to that in Algorithm 36, $w_1, w_1, \ldots, w_{p'}$ is a subset of p' indices

---

Algorithm 37 Function Φ for constructing global basis B from M_{min1}
input*:* $M_{min1} = \{(min1(\beta), idx(\beta))\}$ $(1 \leq \beta < q)$
initialization*:* $B_I = \emptyset$, $B_{II} = \emptyset$
for $\beta = 1$ to $q - 1$
    if $(\beta \notin span\{\alpha'^{(1)}, \alpha'^{(2)}, \ldots, \alpha'^{(|B_I|)}\}$
        $B_I = B_I \bigcup\{(min1(\beta), \beta, idx(\beta))\}$
    else assume $\beta = \alpha'^{(w_1)} \oplus \alpha'^{(w_2)} \oplus \cdots \oplus \alpha'^{(w_{p'})}$ $(1 < p' \leq |B_I|)$
        if $(min1(\beta) < \max\{m_j'^{(w_1)}, m_j'^{(w_2)}, \ldots m_j'^{(w_{p'})}\} = m'^{(w_l)})$
            $B_I = B_I \backslash\{(m'^{(w_l)}, \alpha'^{(w_l)}, I'^{(w_l)})\}$
            $B_I = B_I \bigcup\{(min1(\beta), \beta, idx(\beta))\}$
            $(mt, at, It) = (m'^{(w_l)}, \alpha'^{(w_l)}, I'^{(w_l)})$
        else
            $(mt, at, It) = (min1(\beta), \beta, idx(\beta))$
    -
    if $|B_{II}| < n_b$
        $B_{II} = B_{II} \bigcup\{(mt, at, It)\}$
    else if $(mt < \max\{m''^{(k)}\} = m''^{(l)})$ $(1 \leq k \leq n_b)$
        $B_{II} = B_{II} \backslash\{(m''^{(l)}, \alpha''^{(l)}, I''^{(l)})\}$
        $B_{II} = B_{II} \bigcup\{(mt, at, It)\}$

The steps above the dashed line in Algorithm 37 construct B_I, and those under the dashed line construct B_{II}. One node is read from M_{min1} at a time. The field element of the node, β, is tested first to find if it can be expressed as a combination of the field elements of the nodes that are already in B_I. If not, the node $(min1(\beta), \beta, idx(\beta))$ is put into B_I. Otherwise, assume that $\beta = \alpha'^{(w_1)} \oplus \alpha'^{(w_2)} \oplus \cdots \oplus \alpha'^{(w_{p'})}$ $(1 < p' \leq |B_I|)$. Moreover, among the p' nodes in B_I with these field elements, the one with index w_l has the largest LLR. In the case that $min1(\beta) < m'^{(w_l)}$, $(m'^{(w_l)}, \alpha'^{(w_l)}, I'^{(w_l)})$ should be replaced by $(min1(\beta), \beta, idx(\beta))$. The nodes in B_{II} come from either the nodes of M_{min1}

that have never been inserted into B_I or the nodes that are taken out of B_I. The candidate node is denoted by $(mt, \alpha t, It)$ in Algorithm 37. This candidate is inserted into B_{II} directly if the number of nodes in B_{II} is less than n_b. If not, mt is compared with the maximum LLR in B_{II} to decide if any node in B_{II} should be replaced by this candidate.

The B_I constructed using Algorithm 37 is guaranteed to consist of the p nodes from M_{min1} with the smallest LLRs and independent field elements. A node in B_I can only be replaced by a node with smaller LLR. In addition, the space spanned by the field elements of the nodes in B_I remains unchanged after the node replacement. To simplify later computations, the nodes in B_I are resorted according to their LLRs when an entry is inserted or replaced. Moreover, from Algorithm 37, the nodes in B_{II} have the smallest LLRs among the nodes not belonging to B_I. The nodes in B_{II} are also ordered according to increasing LLR to facilitate later computations in the check node processing.

TABLE 9.8
Example for Function Φ

β	$(min1(\beta), idx(\beta))$	B_I			B_{II}	
1	(0.7,2)	(0.7,1,2)				
2	(0.1,3)	(0.1,2,3)	(0.7,1,2)			
3	(0.4,0)	(0.1,2,3)	(0.4,3,0)		(0.7,1,2)	
4	(0.3,2)	(0.1,2,3)	(0.3,4,2)	(0.4,3,0)	(0.7,1,2)	
5	(0.5,1)	(0.1,2,3)	(0.3,4,2)	(0.4,3,0)	(0.5,5,1)	(0.7,1,2)
6	(0.2,5)	(0.1,2,3)	(0.2,6,5)	(0.4,3,0)	(0.3,4,2)	(0.5,5,1)
7	(1.1,4)	(0.1,2,3)	(0.2,6,5)	(0.4,3,0)	(0.3,4,2)	(0.5,5,1)

Example 48 *Taking the same min1 nodes as those in Table 9.7 as the inputs, the contents of B_I and B_{II} with $n_b = 2$ are listed in Table 9.8 for each iteration of Function Φ. Since the finite field elements of the first two min1 nodes, which are $\beta = 1$ and 2, are independent, $(0.7, 1, 2)$ and $(0.1, 2, 3)$ are put into B_I in the first two iterations. The field element of the third min1 node is $\beta = 3 = 2 \oplus 1$, and $min1(3) = 0.4 < \max\{0.7, 0.1\} = 0.7$. Hence, the third min1 node with its corresponding β, $(0.4, 3, 0)$, replaces $(0.7, 1, 2)$ in B_I. In addition, $(0.7, 1, 2)$ is directly put into B_{II} since $|B_{II}| = 0$ at this time. In the fourth iteration, $\beta = 4$ is independent of the field elements that are already in B_I. Therefore $(0.3, 4, 2)$ is inserted into B_I. In the iteration of $\beta = 5$, although $5 = 2 \oplus 4 \oplus 3$, $(0.5, 5, 1)$ does not replace any node in B_I because $\max\{0.1, 0.3, 0.4\} = 0.4 < min1(5) = 0.5$. Instead, $(0.5, 5, 1)$ is inserted into B_{II}. $6 = 2 \oplus 4$, and $min1(6) = 0.2$ is less than the LLR of the node $(0.3, 4, 2)$. Hence $(0.2, 6, 5)$ replaces $(0.3, 4, 2)$ in B_I in the 6th iteration. Moreover, $(0.3, 4, 2)$ replaces the node with the maximum LLR in B_{II}, which is $(0.7, 1, 2)$. The basis is not changed in the iteration of $\beta = 7$, since $min1(7) = 1.1$ is larger than the maximum LLRs of the nodes in both B_I and*

B_{II}.

Algorithm 38 Function Θ for constructing B_j from B

input: $B_I = \{(m'^{(i)}, \alpha'^{(i)}, I'^{(i)})\}$ $(1 \le i \le p)$
$\qquad B_{II} = \{(m''^{(i)}, \alpha''^{(i)}, I''^{(i)})\}$ $(1 \le i \le n_b)$

initialization: $B_j = B_I = \{(m_j'^{(i)}, \alpha_j'^{(i)}, I_j'^{(i)})\}$

for $i = 1$ *to* p
$\qquad$ *if* $(I_j'^{(i)} = j)$
$\qquad\qquad m_j'^{(i)} = \gamma \cdot max\{m'^{(p)}, m''^{(n_b)}\}$

for $i = 1$ *to* n_b
$\qquad$ *if* $(I''^{(i)} \ne j)$ *&* $(\alpha''^{(i)} = \alpha'^{(w_1)} \oplus \alpha'^{(w_2)} \oplus \cdots \oplus \alpha'^{(w_{p'})})$
$\qquad\qquad l = max_{1 \le k \le p'}\{w_k | I_j'^{(w_k)} = j\}$
$\qquad\qquad B_j = B_j \backslash \{(m_j'^{(l)}, \alpha_j'^{(l)}, I_j'^{(l)})\}$
$\qquad\qquad B_j = B_j \bigcup \{(m''^{(i)}, \alpha''^{(i)}, I''^{(i)})\}$

The minimum basis B_j can be computed from the global basis B according to Function Θ described in Algorithm 38. Although the stage indices of the nodes in B_j are not needed to compute the c2v vector $\tilde{v}_j$, they are added to B_j in Algorithm 38 to facilitate the derivation of B_j from B. B_j is initialized as B_I in the beginning. Then for any node in B_j that is from stage j, the LLR is replaced by a compensation value computed as the maximum of the LLRs in B_I and B_{II} scaled by a constant $\gamma > 1$, whose optimum value is determined from simulations. Such a compensation scheme may over estimate the LLRs and lead to performance loss. To reduce the over-estimation, the nodes in B_{II} that are not from stage j are utilized. These nodes are additional min1 nodes not included in B_I. The field elements of the p nodes in B_I, and accordingly those in B_j after the first loop of Algorithm 38, form a basis of $GF(2^p)$. For each node $(m''^{(i)}, \alpha''^{(i)}, I''^{(i)})$ in B_{II}, the field element $\alpha''^{(i)}$ can be written as the sum of a subset of those basis field elements in B_j. It may happen that multiple nodes of B_j in the subset are from stage j. The last of such nodes is replaced by $(m''^{(i)}, \alpha''^{(i)}, I''^{(i)})$ if $I''^{(i)} \ne j$.

Example 49 *Table 9.9 lists the contents of several minimum bases during the two loops of Algorithm 38 when its input is the global basis $B = B_I \bigcup B_{II}$ constructed in Table 9.8. In the construction of B_0, only the third entry in B_I is from stage 0. The LLR of this entry is replaced by the compensation value $\gamma \max\{m'^{(3)}, m''^{(2)}\} = 0.5\gamma$ in the first loop. In the second loop, $I''^{(1)} = 2 \ne 0$ and $\alpha''^{(1)} = 4 = \alpha_0'^{(1)} \oplus \alpha_0'^{(2)} = 2 \oplus 6$. Neither $I_0'^{(1)}$ nor $I_0'^{(2)}$ is 0, and thus*

TABLE 9.9
Example for Function Θ

		B_I	(0.1,2,3)	(0.2,6,5)	(0.4,3,0)
		B_{II}	(0.3,4,2)	(0.5,5,1)	
B_0	1st loop		(0.1,2,3)	(0.2,6,5)	(0.5γ,3,0)
	2nd loop		(0.1,2,3)	(0.2,6,5)	(0.5,5,1)
B_1	1st loop		(0.1,2,3)	(0.2,6,5)	(0.4,3,0)
	2nd loop		(0.1,2,3)	(0.2,6,5)	(0.4,3,0)
B_5	1st loop		(0.1,2,3)	(0.5γ,6,5)	(0.4,3,0)
	2nd loop		(0.1,2,3)	(0.3,4,2)	(0.4,3,0)

$(m''^{(1)}, \alpha''^{(1)}, I''^{(1)})$ *does not replace any node in* B_0. $I''^{(2)} = 1 \neq 0$, *and* $\alpha''^{(2)} = 5 = \alpha_0'^{(2)} \oplus \alpha_0'^{(3)} = 6 \oplus 3$. *Since* $I_0'^{(3)} = 0$, *the last entry of* B_0 *is replaced by* $(0, 5, 5, 1)$ *at the end of the second loop. It can be observed that this* B_0 *is the same as that derived in Table 9.7. Table 9.9 shows that* $B_1 = B_I$. *This is because none of the nodes in* B_I *is from stage 1. To derive* B_5, $m_5'^{(2)}$ *takes the compensation value in the first loop since* $I'^{(2)} = 5$. *In the second loop,* $I''^{(1)} \neq 5$ *and* $\alpha''^{(1)} = 4 = \alpha_3'^{(1)} \oplus \alpha_3'^{(2)} = 2 \oplus 6$. *Hence,* $(m_5'^{(2)}, \alpha_5'^{(2)}, I_5'^{(2)}) = (0.5\gamma, 6, 5)$ *in* B_5 *is replaced by* $(m''^{(1)}, \alpha''^{(1)}, I''^{(1)}) = (0.3, 4, 2)$. *The second loop continues for the second entry in* B_{II}. $I''^{(2)} \neq 5$ *and* $\alpha''^{(2)} = 5$ *equals the sum of the field elements of the three nodes in the updated* B_5, *which are 2, 4 and 3 at this time. However, since none of the nodes in the updated* B_5 *is from stage 5, no further change is made on* B_5.

Algorithm 39 Simplified Basis-construction Min-max Check Node Processing
input: u_j *and* $\hat{\beta}_j$ *for* $0 \leq j < d_c$; $\hat{\beta}_{sum} = \sum_{0 \leq j < d_c} \hat{\beta}_j$
start:
1: $\tilde{u}_j(\beta) = u_j(\beta + \hat{\beta}_j)$ $(0 \leq j < d_c)$
2: compute $M_{min1} = \{(min1(\beta), idx(\beta))\}$ $(1 \leq \beta < q)$
3: $B = B_I \bigcup B_{II} = \Phi(M_{min1})$
  for $j=0$ to d_c-1
4: $B_j = \Theta(B, j)$
6: $\tilde{v}_j(0) = 0$
  $\tilde{v}_j(\beta) = max\{m_j'^{(w_1)}, m_j'^{(w_2)}, \ldots m_j'^{(w_{p'})}\}$
    if $\beta = \alpha_j'^{(w_1)} \oplus \alpha_j'^{(w_2)} \oplus \cdots \oplus \alpha_j'^{(w_{p'})}$ $(1 \leq p' \leq p)$
7: $v_j(\beta) = \tilde{v}_j(\beta + \hat{\beta}_{sum} + \hat{\beta}_j)$

In summary, the simplified basis-construction Min-max check node processing scheme is given in Algorithm 39. In this algorithm, the global basis, B, is derived using only the min1 nodes, and the min2 nodes are not needed. Only the min1 nodes and a single global basis need to be stored. Hence, the memory requirement is lower than that of the basis-construction check node processing in Algorithm 36. Moreover, instead of recomputing each minimum basis B_j from sorted v2c nodes, this simplified algorithm derives each B_j through choosing nodes from the global basis and making use of compensation values. Therefore, significant logic complexity reduction is also achieved. This simplified scheme has almost the same error-correcting performance as Algorithm 36 unless the finite field is very small, and/or the row weight is low. When the finite field is smaller, there are fewer nodes in the minimum basis. Hence using imprecise LLRs for those nodes would have a larger effect on the computed c2v LLRs. When the row weight is low, the probability that multiple min1 nodes are from the same stage of the trellis becomes higher. Since the min2 nodes are not kept in the simplified scheme, including more min1 nodes from the same stage in the computation of a c2v message leads to a larger deviation. Simulations showed that keeping $n_b = 2$ nodes in B_{II} does not cause noticeable performance loss for codes over $GF(2^5)$ [148]. Larger n_b may be needed for codes over higher-order finite fields.

A CNU implemented according to the simplified basis-construction scheme in Algorithm 39 has five parts: i) a reordering network to derive the reordered v2c message vectors $\tilde{u}_j$ from the original v2c vectors u_j; ii) a sorter for finding the min1 node of each row in the reordered trellis of the v2c messages; iii) a unit that implements the Φ function and accordingly constructs the global basis, B, from the min1 nodes; iv) a minimum basis constructor that derives the minimum basis, B_j, for each c2v vector from B through carrying out the Θ function; v) a basis recovery module that recovers all the c2v messages from the corresponding minimum basis according to steps 6 and 7 in Algorithm 39. As mentioned previously, a reordering network is implemented by p stages of 2^p 2-input multiplexors for a code over $GF(2^p)$. The $2^p - 1$ min1 nodes can be found by using $2^p - 1$ comparator-register feedback loops. These loops take one reordered v2c vector in each clock cycle, and the LLRs of the nonzero field elements in the vector are compared with those intermediate minimum LLRs of the same finite field elements stored in the registers. If a v2c LLR is smaller, the corresponding intermediate minimum value and index are updated. Such a process takes d_c clock cycles to find the min1 nodes. The architectures for implementing the other three parts of the simplified basis-construction CNU are more involved, and they are detailed next.

Global basis constructor The global basis constructor architecture has two parts, one for B_I and the other for B_{II}. It takes in one node from M_{min1} at a time, and derives B_I and B_{II} in $2^p - 1$ clock cycles. The architecture of the part for constructing B_I is illustrated in Fig. 9.21. It has register arrays of length p and corresponding comparators and multiplexors for locating the p nodes with the smallest LLRs and independent field elements. The registers

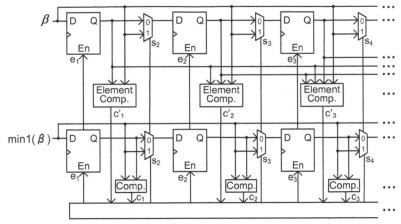

FIGURE 9.21
Global basis construction architecture (modified from [148])

for the LLRs are initialized to the maximum possible value representable by a given word length, and those for the field elements are initialized to zero. In each clock cycle, the LLR of the input min1 node is compared to each of those stored in registers. In addition, the field element of the min1 node is compared to each linear combination of the stored field elements. The results of the comparators are used to decide whether the input min1 node should be inserted into the B_I register array, and if so, where it should be inserted. The select signals of the multiplexors and the enable signals of the registers are generated from the comparison results. There is another register array using the the same select and enable signals for inserting the indices of the nodes. This array is not included in Fig. 9.21 for conciseness.

Let the input min1 node be $(min1(\beta), \beta, idx(\beta))$, and the nodes that are currently in the register arrays are $(m'^{(1)}, \alpha'^{(1)}, I'^{(1)})$, $(m'^{(2)}, \alpha'^{(2)}, I'^{(2)})$, ... Note that the nodes in the register arrays are always in the order of increasing LLR and their associated field elements are always independent of each other. The output of the ith LLR comparator, c_i, is asserted only if $min1(\beta)$ is smaller than the LLR stored in the ith register. The field element comparator after the ith register compares β with the sums of $\alpha'^{(i)}$ and all possible combinations of $\alpha'^{(1)}, \alpha'^{(2)}, \dots, \alpha'^{(i-1)}$. For example, the third field element comparator compares β with $\alpha'^{(3)}$, $\alpha'^{(3)}+\alpha'^{(2)}$, $\alpha'^{(3)}+\alpha'^{(1)}$, and $\alpha'^{(3)}+\alpha'^{(2)}+\alpha'^{(1)}$. The output, c'_i, of the field element comparator is only asserted if β equals to any of the combinations. Since the field elements in the register arrays are linearly independent, at most one of the c'_i signals is asserted.

When the outputs of all element comparators are '0', β is independent of the field elements in the register array. In this case, the location to insert $(min1(\beta), \beta, idx(\beta))$ is decided by the results of all LLR comparators, c_i. If c'_l='1', then β can be represented by a linear combination of the elements in the register arrays, and the largest index of those elements is l. Accordingly, whether to insert $(min1(\beta), \beta, idx(\beta))$ is dependent on the relative values of $min1(\beta)$ and $m'^{(l)}$. $(min1(\beta), \beta, idx(\beta))$ will not be inserted if $min1(\beta) \geq m'^{(l)}$. Otherwise, it is inserted into a location k ($k \leq l$), so that the LLRs in the register arrays are still in nondecreasing order after the insertion. In addition, the registers for $(m'^{(l)}, \alpha'^{(l)}, I'^{(l)})$ will be overwritten. Since the nodes after the lth position should remain the same, the registers storing those nodes are disabled during this process.

To simplify the derivation of the control logic, signals g_i are introduced. g_1 is set to '1', and g_i for $1 < i \leq p$ is defined as

$$g_i = NOT(c'_1 + c'_2 + \cdots + c'_{i-1}).$$

If β can be expressed as a linear combination of the elements stored in the ith register and those before, then $g_j =$ '0' for $i < j \leq p$. Hence these signals are used to disable the registers after the ith position. Since the field elements of the nodes in M_{min1} are nonzero and distinct, $c'_1 =$ '0' and hence $g_2 =$ '1'. Using g_i, the enable signals of the registers, e_i, and the select signals of the multiplexors, s_i, in Fig. 9.21 are generated as

$$\begin{cases} e_i = (c_1 + c_2 + \cdots + c_i)g_i \, (1 \leq i \leq p) \\ s_i = NOT(c_1 + c_2 + \cdots + c_{i-1}) \, (2 \leq i \leq p). \end{cases}$$

A similar architecture is used to derive B_{II}. Nevertheless, this second part of the global basis construction architecture does not have the field element comparators and logic for checking the dependency of the input field elements. It is just a sorter of length n_b. Of course, the nodes in the registers are also kept according to increasing LLR. The input to the second part is either the $(m'^{(l)}, \alpha'^{(l)}, I'^{(l)})$ deleted from B_I or the $(min1(\beta), \beta, idx(\beta))$ from M_{min1}. From Algorithm 37, $(m'^{(l)}, \alpha'^{(l)}, I'^{(l)})$ becomes a candidate of B_{II} only when the node from M_{min1} is inserted into B_I. Therefore, $e_1 + s_2e_2 + \cdots + s_pe_p$ can be used to select the two possible input nodes to B_{II}. Moreover, $\{m'^{(l)}, \alpha'^{(l)}, I'^{(l)}\}$ can be taken from the B_I register array by making use of c'_i ($1 \leq i \leq p$).

Minimum basis constructor The architecture for constructing the minimum basis B_j from the global basis B is depicted in Fig. 9.22. At the beginning, the LLRs and field elements of the nodes in B_I are loaded into register arrays m'_j and α'_j, respectively, and those for B_{II} are stored into m'' and α'' arrays. The indices of these nodes are compared with j, and the output of a comparator is asserted if the index equals j. The comparison results are stored into register arrays f'_j and f'', which are for B_I and B_{II}, respectively. The compensation value $\gamma \cdot max\{m'^{(p)}, m''^{(n_b)}\}$ is computed and written into a separate register. This part is simple, and is not included in Fig. 9.22. The LLRs of the nodes from stage j stored in the m'_j array are not replaced by this compensation value physically. The compensation value is used instead of these LLRs when needed in order to simplify the logic and reduce the latency.

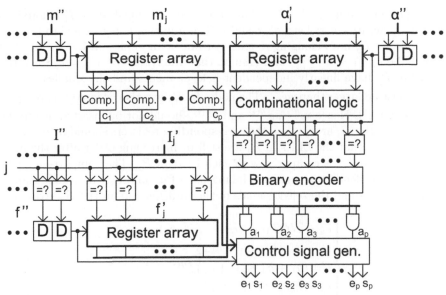

FIGURE 9.22
Minimum basis construction architecture (modified from [148])

Starting from the second clock cycle, the nodes in B_{II} are read out serially according to the order of increasing LLR. From Algorithm 38, each node, $(m'''^{(i)}, \alpha''^{(i)}, I''^{(i)})$, from B_{II} is tested to see if $I''^{(i)} \neq j$ and if $\alpha''^{(i)} = \sum_{k=1}^{p'} \alpha_j'^{(w_k)}$. If both conditions are true, then this node replaces $(m'^{(l)}, \alpha'^{(l)}, I'^{(l)})$ in B_j, where $l = \max_{1 \leq k \leq p'}\{w_k | I_j'^{(w_k)} = j\}$. In addition, the nodes in B_j are still ordered according to increasing LLR after the replacement. The register arrays for B_j, which are the m_j', α_j' and f_j' arrays in Fig. 9.22, have similar architectures as those in Fig. 9.21. They also have enable signals e, and the multiplexors with select signals s pass the correct update values to the registers. However, opposite to that in Fig. 9.21, the top input of each multiplexor comes from the register on its right since the LLR from B_{II} is larger. The node replacement can be done in one clock cycle. As a result, the minimum basis construction takes $n_b + 1$ clock cycles.

The combinational logic block in Fig. 9.22 computes the sums of all possible combinations of the p field elements stored in the α_j' register array. There are $q - 1 = 2^p - 1$ of them excluding the zero sum. Assume that the index k is represented as a p-bit binary tuple $b_{p-1} \cdots b_1 b_0$. This block is designed so that the kth output equals $b_{p-1}\alpha'^{(p)} + \cdots b_1\alpha'^{(2)} + b_0\alpha'^{(1)}$. Since the field elements in the α_j' register array are linearly independent, comparing the sums with a field element from B_{II}, $\alpha''^{(i)}$, would generate a one-hot vector. Accordingly, the output of the binary encoder in Fig. 9.21 is a p-bit vector $b_{p-1} \cdots b_1 b_0$ whose corresponding sum equals $\alpha''^{(i)}$. This p-bit vector is masked by f'_j before it is used to generate control signals e and s. The masking is performed to prevent the nodes of B_j not belonging to stage j from being replaced. Let the masked signals be a_k ($1 \leq k \leq p$), and assume that $\max\{k|a_k =' 1'\} = l$. Since the nodes in B_j have linearly independent field elements and the smallest LLRs, $m''^{(i)}$ must be larger than $m'^{(l)}_j$. The nodes in B_j are also ordered according to increasing LLR. Hence $m''^{(i)}$, $\alpha''^{(i)}$ and $f''^{(i)}$ should not be inserted before the lth nodes in the B_j arrays, and the corresponding registers should be disabled. To make sure that the LLRs in B_j are still in increasing order after the node replacement, $m''^{(i)}$ is compared with each LLR in B_j. The comparison result c_k is asserted if $m''^{(i)}$ is smaller. Making use of a_k and c_k, the enable signals of the registers and the select signals of the multiplexors in the B_j arrays are generated as

$$
\begin{cases}
\check{e}_k = NOT(c_1 + c_2 + \cdots + c_k) \\
\hat{e}_k = NOT(a_{k+1} + a_{k+2} + \ldots) \\
\tilde{e} = (a_1 + a_2 + \cdots + a_p) f''^{(i)} \\
e_k = \check{e}_k \hat{e}_k \tilde{e} \, (1 \leq k \leq p) \\
s_k = c_1 + c_2 + \cdots + c_{k+1} \, (1 \leq k < p).
\end{cases}
$$

The generation of each enable signal, e_k, is simplified by breaking it down into three parts. $\tilde{e}$ decides whether the node $(m''^{(i)}, \alpha''^{(i)}, f''^{(i)})$ should be inserted. If so, $\check{e}_k$ tells which position it should be inserted, and $\hat{e}_k$ disables the registers to the left of the lth position in the B_j arrays if $l = \max\{k|a_k =' 1'\}$.

Basis recovery module Taking $GF(2^3)$ as an example, Fig. 9.23 illustrates the architecture of the basis recovery module implementing steps 6 and 7 of Algorithm 39. This architecture takes the p nodes in the minimum basis B_j, $(m'^{(i)}_j, \alpha'^{(i)}_j, f'^{(i)}_j)$ ($1 \leq i \leq p$), and the compensation value, $\gamma \cdot \max\{m'^{(p)}, m''^{(n_b)}\}$, as the inputs, and generates all messages in the c2v vector v_j in one clock cycle. From step 6, the basis recovery module is basically a selection logic that picks one of the LLRs from B_j, zero or the compensation value to be each $v_j(\beta)$ ($0 \leq \beta < q$). The selection is based on how each β is represented as a linear combination of $\alpha'^{(i)}_j$. As shown in Fig. 9.23 (a), each $v_j(\beta)$ is derived by a $p + 1$-input AND-OR selection tree. To derive all 2^p messages in a c2v vector in parallel, 2^p such selection trees are employed. On the other hand, the control signals for all the selection trees, $s_{i,\beta}$ and $r_{i,\beta}$ ($0 \leq i \leq p, \beta \in GF(2^p)$), can be generated using one copy of the architecture in Fig. 9.23 (b). Remember that the compensation value was actually not written into the register array of B_j during the minimum basis construction process. If a node in B_j is from stage j, then the compensation value is passed through the corresponding multiplexor in Fig. 9.21 (a). Also only one of $s_{i,\beta}$ is '1', and hence only one input value is passed through the AND-OR tree to be $v_j(\beta)$.

To facilitate the generation of the control signals $s_{i,\beta}$ and $r_{i,\beta}$, the combinations of the p field elements in B_j are divided into $p + 1$ groups as shown in

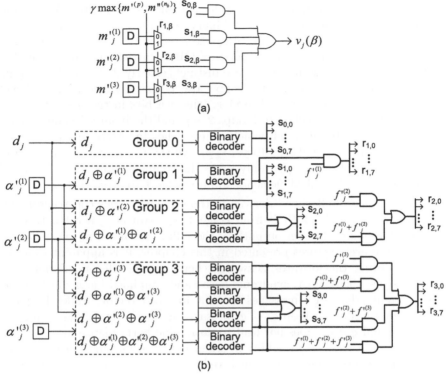

FIGURE 9.23

Architecture of the basis recovery module for $GF(2^3)$ codes (modified from [148])

Fig. 9.23 (b). The *ith* group ($0 < i \leq p$) consists of $\alpha_j'^{(i)}$ plus all possible combinations of $\alpha_j'^{(k)}$ ($1 \leq k < i$). Hence group i has $2^{(i-1)}$ combinations of field elements and the p groups has $2^p - 1$ combinations, whose sums are the $2^p - 1$ nonzero field elements of $GF(2^p)$. As shown in step 7 of Algorithm 39, the reverse reordering of the messages is done as $v_j(\beta) = \tilde{v}_j(\beta + \hat{\beta}_{sum} + \hat{\beta}_j)$. This can be incorporated into the control signal generation by adding $d_j = \hat{\beta}_{sum} + \hat{\beta}_j$ to the sum of the each field element combination. In addition, group 0 has a single element $0 + d_j = d_j$. Note that the nodes in B_j are ordered according to increasing LLR. Therefore, the maximum LLR of the nodes included in each combination belonging to group i is $m_j'^{(i)}$. As a result, if β is equal to the sum of a field element combination in group i, then $m_j'^{(i)}$ should be selected as the value of $v_j(\beta)$ if none of the field elements in the combination is from stage j. Otherwise, $v_j(\beta)$ equals the compensation value since $\gamma > 1$ and hence it is larger than any LLR in B_j.

The sum of each combination of field elements is passed to a binary decoder. The decoder output is a q-bit binary vector in one-hot format. If the decoder input is β, then only the βth output bit is '1'. For the groups that have more than one combination, bit-wise OR is carried out on the one-hot vectors. Since the sums in a group are distinct, exactly $2^{(i-1)}$ of the outputs of the OR gates corresponding to group i are '1', and their positions indicate the sums included in group i. Therefore, the βth bits in the $p-1$ OR gate output vectors corresponding to groups $2 \sim p$ and the binary decoder output vectors for group 0 and 1 tell which group β lies in. Accordingly, they are used as the control signals $s_{i,\beta}$ $(0 \le i \le p)$. All the $r_{i,\beta}$ signals are derived in a similar way. Note that $v_j(\beta)$ only takes the compensation value if some of the nodes in B_j whose field elements add up to be β are from the jth stage of the trellis. Since the flag $f_j'^{(i)}$ derived in the minimum basis constructor tells whether the ith node of B_j is from stage j, $r_{i,\beta}$ can be generated through masking the binary decoder outputs by these flags as shown in Fig. 9.23(b).

From the architectures presented in this chapter, it is apparent that the CNUs adopting the forward-backward scheme is less efficient. Among the other designs, the SMMA, syndrome-based and simplified basis-construction CNUs require that all q messages be kept for each vector, while the path-construction CNU can keep $q' < q$ messages in each vector. The difference between q and q' that does not lead to noticeable performance loss becomes smaller when q is lower. In addition, when not all q messages are kept, the variable node processing is more involved as will be shown in the next subsection. Considering these, the path-construction CNU leads to more efficient decoders when the finite field is large. In the case of moderate or small finite field, the SMMA, syndrome-based and simplified basis-construction CNUs win over the path-construction CNU. Among these three designs, the modified syndrome-based CNU has the lowest complexity for $GF(4)$ codes. The basis-construction CNU has similar error-correcting capability and is more efficient than other designs for finite fields with moderate order. However, simulation results showed that it has performance loss for $GF(4)$ codes. This is because the min2 values contribute more to the c2v messages when the order of the finite field is smaller. However, they come from a compensation value in the basis-construction scheme and are not precise.

9.2.6   Variable node unit architectures

The available CNUs can be divided into two categories according to whether all q messages are kept for each vector. The VNU designs differ for these two categories.

First consider the case that all q messages are kept in each vector. For this case, a VNU architecture that generates all v2c message vectors from the same variable node in parallel for a code with column weight d_v is shown in Fig. 9.24. In non-binary variable node processing, the c2v messages with the same field element are added up. These additions are done by an architecture

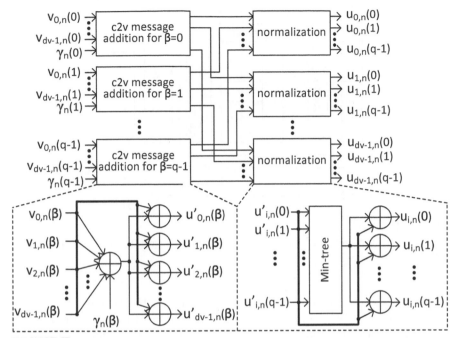

FIGURE 9.24
VNU architecture for NB-LDPC decoders with column weight d_v and all q messages kept

similar to that for binary VNU. Note that sign-magnitude to 2's complement representation conversions are no longer needed since all LLRs in the Min-max algorithm are non-negative. Then normalization is carried out. It finds the smallest LLR in each vector using a tree of 'min' units and subtracts it from each LLR in the vector so that the smallest LLR in each vector is brought back to zero.

When $q' < q$ messages are kept for each vector, the variable node processing still adds up messages with the same field element in the incoming c2v vectors and the channel information vector. However, the q' messages in each vector are the ones with the smallest LLRs, and they do not necessarily correspond to the same set of field elements. Hence, in the additions of the message vectors, compensation values are adopted for those messages not stored [137]. Considering the complex logic for locating the messages with matching field element from different vectors, the addition is carried out over two vectors at a time. Moreover, the CNUs that keep q' messages usually require the messages in each vector to be stored in the order of increasing LLR. Hence, sorting needs to be applied to the sums of messages.

A detailed VNU architecture for the case that $q' < q$ messages are kept

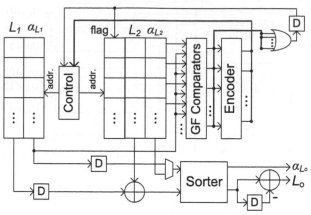

FIGURE 9.25
Architecture for vector addition when $q' < q$ messages are kept (modified from [143])

for each vector is available in [143]. In this design, the largest LLR in each vector is used as the compensation value for those LLRs not stored. Recall that the forward-backward CNU in [143] and trellis-based path construction CNU in [144] may not compute exactly q' valid messages for each c2v vector if a limit is set to the clock cycles can be spent. However, the address of the last computed c2v message, which has the largest LLR, is recorded. Accordingly the compensation value can be read out from the vector using this address. Denote the LLRs and the corresponding field elements of the two input vectors by L_1, α_{L_1} and L_2, α_{L_2}, respectively. The vector addition has two rounds of computations. In the first round, the entries in the L_1 and α_{L_1} vectors are read out serially. In each clock cycle, the field element from α_{L_1} is compared with all those in the α_{L_2} vector. If there is a match, the corresponding LLR from L_2 is added to the entry in L_1. Otherwise, the compensation value of L_2 is used in the addition. In the second round, the LLRs in L_2 whose field elements do not match any of those in α_{L_1} are serially added up with the compensation value of the L_1 vector. In both rounds, the sums of the LLRs are passed to a sorter, which sorts out q' messages with the smallest LLR to become the output vector L_o. Such a vector addition architecture is shown in Fig. 9.25.

In Fig. 9.25, the parts consisting of the registers for L_2 and α_{L_2}, the parallel finite field comparators, and the encoder can be considered as a content-addressable memory. In the first round, the field element from α_{L_1} is compared with each element in α_{L_2} simultaneously. If the comparator outputs are not all zero, which means there is a match, the encoder generates the address of the matching entry in L_2, and the flag corresponding to that entry is set. In the next clock cycle, the LLR from L_2 is read out using this address, and it is added to the LLR from L_1. If all comparator outputs are zero, no flag is set and the compensation LLR of L_2, which is the largest valid LLR in this vector, is read out in the next clock cycle. L_2 and α_{L_2} are stored separately, and hence the field element comparisons and encoding for the next entry of α_{L_1} can be done at the same time as the LLR addition. Therefore, the first round takes $q' + 1$ clock cycles considering the one-stage pipelining latency. In the second round, one LLR is read from L_2 at a time. If the corresponding flag has been set in the first round, nothing needs to be done. Otherwise, this LLR is added to the compensation LLR of L_1. To match the speed of the LLR addition, a parallel sorter is adopted to sort the sums of LLRs in both rounds. Such a sorter is implemented by register arrays and comparators similar to that in Fig. 9.21. It keeps registers of length q', and is able to insert one input into a vector of q' ordered entries in each clock cycle. The q' registers may not be filled with valid entries after the two rounds of additions. In this case, the LLR of the rest entries are set to the largest valid LLR as in the check node processing. This can be also done by storing the number of valid entries. The output vector L_o and α_{L_o} are located in the registers of the sorter at the end. The first LLR, which is also the minimum LLR, is subtracted from each LLR in L_o as they are shifted out to realize the normalization.

When the entries of α_{L_2} are all distinct, at most one of the q' field element comparators outputs '1'. In this case, the encoder is implemented by simple logic. Consider $q' = 16$ as an example. Let the comparator outputs be $c_0, c_1, \cdots, c_{15}$. The output of the encoder is a 4-bit tuple $b_3 b_2 b_1 b_0$ whose decimal value is i if $c_i =' 1'$. It can be derived that

$$
\begin{cases}
d_0 = c_1 \oplus c_3 \oplus c_5 \oplus c_7 \oplus c_9 \oplus c_{11} \oplus c_{13} \oplus c_{15} \\
d_1 = c_2 \oplus c_3 \oplus c_6 \oplus c_7 \oplus c_{10} \oplus c_{11} \oplus c_{14} \oplus c_{15} \\
d_2 = c_4 \oplus c_5 \oplus c_6 \oplus c_7 \oplus c_{12} \oplus c_{13} \oplus c_{14} \oplus c_{15} \\
d_3 = c_8 \oplus c_9 \oplus c_{10} \oplus c_{11} \oplus c_{12} \oplus c_{13} \oplus c_{14} \oplus c_{15}
\end{cases}
$$

It is possible that fewer than q' valid entries have been computed for the L_2 vector from the check node processing in a given number of clock cycles. In this case, the rest of the entries of the α_{L_2} vector are the initial value, such as zero. A valid element of α_{L_2} may also be zero. As a result, more than one field element comparators may output '1'. This would demand a much more complicated encoder. Alternatively, the address of the first zero element in the α_{L_2} vector can be recorded either during the check node processing or when the messages are loaded into the registers of the vector addition architecture. When the field element from α_{L_1} in the first round is zero, the recorded address is used to directly read the corresponding LLR from L_2.

9.2.7 Overall NB-LDPC decoder architectures

The same decoding schemes for binary LDPC decoders discussed in Chapter 8, namely the flooding, sliced message-passing, row-layered, and shuffled schemes, can be applied to NB-LDPC decoders. However, in NB-LDPC decoders, the check node processing is much more complicated than variable node processing. A variety of CNU architectures for NB-LDPC decoders have been presented in this chapter. They have different latency and store different intermediate messages. As a result, the scheduling of the computations in NB-LDPC decoding is more involved and affects the memory requirement of the overall decoder. The sliced message-passing and row-layered decoders have higher hardware efficiency than decoders adopting the flooding scheme. The shuffled scheme leads to faster decoding convergence as the row-layered scheme. However, a perfect shuffled scheme requires the c2v messages to be recomputed every time a v2c message is updated. This can not be easily done in forward-backward CNUs, and requires more complicated calculations in those CNUs making use of min1 and min2 nodes. In addition, similar to that in shuffled binary LDPC decoders, updating the min1 and min2 nodes using only the updated v2c messages and previous min1 and min2 nodes leads to error-correcting performance loss. Therefore, QCNB-LDPC decoders adopting the sliced message-passing and layered schemes are the focus of the discussions next.

As explained in Chapter 8, a sliced message-passing decoder carries out check node processing for all rows of H simultaneously. Usually, one block column of H is processed at a time, and hence the number of VNUs equals the size of the submatrices in a quasi-cyclic H matrix. Each non-binary VNU is capable of taking d_v c2v message vectors and generating d_v v2c message vectors for a column of H in parallel. In addition, the variable node processing is overlapped with the check node processing of the next decoding iteration.

Compared to other CNUs, the SMMA, basis-construction and syndrome-based CNUs can better take advantage of the sliced message-passing scheme. The common features of these CNUs are i) all q messages are kept in each vector; ii) Once after the min1 nodes, min2 nodes, and syndromes of the trellis are derived, the c2v messages are computed using simple logic without referring to the v2c messages. These features help to reduce the memory requirement and balance the latency of the variable and check node processing. Using these CNUs, the actual c2v message vectors do not need to be stored. Instead, the min1, min2 and syndromes are recorded as compressed messages, and the c2v messages are computed from them when needed. The v2c messages do not need to be stored either, if they are consumed in the min1 and min2 computations right afterwards. This would require the v2c messages to be generated at the same pace as they are consumed in the min1 and min2 computations. When the check and variable node processing is overlapped, ideally, the CNUs and VNUs should have similar critical paths and they should take about the same number of clock cycles in order to maximize the hardware efficiency. Each CNU processes one v2c message to update the min1 and min2 values in each clock cycle. When all q messages are kept in each vector, the VNU architecture in Fig. 9.24 can be used to compute all v2c messages from a variable node in one clock cycle.

When $q' < q$ messages are kept for each vector, the v2c messages are

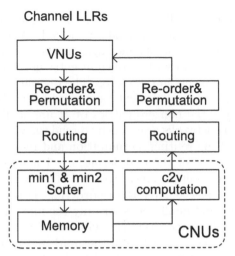

FIGURE 9.26
Overall architecture for sliced message-passing QCNB-LDPC decoders

generated at a lower throughput, and hence the check node processing needs to slow down to wait for the v2c messages. This is the reason that the sliced message-passing scheme is not a good choice when the trellis-based path-construction CNU is used, although it can also store the $1.5q'$ nodes of the smallest LLR as compressed c2v messages. When $q' < q$ messages are kept in each vector, the vector addition needed in the variable node processing is done over only two vectors at a time. Using the architecture in Fig. 9.25, each such vector addition requires two rounds of computations. Moreover, the computations can not be pipelined because of data feedback, and the sum vector is not available until the end of the second round. As a result, the addition of each pair of vectors takes about $2q'$ clock cycles using the architecture in Fig. 9.25, and hence the computation of each v2c vector would need around $2q'\lceil \log_2 d_v \rceil$ clock cycles using a tree consisting of $d_v - 1$ vector adders. Even if a serial sorter such as that in Fig. 9.13 is employed to find the $1.5q'$ messages with the smallest LLR, processing a v2c vector only takes $1.5q'$ clock cycles in the trellis-based path-construction CNU. Considering the large discrepancy between the latencies, overlapping the variable node processing and trellis-based path-construction check node processing as needed in the sliced message-passing scheme leads to inefficient decoders.

The forward-backward CNUs can not take full advantage of the sliced message-passing scheme either. The storage of the v2c messages can not be eliminated when forward-backward CNUs are used, even if all q messages are kept in each vector. The forward and backward processes of a decoding iteration can be switched. However, both of them need to read v2c messages, and hence the v2c messages need to be held until both processes are finished.

Fig. 9.26 shows the overall sliced message-passing decoder architecture for QCNB-LDPC codes when the CNUs adopt the SMMA, basis-construction or syndrome-based design. When all q messages in a vector are kept, multiplying the field elements in a vector with a constant entry in the H matrix is done by permuting the field elements in the vector. The re-ordering of the messages in a vector to make the zero field element have zero LLR is also done by columns of multiplexors. Since the nonzero entries of H are at different locations in each column, routing networks are needed to send the v2c messages to correct CNUs. The order of the routing, permutation and re-ordering functions can be switched. However, after the re-ordering, $q - 1$ nonzero LLRs need to be recorded. Hence, fewer multiplexors are needed in the permutation and routing functions if they are placed after the re-ordering. Depending on the CNUs used, both the min1 and min2 nodes, or only the min1 nodes are sorted out from the v2c vectors. In the case that the syndrome-based CNUs are employed, the syndromes are also computed from the min1 and min2 nodes. This part is not included in Fig. 9.26 for clarity. The min1 nodes, min2 nodes, and syndromes are loaded into memory to be used for computing c2v messages, right after which the sorter starts the check node processing for the next decoding iteration. The computed c2v vectors go through reverse routing, permutation, and re-ordering before they are added up in the VNUs.

Since the v2c messages are sent to the sorter right after they are computed by the VNUs, the min1 and min2 nodes are available shortly after the variable node processing of an iteration is completed in the sliced message-passing scheme. The c2v message computation methods vary in different CNUs, and they have different latencies. From the min1 and min2 nodes, if the first c2v vector can be made available in a few clock cycles, and the number of clock cycles needed for the check node processing is similar to that of the variable node processing, then the overall computation scheduling of the sliced message-passing NB-LDPC decoder is similar to that shown in Fig. 8.6 for the binary decoder. The SMMA and modified syndrome-based CNUs belong to this case. In another case, such as in the basis-construction CNU, more steps with longer latency are needed to derive the first c2v vector. To balance the latency of the variable and check node processing, the numbers of VNUs and hardware engines involved in each step of the CNUs need to be adjusted. Moreover, the decoding of multiple blocks can be interleaved in a similar manner as shown in Fig. 8.5 to better utilize the hardware units if the CNU latency is significantly longer.

As explained in Chapter 8, in layered decoding, the H matrix is divided into blocks of rows, also called layers. The c2v messages derived from the decoding of a layer are used to update the v2c messages to be used in the decoding of the next layer. To simplify the v2c message updating rule and speed up the decoding convergence, each column in a layer should have at most one nonzero entry. This means that, for QCNB-LDPC codes, each layer should consist of one block row of submatrices. If higher throughput is desired, more columns in a layer can be processed in parallel if the H matrix can be

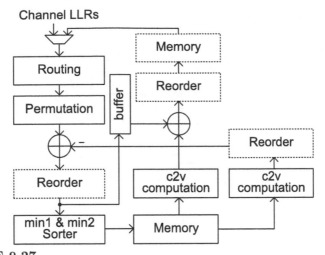

FIGURE 9.27

Layered NB-LDPC decoder architecture when CNUs adopting min1 and min2 nodes are used

designed to facilitate this. It has been shown that single-block-row layered decoders usually require about half the iterations to converge compared to sliced message-passing decoders. On the other hand, the pipelining latency of the VNUs and CNUs has a higher impact on the number of clock cycles needed by layered decoders due to the data dependency among the decoding of adjacent layers. If there are ξ pipelining stages, the pipelining latency adds $d_v\xi$ to the number of clock cycles needed by one iteration of layered decoding, while it increases the clock cycle number by ξ in one iteration of sliced message-passing decoding.

Fig. 9.27 depicts the architecture for a layered NB-LDPC decoder when CNUs making use of the min1 and min2 nodes are employed. These CNUs include the SMMA, trellis-based path-construction, basis-construction, and syndrome-based CNUs. The channel LLRs are used as the initial v2c messages for the decoding of the first layer, and the v2c message updating rule is the same as that in layered binary decoding shown in (8.10). There is no explicit VNU in the layered decoder. Instead, the most updated v2c messages are derived using the adder and subtractor. Although it is not included in Fig. 9.27, vector normalization needs to be done at the output of the subtractor. The c2v messages are added up with buffered v2c messages to derive intrinsic messages. The size of the buffer is decided by the parallelism of the variable node processing and the latency of the sorter and c2v message computation. Similar to that in layered binary decoders, if the submatrix in the same block column of the next layer is nonzero, the intrinsic messages are consumed in the check node processing for the next layer right away, and do not need to be stored. To compute the extrinsic v2c messages, the c2v messages of the

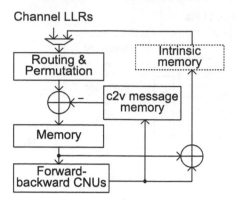

FIGURE 9.28
Layered NB-LDPC decoder architecture when forward-backward CNUs are used

same layer derived in the previous decoding iteration need to be subtracted. Instead of storing the c2v message vectors, the memory requirement is greatly reduced by recording the min1 nodes, min2 nodes, and syndromes as compressed messages. A second copy of the c2v message computation unit is used to derive the c2v messages of the previous iteration. The overhead brought by this second unit is usually smaller than the saving achieved by the reduced memory requirement unless d_c is very small. The compressed c2v messages belonging to different layers or iterations are stored in different memory blocks or registers, and they can be accessed independently for computing the c2v messages of the current and previous iterations at the same time. When the messages of a layer are no longer needed for later computations, the corresponding memory is reused to store the newly computed min1 nodes *etc.* The SMMA, basis-construction, and syndrome-based CNUs make use of the re-ordered trellis. At the input of the sorter for check node processing, the zero field element must have zero LLR in each v2c vector. Therefore, the re-ordering is done right before the min1 and min2 sorting. Originally, the intrinsic messages should go through reverse permutation and reverse routing. However, the outputs of these two blocks are sent to the routing and permutation for the next layer. Since the H matrix is fixed, the reverse permutation (routing) is combined with permutation (routing). Nevertheless, since the $\hat{\beta}$ that controls the re-ordering varies as the decoding proceeds, the two inputs of the subtractor may have been re-ordered differently. Therefore, the re-ordering and its reverse can not be combined.

When forward-backward CNUs are employed, the v2c messages need to be held during the forward and backward processes. Moreover, no compressed format of the c2v messages is available. Because of these reasons, the layered decoder architecture is modified to that in Fig. 9.28 when forward-backward CNUs are employed. $d_c d_v$ c2v message vectors of the previous decoding iterations need to be stored. Moreover, large memories are required to hold the inputs to the CNUs and the intermediate vectors computed during the forward and backward processes. For example, using the Scheme B in Fig. 9.10 to schedule the forward and backward processes, the memory at the input of the CNUs needs to store $2d_c$ message vectors in order to enable the overlapping of the computations from adjacent layers. Moreover, $(d_c - 2)$ intermediate message vectors from the forward and backward computations need to be recorded. Also, due to the iterative forward and backward processes, the achievable throughput of the decoder is much lower than those making use of min1 and min2 nodes, despite similar or larger gate count.

In summary, among available Min-max QCNB-LDPC decoder architectures, those adopting forward-backward CNUs are the least efficient, and their achievable throughput is low. When the finite field is small, all q messages can be kept to avoid the more complicated vector additions needed in the variable node processing. In addition, when $q' < q$ LLRs are kept for each vector, their associated field elements also need to be stored. On the other hand, the field elements do not need to be explicitly recorded when all q messages in a vector are stored according to a predetermined order. When q is small, such as 4 or 8, the absolute difference between q and q' that does not lead to noticeable performance loss is small. As a result, storing q' messages in each vector does not necessarily lead to memory saving. For $q = 4$, the modified syndrome-based CNU has the lowest complexity. Although it requires fewer comparators than the SMMA CNU, the complexities of both of them are higher than that of basis-construction CNU for moderate q. Nevertheless, the basis-construction CNU has performance loss for $GF(4)$ codes because the min2 nodes are not recorded and their compensation values are more biased. When q is large, keeping $q' < q$ messages in each vector leads to more significant saving on the memory, which can contribute to the major part of the overall decoder complexity. Hence, the trellis-based path construction decoder is more efficient when q is large.

Compared to layered decoders, sliced message-passing decoders can achieve higher throughput. If one block column of H is processed at a time, the number of clock cycles needed in an iteration of the sliced message-passing decoding is decided by the number of block columns. On the other hand, the number of clock cycles needed for an iteration of layered decoding is proportional to the row weight and number of layers. As mentioned previously, including multiple block rows of H in a layer complicates the v2c message updating rule and slows down the decoding convergence. In addition, in a given H, if the nonzero submatrices are located at irregular positions, it may not be possible to read the messages associated with multiple nonzero submatrices from the memory without causing conflicts. These obstacles prevent layered decoders from achieving as high throughput as sliced message-passing decoders. Nevertheless, a sliced message-passing decoder processes all rows of H simultaneously, and hence needs a large number of CNUs for codes that are not short, even if the code rate is high. Which type of these decoders is more efficient for a given application depends on the throughput requirement, area and power budgets, and code parameters, such as row weight, column weight, sub-matrix size, and finite field order.

9.3   Extended Min-sum decoder architectures

The difference between the Min-max and EMS decoding algorithms only lies in the check node processing. The 'max' in (9.3) is replaced by sum in the EMS algorithm. Therefore, some of the CNU architectures for the Min-max algorithm can be easily modified to implement the EMS algorithm. For example, if the 'max' units in Figs. 9.3, 9.5, 9.6 and 9.7 are replaced by adders, elementary step architectures for forward-backward EMS CNUs keeping all q messages are derived. Moreover, the 'max' computations in the SMMA and syndrome-based Min-max check node processing can be also replaced by adders to implement the EMS algorithm. Nevertheless, the elementary step architecture with $q' < q$ messages in Fig. 9.9, and the trellis-based path construction [144] and basis-construction [148] Min-max CNUs are developed by making use of the property that the output of a 'max' unit equals one of its inputs. These designs do not directly extend for the EMS algorithm. In the following, an EMS forward-backward elementary step architectures with $q' < q$ messages and a trellis-based path construction EMS CNU [151] are presented.

9.3.1   Extended Min-sum elementary step architecture keeping $q' < q$ messages

Each elementary step of the forward-backward EMS check node processing calculates the sums of pairs of messages from the two input vectors. Then the q' smallest sums corresponding to distinct field elements become the entries in the output vector. Since the field elements of the q' messages kept in each vector are different, the number of message pairs whose field elements add up to be the same varies. As a result, it is very difficult to carry out the min and sum computations over the entries of two vectors in a parallel way. Moreover, choosing the smallest sums involve more complicated logic than selecting the maximum values.

Denote the two LLR and field element input vectors to the elementary step by L_1, L_2, and α_{L_1}, α_{L_2}, respectively. The sum of each LLR pair can be listed in a two-dimension array, and the entry in the i row and jth column, $l_{i,j}$, is equal to $L_1(i) + L_2(j)$. Fig. 9.29 shows the array for a pair of example input vectors. The LLRs in the output vector are the smallest entries in this array. They are $0, 5, 6, 11, 13, 14 \ldots$. Starting from 0, these entries are linked by the dashed lines in non-decreasing order in Fig. 9.29. It can be observed

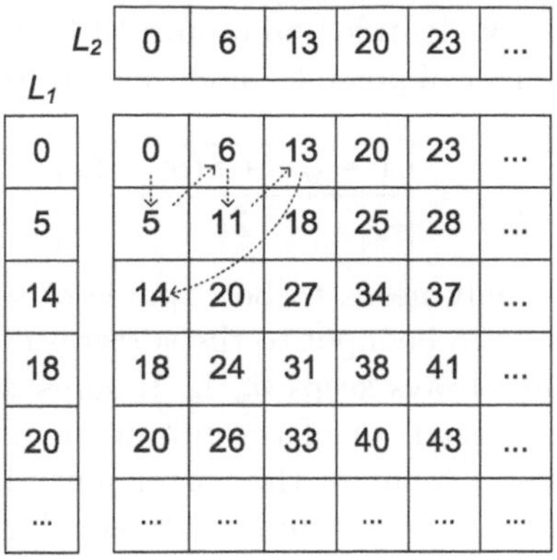

L_2	0	6	13	20	23	...

L_1						
0	0	6	13	20	23	...
5	5	11	18	25	28	...
14	14	20	27	34	37	...
18	18	24	31	38	41	...
20	20	26	33	40	43	...
...	...	...	...	...	...	...

FIGURE 9.29
Array of LLR sums for two example input vectors

that, due to the 'sum' computation, the next smallest entry may not always be a neighbor of the current smallest entry as in the Min-max algorithm. Hence, sorting needs to be applied to the sums to find the smallest entries. When the next smallest entry is identified, its field element is compared with those of previous entries. Only when the field element is different, the next smallest entry is inserted into the output vector.

The design in [149] makes use of a sorter that is able to insert an LLR into a vector of q' sorted LLRs in one clock cycle. Such a sorter can be implemented by a register-multiplexor-comparator array architecture similar to that in the bottom part of Fig. 9.21. It records q' LLRs, and the minimum LLR is shifted out to be stored in the output vector after every insertion. In addition, the input is inserted properly so that the q' LLRs in the registers remain in nondecreasing order. The registers in the sorter are initialized to the LLRs in L_1, which are also the entries in the first column of the sum array in Fig. 9.29. The sum input to the sorter in the next clock cycle is the right-hand-side neighbor of the minimum-LLR that is shifted out of the sorter in the current clock cycle. Taking the sum array in Fig. 9.29 as an example, when LLR=0 is shifted out, the next sum to be input to the sorter is 6. The minimum LLR shifted out in the next clock cycle is 5. Hence, the sorter input in the clock cycle after is 11.

A bubble check scheme was developed in [150] to reduce the complexity of the sorting. Although q' messages need to be found for the output vector, the sorting does not have to involve q' LLRs. Since the LLRs in the two input vectors are in the order of increasing LLR, the entries of the sum array satisfy $l_{i,j} \le l_{i,j+1}$ and $l_{i,j} \le l_{i+1,j}$. Therefore, the possible candidates for the next smallest LLR are the immediate neighbors of the minimum LLRs that have already been identified. For example, after the minimum LLRs $0, 5, 6, 11$ are found in the array of Fig. 9.29, the next smallest entry can be only from $l_{0,2}$, $l_{1,2}$, $l_{2,0}$ and $l_{2,1}$. It has been analyzed in [150] that in the worst case, the locations of the previously identified minimum LLRs form a triangle, and the upper bound of the number of possible candidates to be searched for the next smallest LLR is

$$n_u = \left\lceil \frac{1 + \sqrt{1 + 8(q' - 1)}}{2} \right\rceil . \tag{9.14}$$

Accordingly, the sorter only needs to be capable of sorting n_u instead of q' LLRs. The next LLR sum to be input to the sorter is decided using Algorithm 40 [150]. This algorithm always keeps n_u LLR sums at the sorter. If flag=1, then the right-hand-side neighbor of the minimum LLR that has just been shifted out of the sorter is input to the sorter next. If flag=0, the neighbor below is selected. When $a = n_u - 1$ and $b = 0$, the flag is flipped to 0. Before that, the LLR sums to be fed to the sorter are limited to those in the first n_u rows of the sum array.

Algorithm 40 Bubble Check Sorting for EMS Elementary Step
input: $L_1, \alpha_{L_1}, L_2, \alpha_{L_2}$
initialization: *copy the first n_u LLRs in L_1 to the registers in the sorter*
repeat *for I_{max} iterations or until q' messages are found for output vector*
        shift out the smallest entry in the sorter; assume it is $l_{i,j}$
        if $\alpha_{L_1}(i) + \alpha_{L_2}(j)$ is not in output vector
            store $\{l_{i,j}, \alpha_{L_1}(i) + \alpha_{L_2}(j)\}$ into output vector
        if $(i = 0)$
            flag=1
        if $(j = 0)$ & $(i = n_u - 1)$
            flag=0
        if $l_{i+(1-flag),j+flag}$ has not been introduced in the sorter before
            input $l_{i+(1-flag),j+flag}$ to the sorter
        else
            input $l_{i+flag,j+(1-flag)}$ to the sorter

Example 50 *Fig. 9.30 gives an example of carrying out Algorithm 40 on the two input vectors shown in Fig. 9.29. When $q' = 7$, it is computed from (9.14) that at most $n_u = 4$ sums need to be considered as the candidates of the next smallest LLR in the output vector. In Fig. 9.30, the minimum LLRs that have already been identified at the beginning of each iteration of Algorithm 40 are denoted by shaded circles in each matrix. The candidates for the next smallest LLR are represented by clear circles.*

Simulation results in [150] showed that the length of the sorter can be made smaller than the n_u computed from (9.14) without causing error-correcting

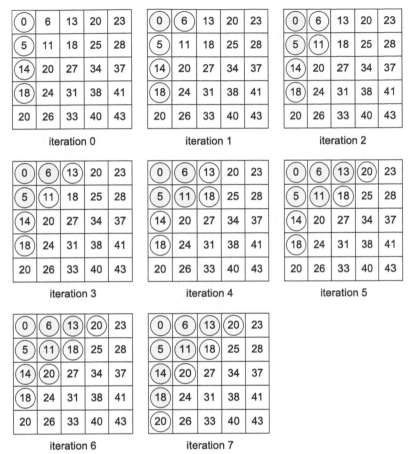

FIGURE 9.30

Example for bubble check sorting EMS elementary step

performance loss. For a code over $GF(64)$ and $q' = 16$, $n_u = 6$. However, using sorters of length 4 or 5 leads to the same performance. Accordingly, an L-bubble check algorithm was proposed in [150] to further reduce the complexity of the elementary steps in the EMS check node processing. In the L-bubble scheme, the length of the sorter is fixed to 4. Moreover, the next candidate to be input to the sorter is selected using a simpler approach. If $i = 0$ or 1, the next candidate is always the right-hand-side neighbor of $l_{i,j}$. If $l_{i,j} = l_{2,0}$, then the next candidate is $l_{2,1}$, and the candidates after $l_{2,1}$ are $l_{3,1}, l_{4,1}, l_{5,1} \ldots$. In the case that $l_{i,j} = l_{3,0}$, then the next candidates are $l_{4,0}, l_{5,0}, l_{6,0} \ldots$. As a result, only the LLRs in the first two rows and first two columns of the sum array can be the candidates of the sorter input. The L-bubble scheme can be implemented with the same sorter architecture and simpler control for selecting the next sorter input.

9.3.2   Trellis-based   path-construction   extended   Min-sum check node processing

A trellis-based path-construction EMS decoder has been introduced in [151]. Similar to the path-construction scheme for the Min-max algorithm, the $1.5q'$ nodes in the v2c message trellis with the smallest nonzero LLRs are first sorted out, and a similar process is adopted to construct paths in the trellis. However, for the EMS algorithm, the LLR of a path is defined as the sum of the LLRs of the nodes in the path. Although the nodes are still used for path construction in the order of increasing LLR, it may happen that a newly constructed path has smaller LLR than a previous path. For example, denote the smallest nonzero LLRs of the nodes by $x(1), x(2), x(3) \ldots$, and they belong to nodes $O(1), O(2), O(3), \ldots$. Also assume that $O(1)$ and $O(2)$ are not in the same stage. Hence, there will be a path with two nonzero-LLR nodes $O(1)$ and $O(2)$. The path with only O_3 as the nonzero-LLR node is constructed after this path. If $x(3) < x(1) + x(2)$, then the path with $O(3)$ has smaller LLR than the path with $O(1)$ and $O(2)$, which was previously constructed. As a result, sorting needs to be applied in order to insert the LLR of the newly-constructed path into the output c2v message vector. Similarly, if the new path has the same field element as that of a previous path but larger LLR, then it should be not inserted.

Let the field elements of the nodes with $O(1), O(2), O(3) \cdots$ be $\alpha(1), \alpha(2), \alpha(3), \cdots$, and the corresponding variable node indices be $e(1), e(2), e(3) \cdots$. Assume that the field element of the zero-LLR node in stage j is $z(j)$ and $\alpha_{sum} = \sum_{0 \le j < d_c} z(j)$. Then the trellis-based path construction for the EMS algorithm can be carried out according to the pseudo codes in Algorithm 41 to compute the c2v LLR vector L_n and the corresponding finite field element vector α_{L_n} for variable node n. In this algorithm, i tracks the sorted nodes of minimum LLRs considered in the path construction, and num tells the number of valid entries that have been derived for the output vector. d denotes the index of the entry in α_{L_n} such that $\alpha_{L_n}(d) = \delta$. Similar to that for the Min-max algorithm, when $O(i)$ is considered, new paths are constructed by replacing the nodes of stage $e(i)$ in each previous path by $O(i)$. Also if a previous path has a nonzero-LLR node in stage $e(i)$, then the path construction is skipped because it leads to a duplicated path. Different from that for the Min-max algorithm, a newly constructed path may have smaller LLR than previously constructed paths. Therefore, when the new path is inserted into the output vector according to increasing LLR, the order of the paths may be changed. This makes it difficult to track which previous paths have been used in constructing new paths. To solve this issue, L_n and α_{L_n} are copied to L'_n and α'_{L_n}, respectively, at the beginning of the path construction using a new node. New paths are constructed by substituting the new node into the paths corresponding to L'_n and α'_{L_n}. The computations of the field element and LLR of the new path are simplified by taking care of only the discrepancies.

Algorithm 41 EMS Path Construction Algorithm
input: $x(i), \alpha(i), e(i)$ $i = 1, 2, 3, \ldots$
initialization: $L_n(0) = 0$, $\alpha_{L_n}(0) = \alpha_{sum} + z(n)$
$\qquad\qquad\quad i = 1$, $num = 1$

loop: if $(e(i) \neq n)$
$\qquad\qquad \kappa = num$, $L'_n = L_n$, $\alpha'_{L_n} = \alpha_{L_n}$
$\qquad\qquad$ for $k = 0$ to $\kappa - 1$
$\qquad\qquad\qquad \delta = \alpha'_{L_n}(k) \oplus z(e(i)) \oplus \alpha(i)$
$\qquad\qquad\qquad y = x(i) + L'_n(k)$
$\qquad\qquad\qquad$ if (path k has a zero-LLR node in stage $e(i)$)&
$\qquad\qquad\qquad\qquad\qquad ((\delta \notin \alpha_{L_n}) | ((\delta = \alpha_{L_n}(d))$ & $(y < L_n(d))))$
$\qquad\qquad\qquad\qquad$ insert y into L_n, insert δ into α_{L_n}
$\qquad\qquad\qquad\qquad num = num + 1$
$\qquad\qquad\qquad\qquad$ if $(num = q')$, stop
$\qquad\quad i = i + 1$; goto loop

A maximum iteration number can be also set to the computations in Algorithm 41 to limit the worst-case latency. If less than q' valid entries have been derived for the output vector at the end, the rest of the entries are compensated in the same way as those for the Min-max algorithm. The LLRs are set to the last computed LLR and the field elements are set to zero. New paths are only constructed by replacing the nodes in each of the previously recorded path. Since the path LLR in the EMS algorithm is the sum of the LLRs of the nodes in the path, it is more likely that some valid paths that should contribute to the output vector are not constructed by Algorithm 41 compared to Algorithm 34 for the Min-max algorithm. To address this issue, a set of supplementary paths are constructed by replacing the nodes in the first two paths with $O(i), O(i+1), \ldots$ for a larger i. These nodes may not have been used in the path construction when q' valid entries are found for the output vector or the maximum iteration is reached. The paths constructed from Algorithm 41 and the supplementary paths are merged to generate the final output vector. Experiments in [151] showed that for a NB-LDPC code over $GF(2^5)$, when $q' = 16$, Algorithm 41 with $2q'$ iterations leads to noticeable performance loss. The performance gap is eliminated after including supplementary paths constructed using $x(5), x(6), \ldots, x(12)$.

The nodes of the minimum LLRs and α_{sum} can be computed by the architecture in Fig. 9.13. Fig. 9.31 illustrates an architecture for implementing the main path construction according to Algorithm 41. The sorted nodes are read from memory serially in the order of increasing LLR. A d_c-bit binary path

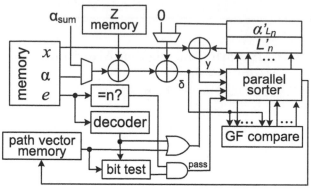

FIGURE 9.31

Main path construction architecture for the EMS algorithm (modified from
[151])

vector is also kept for each path. The vector is '1' in the jth bit if the path
has a nonzero LLR node in stage j. The binary decoder in Fig. 9.31 converts
the stage index into an active-high one-hot vector. The bit-test block carries
out bit-wise AND over the two input vectors, and the AND outputs are sent
to a NOR gate. Its output is '1' when the path has a zero-LLR node in stage
$e(i)$. Hence, if the signal 'pass' in Fig. 9.31 is asserted, then $e(i) \neq n$ and the
previous path has a zero-LLR node in stage $e(i)$. Accordingly a new path can
be constructed. Then δ and y are sent to the parallel sorter and field element
comparators to find if the newly constructed path should be inserted into the
output vector. The output of the ith field element comparator, g_i, is '1' if δ
matches the ith field element in the output vector. The architecture of the
parallel sorter is similar to that in Fig. 9.21, except that the select signals for
the multiplexor and the enable signals for the registers are modified to

$$\begin{cases} s_i = NOT(c_0 + c_1 + \cdots, +c_{i-1}), (0 \leq i \leq q' - 1) \\ e_i = en_i \cdot pass \cdot NOT(g_0 + g_1 + \cdots + g_{i-1}), (1 \leq i \leq q' - 1) \end{cases}$$

where c_i is the output of the ith LLR comparator, and $en_i = c_0 + c_1 + \cdots + c_i$.
Similarly, the field elements in the output vector and the path vectors are
reordered using these control signals. At the end of constructing paths using
a new node, the vectors located in the parallel sorter are copied to the L'_n,
α'_{L_n}, and path vector memory in Fig. 9.31.

    There are two groups of supplementary paths: the ones derived by replacing
the nodes in path 0 and those formed by replacing the nodes in path 1. For the
path construction of each group, the nodes corresponding to $x(i), x(i+1), \cdots$
are read out serially, and one new path is constructed each time. The LLR and
field element of each path are computed using one finite field adder and one
integer adder. Of course, necessary testing is done to exclude nodes from stage
n and avoid constructing duplicated paths. Since the LLRs of the paths in each

group is non-decreasing, the two groups of paths can be merged using a design similar to the serial sorter in Fig. 9.13. Nevertheless, field element comparators need to be adopted to exclude paths with the same field elements. At the end, the main and supplementary paths are merged using the same serial sorter architecture.

9.4 Iterative majority-logic decoder architectures

The details of the ISRB and IHRB-MLGD algorithms can be found in Algorithm 30 and 31, respectively, presented in the beginning of this chapter. These algorithms do not require complicated check node processing as in the Min-max or EMS algorithms. Instead, only simple check sums need to be computed. Moreover, the messages from a variable node to all connected check nodes are the same, and hence only one reliability message vector needs to be stored for each variable node. This significantly reduces the memory requirement. As a result, these MLGD algorithms achieve efficient performance-complexity tradeoff. When the column weight is large, the ISRB algorithm can even achieve similar performance as the Min-max algorithm [152]. Compared to the ISRB algorithm, the IHRB algorithm has lower hardware complexity. However, the savings come at the cost of performance loss. To bridge the performance gap between the IHRB and ISRB algorithms, two enhancement schemes were developed for the IHRB algorithm in [152]. The first scheme is to incorporate the soft information from the channel into the message initialization. The second is to exclude the contribution from the same check node and only use extrinsic information in the v2c message computation.

Different from IHRB decoders, ISRB decoders accumulate soft reliability measures. Hence ISRB decoders not only require longer word length to represent the reliability measures, but also need more complicated VNUs. As a result, the hardware complexity of ISRB decoders is much higher than that of IHRB decoders. As it will be shown in this section, the two enhancement schemes to the IHRB algorithm can be implemented by adding small hardware overhead to the IHRB decoders, although the error-correcting performance improvement brought by adopting soft information in the message initialization is more significant than employing extrinsic messages. Since the performance of the enhanced (E)-IHRB algorithm is similar or quite close to that of the ISRB algorithm, the implementation of ISRB decoders is not further discussed in this section. The interested reader is referred to [153] for more details.

As mentioned previously, MLGD algorithms perform well when the column weight of the code is not small. Besides QC codes, another type of hardware-friendly LDPC codes is cyclic LDPC codes, such as those constructed based on Euclidean Geometry (EG) [103]. These cyclic codes usually have high column weight. The H matrix of a cyclic NB-LDPC code consists of a single circulant matrix whose entries are elements of $GF(q)$ or a column of circulant matrices. For a $(255,175)$ EG-LDPC code over $GF(2^8)$ with $d_v = 16$, the ISRB algorithm and E-IHRB algorithm with both enhancement schemes can achieve similar error-correcting performance as the Min-max algorithm. It should be noted that the performance of these two algorithms become inferior when the column weight is lower for both cyclic and QC codes. The size of the circulant matrix of a cyclic EG-LDPC code over $GF(q)$ is $q - 1$. Hence, large finite fields are involved to construct cyclic LDPC codes with reasonable codeword length. Besides the cyclic H matrix structure, the large finite field is another factor that makes the cyclic NB-LDPC decoder architecture different from that for QCNB-LDPC codes. The remaining of this section first presents the IHRB decoder architectures for both QC and cyclic codes. Then the E-IHRB decoding algorithms and architectures are introduced.

9.4.1   IHRB decoders for QCNB-LDPC codes

Partial-parallel decoders are the focus of the design since they achieve speed-area tradeoff compared to fully-parallel and serial decoders. Assume that the H matrix of a QCNB-LDPC code consists of submatrices of dimension $(q-1) \times (q-1)$, and the row weight is d_c. Additionally, one block row of submatrices is processed at a time. In this case, each column in a block row has at most one nonzero entry. If only one nonzero entry is processed for each column at a time, then the reliability measures of each variable node is only increased by one each time. Accordingly, the IHRB algorithm is re-organized as in Algorithm 42 to simplify the hardware architectures.

Algorithm 42 Re-organized IHRB-MLGD for NB-LDPC Codes
input: z_j
initialization: $z_j^{(0)} = z_j$; $R_j^{(0)}(\beta) = \gamma$ if $\beta = z_j^{(0)}$
$$R_j^{(0)}(\beta) = 0 \ \text{if} \ \beta \neq z_j^{(0)}$$
for $k = 1 : I_{max}$
  stop if $z^{(k-1)} H^T = 0$
  for each check node i
    for each $j \in S_v(i)$
$$\sigma_{i,j} = h_{i,j}^{-1} \sum_{u \in S_v(i) \backslash j} z_u^{(k-1)} h_{i,u}$$
      if $(\sigma_{i,j} = \beta)$
$$R_j^{(k-1)}(\beta) \Leftarrow R_j^{(k-1)}(\beta) + 1$$
  for each variable node j
$$R_j^{(k)}(\beta) = R_j^{(k-1)}(\beta)$$
$$z_j^{(k)} = \arg\max_{\beta} R_j^{(k)}(\beta)$$

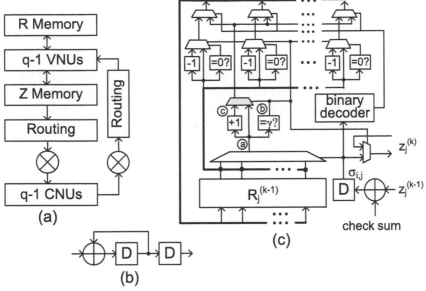

FIGURE 9.32
IHRB decoder architecture for QCNB-LDPC codes. (a) overall decoder architecture; (b) CNU architecture; (c) VNU architecture (modified from [152])

In the beginning of Algorithm 42, each reliability measure, $R_j^{(0)}(\beta)$, is initialized as either zero or γ, which is the largest positive integer allowed for reliability measures. When a message exceeds γ, the entire vector is clipped. The clipping is done as offsetting each message in the vector by the same value so that the largest message remains γ. If a message becomes negative as a result of the offsetting, it is set to zero. As a result, all messages remain in the range of $[0, \gamma]$ during the IHRB decoding. This clipping process is not shown in Algorithm 42 for brevity purpose.

Fig. 9.32(a) shows the overall architecture of the IHRB decoder for QCNB-LDPC codes. To process one block row at a time, $q-1$ CNUs are adopted. Also $q-1$ VNUs are employed to handel the reliability measure vector updating for one block column in parallel. To accommodate this, the $q-1$ vectors belonging to the same block column need to be accessed simultaneously from the R memory. The Z memory stores the hard-decision symbols for two consecutive iterations, $z^{(k-1)}$ and $z^{(k)}$, in a similar manner. The $z^{(k-1)}$ for one block column are fed to the $q-1$ CNUs at a time. The routing network sends these hard-decision symbols to the correct CNUs according to the locations of the nonzero entries in H, and the multipliers take care of the multiplications with the nonzero entries. To reduce the complexity, $\sigma_{i,j}$ is computed instead as

$$\sigma_{i,j} = z_j^{(k-1)} + h_{i,j}^{-1} \sum_{u \in S_v(i)} z_u^{(k-1)} h_{i,u}.$$

The check sum $\sum_{u \in S_v(i)} z_u^{(k-1)} h_{i,u}$ only needs to be computed once for each check node, and it is reused to calculate $\sigma_{i,j}$ with different j. Since there is only one nonzero entry in each column of a nonzero submatrix, a CNU can be implemented as a simple feedback loop with one XOR gate in the critical path as shown in Fig. 9.32(b). The check sum is available after d_c clock cycles. Then it is loaded into the register outside of the register loop, so that the check sum computation for the next block row of H starts right after.

When all q messages are kept in a vector, an efficient architecture for implementing the VNUs for the IHRB decoder is shown in Fig. 9.32(c). The check sum is multiplied by $h_{i,j}^{-1}$ and goes through a reverse routing network before it is added up with $z_j^{(k-1)}$ to compute $\sigma_{i,j}$ in the VNU. As mentioned previously, only one measure in each $R_j^{(k-1)}$ vector is increased by one at a time. If $\sigma_{i,j} = \beta$, $R_j^{(k-1)}(\beta)$ is selected by the multiplexor to be added by one. The clipping is done jointly with the reliability measure updating. In the case that the selected $R_j^{(k-1)}(\beta)$ is already γ, then it should remain at γ instead and each of the other nonzero measures in the same vector should be subtracted by one to incorporate the clipping. The updated and clipped reliability measures in the vector, except $R_j^{(k-1)}(\beta)$, are selected by the second-row multiplexors in the top part of the VNU architecture in Fig. 9.32(c). The updated $R_j^{(k-1)}(\beta)$ is available at the output of the multiplexor in grey color. To pass this value as the βth updated reliability measure, a binary decoder is adopted to generate the select signals for the top-row multiplexors in the VNU.

The hard decision, $z_j^{(k)}$, needs to be made at the end of decoding iteration k. The hard decision is the finite field element corresponding to the largest reliability measure in the vector. Finding the largest measure in a vector requires an expensive comparator tree if it needs to be done as fast as possible to avoid bringing extra latency to each decoding iteration. Alternatively, $z_j^{(k)}$ can be derived through tracking the reliability measure updating. Using the initialization in Algorithm 42 and adopting the message clipping, the jth hard-decision symbol can only be replaced by $\sigma_{i,j} = \beta$ when $R_j^{(k-1)}(\beta)$ is already γ before it is increased by one. Therefore, $z_j^{(k)}$ is updated using a multiplexor during the processing of each block row of H. Since the reliability measure vectors and hard-decision symbols for one block column are updated at a time using $q - 1$ VNUs, it takes d_c clock cycles to finish the variable node processing for one block row. These computations are overlapped with the check node processing of the next block row to better utilize the hardware units and increase the throughput. Accordingly, the IHRB decoder for QCNB-LDPC codes takes around $(1 + rI_{max})d_c$ clock cycles to complete I_{max} decoding iterations, if there are r block rows in H.

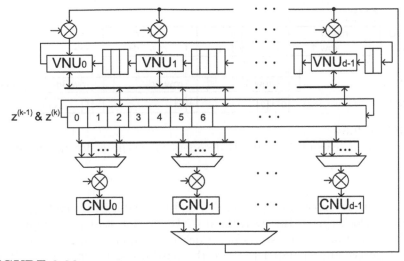

FIGURE 9.33
IHRB decoder architecture for cyclic NB-LDPC codes (modified from [152])

9.4.2 IHRB decoders for cyclic NB-LDPC codes

The decoder architecture for cyclic NB-LDPC codes whose H matrix consists of a single circulant matrix is discussed next. It can be easily extended for codes having a column of circulant matrixes in H. Using EG or projective geometry-based construction methods in [103], each row of the $(q-1) \times (q-1)$ cyclic H matrix is the previous row shifted by one position and multiplied by α, which is a primitive element of $GF(q)$. Different from the submatrices in the H matrix of a QC code, each row of the cyclic H has multiple nonzero entries and they appear at irregular locations. Hence, a cyclic H can not be partitioned into block rows and columns that have fixed number of nonzero entries in each block. As a result, previous decoder architectures for QC codes can not achieve efficient partial-parallel processing when applied to cyclic codes. Moreover, the codeword length is $q - 1$ for cyclic codes constructed over $GF(q)$. Hence large finite field is used to construct codes with reasonable codeword length. Storing all q messages in a vector leads to very large memory requirement in this case. To address these issues, a shift-message decoder architecture was developed in [152] to achieve efficient partial-parallel processing for cyclic NB-LDPC codes, and an efficient VNU was designed to keep only the $q' < q$ most reliable measures in each vector without causing noticeable error-correcting performance loss.

The shift-message decoder architecture is depicted in Fig. 9.33. Assume that each row of the cyclic H has d nonzero entries, and they are located at positions $p_0, p_1, \cdots, p_{d-1}$ in the first row. This decoder has d CNUs and d VNUs. Each CNU has a simple feedback loop with an XOR gate in the

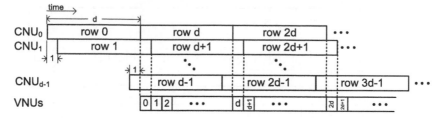

FIGURE 9.34

Computation scheduling in IHRB decoders for cyclic codes (modified from [152])

critical path as shown in Fig. 9.32 (b), and is capable of computing a check sum in d clock cycles. However, the d CNUs do not start the computations at the same time. CNU_i starts one clock cycle after CNU_{i-1}. Two sets of shift registers are employed to store the hard-decision symbols $z^{(k-1)}$ and $z^{(k)}$. In each clock cycle, they are cyclically shifted by one position to the left. As a result, all CNUs read from the same d registers of $z^{(k-1)}$ at locations $p_0, p_1 - 1, \cdots, p_{d-1} - (d-1)$ in each clock cycle, and hence the routing between the registers and CNUs is simplified. For an example cyclic code with $p_0 = 0, p_1 = 3, p_2 = 7, \cdots$, the connections from the $z^{(k-1)}$ registers are shown in Fig. 9.33. Each multiplexor in the middle selects one of the d hard-decision symbols, and its select signal is generated by a counter. The counter for the ith multiplexor starts one clock cycle after that for the $i - 1th$ multiplexor. The selected hard-decision symbols are multiplied with the entries of H before they are sent to the CNUs. The first check sum is computed by CNU_0 in d clock cycles. After that, one additional check sum is available in each clock cycle from $CNU_1, CNU_2, \cdots$. After a CNU finishes computing the check sum for row i, it starts to calculate the check sum for row $i + d$ right after. Fig. 9.34 shows the scheduling of the check node processing.

Once the check sum for row i is computed, it is multiplied with $h_{i,j}^{-1}$ for every $j \in S_v(i)$ and then sent to the d VNUs. The updating of the reliability measure and hard-decision symbol for the jth nonzero column in each row of H is taken care of by VNU_j. The variable node processing for all nonzero entries in a row is carried out by the d VNUs in parallel. As shown in Fig. 9.34, after the initial latency of d clock cycles, one check sum is available in each clock cycle. Hence only one reliability measure is increased by one at a time in each VNU, and an approach similar to that in the partial-parallel IHRB decoder for QC codes can be adopted to update the reliability measure and hard-decision symbols. The relevant updating is also done in one clock cycle, and the scheduling of the variable node processing is illustrated in Fig. 9.34. The row number for which the variable node processing is carried out in each clock cycle is labeled in this figure. Taking into account the initial check node processing latency of d clock cycles, the decoding of each iteration requires

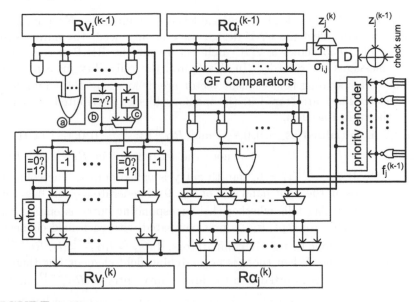

FIGURE 9.35
IHRB VNU architecture when $q' < q$ messages are kept in each vector (modified from [152])

$n + d$ clock cycles, where n is the number of block columns in H and is equal to $q - 1$ for cyclic codes.

The VNUs need to read from the $z^{(k-1)}$ shift registers in order to compute $\sigma_{i,j}$. Also the updated hard-decision symbols are written back to the $z^{(k)}$ shift registers. To reduce the routing complexity, the reliability measures are shifted together with $z^{(k-1)}$ and $z^{(k)}$. In this case, the VNUs can also be connected to fixed locations of the $z^{(k-1)}$ and $z^{(k)}$ shift registers. Since the check node processing has a latency of d clock cycles, when the check sum for a row of H is available, the hard-decision symbols involved in that check sum have been shifted by d positions in the registers. As a result, VNU$_j$ needs to be connected to register $(p_j - d) \mod n$ in the $z^{(k-1)}$ and $z^{(k)}$ arrays. The reliability measures are stored in memory blocks distributed among the VNUs as shown in Fig. 9.33. The memory block between VNU$_j$ and VNU$_{j+1}$ stores $p_{j+1} - p_j$ reliability vectors. Every time, a VNU reads messages from the memory on its right, and writes the updated messages to the memory on its left. Hence, the VNUs also serve as connecting ports to shift messages from one memory block to another.

As aforementioned, high-order finite fields are employed to construct cyclic NB-LDPC codes. To reduce the memory requirement, only the $q' < q$ largest entries are stored for each reliability measure vector. It has been shown in [152] that keeping $q' = 16$ messages in each vector for a cyclic code constructed over $GF(256)$ does not lead to noticeable performance loss. Keeping $q' < q$ entries does not affect the CNU design since it only needs the hard-decision symbols. Fig. 9.35 shows a VNU architecture for the IHRB decoder when $q' < q$ messages are recorded. In this figure, the reliability measure vector and the corresponding field element vector are denoted by $Rv_j^{(k)}$ and $Ra_j^{(k)}$, respectively. They are stored in the memory blocks distributed among the VNUs shown in Fig. 9.35.

$Rv_j^{(0)}$ is initialized as $[\gamma, 0, 0, \cdots]$ and $Ra_j^{(0)} = [z_j^{(0)}, 0, 0, \cdots]$ at the beginning. $Rv_j^{(k)}(l)$ and $Ra_j^{(k)}(l)$ denote the lth entries in $Rv_j^{(k)}$ and $Ra_j^{(0)}$, respectively. If $Rv_j^{(k-1)}(l) = 0$, the lth entry is called empty. The column of the NOR gates on the right-hand side of Fig. 9.35 generate a q'-bit flag vector $f_j^{(k-1)}$, and each flag tells whether the corresponding entry is empty. Since not all q messages are kept, $\sigma_{i,j}$ is first compared with each field element in $Ra_j^{(k-1)}$. This is done by the GF comparators. If $\sigma_{i,j} = Ra_j^{(k-1)}(l)$, $Rv_j^{(k-1)}(l)$ is increased by one. When there is no matching field element but there are empty entries, $\sigma_{i,j}$ is inserted into the first empty entry of $Ra_j^{(k-1)}$, and the corresponding entry from $Rv_j^{(k-1)}$ is increased from zero to one. Otherwise, no further action is taken and $\sigma_{i,j}$ is discarded. It is possible that $\sigma_{i,j}$ equals the field elements in those empty entries, and the outputs of the corresponding GF comparators are asserted. However, the comparator outputs are masked by the flags from the empty entry testing. Hence, the output of the OR gate in the right middle part is only asserted if there is a real match. Taking the flags as inputs, the priority encoder generates a q'-bit one-hot binary string, in which the lth bit is '1' if the lth entry is the first empty entry. According to whether there is a real match, either the field elements from $Ra_j^{(k-1)}$ are copied to $Ra_j^{(k)}$, or $\sigma_{i,j}$ is inserted into the first empty entry of $Ra_j^{(k)}$. This is achieved by using either the masked GF comparator outputs or the priority encoder outputs as the select signals of the multiplexor row right above the $Ra_j^{(k)}$ registers.

The output of the AND-OR tree in the top left corner of Fig. 9.35 is zero if none of the elements in $Ra_j^{(k-1)}$ matches $\sigma_{i,j}$. Otherwise, it is the reliability measure whose field element equals $\sigma_{i,j}$. To reduce the critical path, the GF comparator outputs are used instead of the masked comparator outputs as the select signals of this tree. The reason is that even if some GF comparator outputs are falsely asserted, the corresponding reliability measures are zero, and hence the outputs of the corresponding AND gates in the tree do not contribute to the OR gate output. The AND-OR tree output is added by one, which is needed for both increasing the reliability by one when there is a match, and setting the reliability measure to one for the newly inserted entry when there is no match. The message clipping is done following the same scheme used in the VNU architecture keeping all q messages. Hence, Fig. 9.35 has similar units used for this purpose, such as the blocks for subtracting one and zero testers followed by multiplexors. The updated reliability measures are also routed through the multiplexors to the $Rv_j^{(k)}$ registers in a way similar to that for $Ra_j^{(k)}$. It may happen that the decoding converges to different symbols in later iterations, but there is no empty entries to insert the different symbols. Ignoring this case leads to performance loss. One way to address this issue is to clear the entries with the lowest reliability measure, which equals one, at the end of each decoding iteration to clean up some space. This scheme can be implemented easily by adding an equal-to-one tester to each entry in $Rv_j^{(k-1)}$. The control block in Fig. 9.35 generates multiplexor select signals using the tester outputs and according to whether the row being processed is one of the last d rows of H.

9.4.3 Enhanced IHRB decoding scheme and architectures

The major reason that the ISRB algorithm performs better than the IHRB algorithm is that it makes use of soft reliability information from the channel. On the other hand, the accumulation of the soft reliability measure requires longer word length and makes the message clipping much more complicated. When soft information is available, it can also be incorporated into the message initialization of the IHRB algorithm to improve the performance. However, unlike the message initialization for the ISRB algorithm, the maximum value for each reliability measure is limited to a constant γ in the E-IHRB algorithm [152] in order to simplify the message clipping. To achieve such initialization, $\varphi_j(\beta)$, the probability information from the channel that the jth received symbol equals β, is first multiplied by a constant scaler ρ. The role of ρ is similar to that of λ in Algorithm 30 for the ISRB decoding, and its optimal value can be found from simulations. After the scaler multiplications, each reliability measure vector is clipped so that the largest measure in each vector is γ and all the measures are in the range of $[0, \gamma]$.

In the IHRB algorithm, the messages sent from a variable node to all connected check nodes are the same. They are the hard-decision symbol of the variable node. Hence, the v2c messages in the IHRB algorithm are actually not *extrinsic* information. To further improve the performance, the contribution from a check node can be excluded from the v2c message sent to this check node as in the Min-max and other BP-based decoding algorithms. Fortunately, each $\sigma_{i,j}$ only increases one of the reliability measures by one in the IHRB decoding shown in Algorithm 42. Making use of this property, the contribution from a check node can be excluded in an easier way. Let $R_j^{(k)max}$ and $R_j^{(k)max2}$ be the largest and second largest reliability measures, respectively, in the vector $R_j^{(k)}$. Apparently, $z_j^{(k)}$ is the field element corresponding to $R_j^{(k)max}$. Let $z'^{(k)}_j$ be the field element corresponding to $R_j^{(k)max2}$. Denote the extrinsic message from variable node j to check node i by $z_{i,j}^{(k)}$. In the case that $\sigma_{i,j}^{(k-1)} \neq z_j^{(k)}$, $R_j^{(k)max}$ does not change after the contribution from

$\sigma_{i,j}^{(k-1)}$ is excluded, and hence $z_{i,j}^{(k)} = z_j^{(k)}$. If $\sigma_{i,j}^{(k-1)} = z_j^{(k)}$, there are three cases to consider.

Case 1: $R_j^{(k)max} = R_j^{(k)max2}$. $R_j^{(k)max}$ would be one less and smaller than $R_j^{(k)max2}$ without the contribution of $\sigma_{i,j}^{(k-1)}$. Hence, $z_{i,j}^{(k)}$ should be $z_j'^{(k)}$.

Case 2: $R_j^{(k)max} = R_j^{(k)max2} + 1$. If the contribution of $\sigma_{i,j}^{(k-1)}$ is excluded, $R_j^{(k)max} = R_j^{(k)max2}$ and $z_{i,j}^{(k)}$ can be either $z_j^{(k)}$ or $z_j'^{(k)}$. Nevertheless, from simulations, the decoder performance would be better if $z_{i,j}^{(k)}$ is set to $z_j'^{(k)}$. One possible explanation is that using $z_j'^{(k)}$ may introduce some disturbance to help the decoder jump out of trapping sets.

Case 3: $R_j^{(k)max} > R_j^{(k)max2} + 1$. $R_j^{(k)max}$ is still larger than $R_j^{(k)max2}$ after the contribution of $\sigma_{i,j}^{(k-1)}$ is removed. Accordingly, $z_{i,j}^{(k)}$ should be $z_j^{(k)}$.

From these analyses, the E-IHRB algorithm using both soft initialization and extrinsic v2c messages is summarized in Algorithm 43.

Algorithm 43 E-IHRB-MLGD Algorithm for NB-LDPC Codes

input: φ_j

initialization: $R_j^{(0)}(\beta) = \max(\lfloor \rho\varphi_j(\beta) \rfloor + \gamma - \max_\beta(\lfloor \rho\varphi_j(\beta) \rfloor), 0)$

$\qquad\qquad z_{i,j}^{(0)} = \arg\max_\beta(\varphi_j(\beta))$

for $k = 1 : I_{max}$

$\qquad$ *stop if* $z^{(k-1)}H^T = 0$

$\qquad$ *for each check node* i

$\qquad\qquad$ *for each* $j \in S_v(i)$

$\qquad\qquad\qquad \sigma_{i,j}^{(k-1)} = h_{i,j}^{-1} \sum_{u \in S_v(i) \backslash j} z_{i,u}^{(k-1)} h_{i,u}$

$\qquad\qquad\qquad$ *if* $(\sigma_{i,j}^{(k-1)} = \beta)$

$\qquad\qquad\qquad\qquad R_j^{(k-1)}(\beta) \Leftarrow R_j^{(k-1)}(\beta) + 1$

$\qquad$ *for each variable node* j

$\qquad\qquad R_j^{(k)}(\beta) = R_j^{(k-1)}(\beta)$

$\qquad\qquad R_j^{(k)max} = \max_\beta R_j^{(k)}(\beta)$

$\qquad\qquad z_j^{(k)} = field\ element\ of\ R_j^{(k)max}$

$\qquad\qquad R_j^{(k)max2} = second\ largest\ R_j^{(k)}(\beta)$

$\qquad\qquad z_j'^{(k)} = field\ element\ of\ R_j^{(k)max2}$

$\qquad\qquad$ *for each* $i \in S_c(j)$

$\qquad\qquad\qquad$ *if* $(\sigma_{i,j}^{(k-1)} = z_j^{(k)})\&(R_j^{(k)max} \leq R_j^{(k)max2} + 1)$

$\qquad\qquad\qquad\qquad z_{i,j}^{(k)} = z_j'^{(k)}$

$\qquad\qquad\qquad$ *else*

$\qquad\qquad\qquad\qquad z_{i,j}^{(k)} = z_j^{(k)}$

Although the E-IHRB algorithm adopts soft information in the initialization, the reliability measure updating is done in the same way as in the IHRB algorithm. Unless γ is changed, using different initial values for the reliability measures does not change the decoder architecture. When γ is larger, more bits are needed to represent each reliability measure, and hence the memory requirement is larger. In the case that $q' < q$ messages are kept in each vector, the least reliable entries are cleared from each vector at the end of each decoding iteration to free up space for symbols that the decoding may converge to in later iterations. If γ is larger, the entries with larger measures, for example, those equal to two, may also need to be cleared. Accordingly, more testers are needed for each entry of the $Rv_j^{(k)}$ vector in the VNU architecture of Fig. 9.35. On the other hand, more significant modifications need to be made on the IHRB decoders to generate the extrinsic v2c messages $z_{i,j}^{(k)}$.

To reduce the memory requirement, $z_{i,j}^{(k)}$ are not stored. Instead, it is chosen from $z_j^{(k)}$ and $z_j'^{(k)}$ based on the value of $\sigma_{i,j}^{(k-1)}$ and $R_j^{(k)max2}$. Using the initialization in Algorithm 43 and adopting the message clipping, $R_j^{(k)max}$ is always γ and does not need to be stored explicitly. Similar to that in the IHRB decoder, $\sigma_{i,j}^{(k-1)}$ is computed by adding $z_{i,j}^{(k-1)}$ to the check sum $\sum_{u \in S_v(i)} z_{i,u}^{(k-1)} h_{i,u}$. $z_{i,j}^{(k-1)}$ also needs to be selected from $z_j^{(k-1)}$ and $z_j'^{(k-1)}$. To stop this selection process with infinite depth, flags are stored to indicate whether $z_j^{(k-1)}$ or $z_j'^{(k-1)}$ has been picked as $z_{i,j}^{(k-1)}$. These flags are denoted by $f_{i,j}^z$, and are updated with the message selection in the last four rows of Algorithm 43 in each iteration. $z_j^{(k+1)}$, $z_j'^{(k+1)}$ and $R_j^{(k+1)max2}$ can be also updated with each addition to the reliability measure in iteration k, and they need to be stored. Therefore, the hard decisions and second most likely symbols need to be kept for three consecutive iterations. Nevertheless, the condition test for $z_{i,j}^{(k)}$ selection at the end of Algorithm 43 only needs to know if $R_j^{(k)max2} \geq R_j^{(k)max} - 1 = \gamma - 1$. To further save memory, a single-bit flag, f_j^r, is stored to tell if this condition is satisfied instead of recording $R_j^{(k)max2}$.

For both QC and cyclic codes, E-IHRB decoder architectures can be designed based on IHRB decoder architectures with small modifications as listed in the following.

1) *Extra storage.* Compared to those in the IHRB decoder, the additional variables need to be stored in the E-IHRB decoder are $z^{(k+1)}$, $z'^{(k-1)}$, $z'^{(k)}$, $z'^{(k+1)}$, $R^{(k+1)max2}$, the check sums, and the flags f^z and f^r. Copies of the Z memories in Fig. 9.32(a) for QC codes and shift registers in Fig. 9.33 for cyclic codes need to be added for storing the extra z and z'. f^r is stored in the same manner as z and z'. One more entry is included in the memory for each R_j vector to store R_j^{max2}. In addition, the f^z flags are stored in separate memories located in the selection logic detailed next.

2) *Extra extrinsic v2c message selection logic.* A selector is required to decide $z_{i,j}^{(k)}$ according to the last four rows in Algorithm 43 before the check

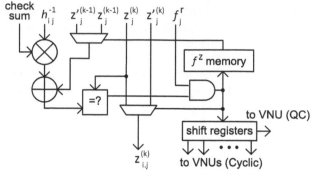

FIGURE 9.36

Extrinsic v2c message selector architecture (modified from [152])

TABLE 9.10

Updating of extra values in the E-IHRB decoding [152]

			updated values		
$R = \gamma$?	$R = R^{max2}$?	$\sigma = z$?	z	z'	R^{max2}
yes	yes	yes	σ (or z)	z'	$R^{max2} - 1$
yes	no	yes	σ (or z)	z'	$R^{max2} - 1$
yes	yes	no	σ (or z)	z (or z')	$R^{max2} - 1$
yes	no	no	σ	z	$R^{max2} - 1$
no	yes	-	z	σ	$R^{max2} + 1$
no	no	-	z	z'	R^{max2}

node processing for the next decoding iteration is carried out. One selector is needed for each CNU. The architecture of the selector is shown in Fig. 9.36. In this figure, the output of the equality tester is asserted if $\sigma_{i,j}^{(k-1)} = z_j^{(k)}$, and $f_j^r = 1$ if $R_j^{(k)max2} \geq \gamma - 1$. Hence the output of the AND gate is used to select the value for $z_{i,j}^{(k)}$. It is also the updated $f_{i,j}^z$ flag, which is stored and used to select the value for $z_{i,j}^{(k)}$ again in the next decoding iteration. $z_{i,j}^{(k)}$ is also added up with the check sum in the VNUs to derive $\sigma_{i,j}^{(k)}$, and the computations in the VNUs lag behind those in the CNUs. Hence $f_{i,j}^z$ is delayed by a shift register before it is sent to the corresponding VNU in the decoders for QC codes. In cyclic decoders, the VNUs need all $z_{i,j}^{(k)}$ with $j \in S_v(i)$ at the same time. For this case, the shift registers in Fig. 9.36 are used as a serial-to-parallel converter. All the updated $f_{i,j}^z$ with $j \in S_v(i)$ are shifted in serially before they are sent to the VNUs in parallel.

3) *VNU modifications.*

During the *k*th iteration of the E-IHRB decoding, $z_j^{(k)}$, $z_j'^{(k)}$, and $R_j^{(k)max2}$ need to be updated during the variable node processing according to each $\sigma_{i,j}^{(k-1)}$ and corresponding reliability measure $R_j^{(k)}(\beta)$. $R_j^{(k)}(\beta)$ is available at

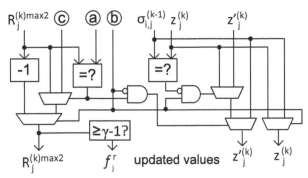

FIGURE 9.37
VNU architecture modifications for implementing E-IHRB decoding (modified from [152])

the point labeled as 'a' in Fig. 9.32(c) and Fig. 9.35. These updating can be carried out according to Table 9.10. For the purpose of clarity, some super-scripts and subscripts are omitted from the variables in this table. $R_j^{(k)max}$ is always γ in the E-IHRB algorithm. The updated value of $R_j^{(k)max2}$ is only dependent on if $R_j^{(k)}(\beta)$ equals γ and/or current $R_j^{(k)max2}$. $R_j^{(k)max2}$ may be also equal to γ. Hence the testings of whether $R_j^{(k)}(\beta)$ equals γ and $R_j^{(k)max2}$ are both necessary. When $R_j^{(k)}(\beta) = \gamma$, clipping needs to be done and hence $R_j^{(k)max2}$ is decreased by one. To ensure that the updated $z_j^{(k)}$ and $z_j'^{(k)}$ are different, the testing of whether $\sigma_{i,j}^{(k-1)}$ equals the current $z_j^{(k)}$ is required.

The updating function in Table 9.10 is implemented by the architecture depicted in Fig. 9.37. This architecture replaces the multiplexors for $z_j^{(k)}$ up-dating in the VNU architectures shown in Fig. 9.32(c) and Fig. 9.35 to im-plement the E-IHRB algorithm. The 'a', 'b', and 'c' inputs in this figure are $R_j^{(k)}(\beta)$, whether $R_j^{(k)}(\beta) = \gamma$, and $R_j^{(k)}(\beta)+1$, respectively. They are available from the signals with the same labels in Fig. 9.32(c) and Fig. 9.35. Moreover, the architecture in Fig. 9.37 also generates the flag f_j^r that tells whether the updated $R_j^{(k)max2}$ is equal to or larger than $\gamma - 1$.

It has been shown in [152] that the IHRB and E-IHRB decoders can achieve more than ten times higher hardware efficiency than the Min-max decoders in terms of throughput-over-area ratio. The major saving comes from the simple check and variable node updating. Moreover, the memory requirement is sig-nificantly lower since the messages sent from a variable node to all connected check nodes are the same in the IHRB decoder. Although the E-IHRB algo-rithm utilizes extrinsic v2c messages to improve the performance, those mes-sages are not stored and are generated from a few vectors and single-bit flags. As a result, the memory requirement of the E-IHRB algorithm is only slightly higher than that of the IHRB decoder. Nevertheless, the error-correcting per-

formance of these majority-logic-based decoders deteriorates when the column
weight of the code is small.

Bibliography

[1] R. Lidl and H. Niederreiter, *Finite Fields*, Addison-Wesley, 1983.

[2] R. J. McEliece, *Finite Fields for Computer Scientists and Engineers*, Kluwer, Boston, 1987.

[3] A. J. Menezes (Ed.), *Application of Finite Fields*, Kluwer, 1993.

[4] E. R. Berlekamp, *Algebraic Coding Theory*, McGraw-Hill, New York, 1968.

[5] S. B. Wicker, *Error Control Systems for Digital Communication and Storage*, Prentice Hall, Upper Saddle River, New Jersey, 1995.

[6] L. Shu and D. J. Costello, *Error Control Coding : Fundamentals and Applications*, 2nd Ed., Pearson-Prentice Hall, Upper Saddle River, 2004.

[7] X. Zhang and K. K. Parhi, "Fast factorization architecture in soft-decision Reed-Solomon decoding," *IEEE Trans. VLSI Syst.*, vol. 13, no. 4, pp. 413-426, Apr. 2005.

[8] S. K. Jain and K. K. Parhi, "Low latency standard basis $GF(2^M)$ multiplier and squarer architectures," *Proc. of Intl. Conf. on Acoustics, Speech, and Signal Processing*, pp. 2747-2750, Washington, May 1995.

[9] S. K. Jain, L. Song and K. K. Parhi, "Efficient semi-systolic architectures for finite field arithmetic," *IEEE Trans. on VLSI Syst.*, vol. 6, pp. 101-113, Mar. 1998.

[10] L. Gao and K. K. Parhi, "Custom VLSI design of efficient low latency and low power finite field multiplier for Reed-Solomon codec," *Proc. of IEEE Intl. Symp. on Circuits and Syst,*, pp. 574-577, Sydney, Australia, May 2001.

[11] W. Kim, S. Kim H. Cho and K. Lee, "A fast-serial finite field multiplier without increasing the number of registers," *Proc. of IEEE Intl. Symp. on Circuits and Syst.*, pp. 157-160. Bangkok, Thailand, May 2003.

[12] E. D. Mastrovito, "On fast Galois-Field multiplication," *Proc. of IEEE Symp. on Info. Theory*, pp. 348-348, Hungary, Jun. 1991.

[13] B. Sunar and C. K. Koc, "Mastrovito multiplier for all trinomials," *IEEE Trans. on Computers*, vol. 48, no. 5, pp. 522-527, May 1999.

[14] L. Song and K. K. Parhi, "Low-complexity modified Mastrovito multipliers over finite fields $GF(2^M)$," *Proc. of IEEE Intl. Symp. on Circuits and Syst.*, pp. 508-512, Florida, May 1999.

[15] A. Halbutogullari and C. K. Koc, "Mastrovito multiplier for general irreducible polynomials," *IEEE Trans. on Computers*, vol. 49, no. 5, pp. 503-518, May 2000.

[16] C. C. Wang, T. K. Troung, H. M. Shao, L. J. Deutsch, J. K. Omura and I. S. Reed, "VLSI architectures for computing multiplications and inverses in $GF(2^m)$," *IEEE Trans. on Computers*, vol. 34, no. 8, pp. 709-717, Aug. 1985.

[17] M. A. Hasan, M. A. Wang and V. K. Bhargava, "A modified Massey-Omura parallel multiplier for a class of finite fields," *IEEE Trans. on Computers*, vol. 42. no. 10, pp. 1278-1280, Oct. 1993.

[18] C. K. Koc and B. Sunar, "Low-complexity bit-parallel canonical and normal basis multipliers for a class of finite fields," *IEEE Trans. Computers*, vol. 47, no. 3, pp. 353-356, Mar. 1998.

[19] A. Reyhani-Masoleh and M. A. Hasan, "A new construction of Massey-Omura parallel multiplier over $GF(2^m)$," *IEEE Trans. on Computers*, vol. 51, no. 5, pp. 511-520, May 2002.

[20] A. Reyhani-Masoleh and M. A. Hasan, "Low complexity word-level sequential normal basis multipliers," *IEEE Trans. on Computers*, vol. 54, no. 2, pp. 98-110, Feb. 2005.

[21] K. K. Parhi, *VLSI Digital Signal Processing Systems: Design and Implementation*, John Wiley & Sons, 1999.

[22] "Advanced Encryption Standard (AES)," *Federal Information Processing Standards publication 197*, Nov. 26, 2001.

[23] I. S. Reed and G. Solomon, "Polynomial codes over certain finite fields," *SIAM Journ. on Applied Mathematics*, vol. 8, pp. 300-304, 1960.

[24] M. H. Jing, Y. H. Chen, Y. T. Chang and C. H. Hsu, "The design of a fast inverse module in AES," *Proc. of Intl. Conf on Info-tech and Info-net*, vol. 3, pp. 298-303, Beijing, China, Nov. 2001.

[25] T. Itoh and S. Tsujii, "A fast algorithm for computing multiplicative inverses in $GF(2^m)$ using normal bases," *Information and Computations*, vol. 78, pp. 171-177, 1998.

[26] C. Paar, *Efficient VLSI Architecture for Bit-Parallel Computations in Galois Field*, Ph.D thesis, Institute for Experimental Mathematics, University of Essen, Germany, 1994.

[27] S. H. Gerez, S. M. Heemstra De Groot, and O. E. Herrmann, "A polynomial-time algorithm for the computation of the iteration-period bound in recursive data-flow graphs," *IEEE Trans. on Circuits and Syst.-I*, vol. 39, no. 1, pp. 49-52, Jan. 1992.

[28] K. Ito, and K. K. Parhi, "Determining the minimum iteration period of an algorithm," *Journ. of Signal Processing Syst.*, vol. 11, pp. 629-634, Sep. 1993.

[29] K. Lee, H.-G. Kang, J.-I. Park and H. Lee, "A high-speed low-complexity concatenated BCH decoder architecture for 100Gb/s optical communications," *Journ. Signal Processing Syst.*, vol. 6, no. 1, pp. 43-55, Jan. 2012.

[30] Y. Chen and K. K. Parhi, "Small area parallel Chien search architecture for long BCH codes," *IEEE Trans. VLSI Syst.*, vol. 12, no. 5, pp. 545-549, May, 2004.

[31] J. Ma, A. Vardy and Z. Wang, "Low-latency factorization architecture for algebraic soft-decision decoding of Reed-Solomon codes," *IEEE Trans. VLSI Syst.*, vol. 15, no. 11, pp. 1225-1238, Nov. 2007.

[32] X. Zhang and Z. Wang, "A low-complexity three-error-correcting BCH decoder for optical transport network," *IEEE Trans. on Circuits and Systems-II*, vol. 59, no. 10, pp. 663-667, Oct. 2012.

[33] H. Okano and H. Imai, "A construction method of high-speed decoders using ROM's for Bose-Chaudhuri-Hocquenghem and Reed-Solomon codes," *IEEE Trans. on Computers*, vol. C.-36, no. 10, pp. 1165-1171, Oct. 1987.

[34] M. E. O'Sullivan and X. Zhang, "A new method for decoding 3-error-correcting BCH codes for optical transport network," preprint.

[35] I. S. Reed and G. Solomon, "Polynomial codes over certain finite fields," *Journ. Society Indust. and Appl. Math*, vol. 8, pp. 300-304, Jun. 1960.

[36] W. Zhang, X. Zhang and H. Wang, "Increasing the energy efficiency of WSNs using algebraic soft-decision Reed-Solomon decoders," *Proc. of IEEE Asia Pacific Conf. on Circuits and Syst.*, pp. 49-52, Taiwan, Dec. 2012.

[37] J. H. Baek and M. H. Sunwoo, "Simplified degree computationless Euclid's algorithm and its architecture," *Proc. IEEE Intl. Symp. Circuits and Syst.*, pp. 905-908, May 2007.

[38] L. R. Welch and E. R. Berlekmap, "Error correction for algebraic block codes," U.S. patent no. 4,633,470, Dec. 1986.

[39] W. W. Peterson, "Encoding and error-correction procedures for the Bose-Chaudhuri codes," *IRE Trans. on Info. Theory*, vol. 6, pp. 459-470, Sep. 1960.

[40] D. Gorenstein and N. Zierler, "A class of error correcting codes in p^m symbols," *Journ. Society Indust. and Appl. Math*, vol. 9, pp. 207-214, Jun. 1961.

[41] J. L. Massey, "Shift register synthesis and BCH decoding," *IEEE Trans. on Info. Theory*, vol. 15, no. 1, pp. 122-127, Jan. 1969.

[42] D. G. Forney, "On decoding of BCH codes," *IEEE Trans. on Info. Theory*, vol. IT-11, pp. 549-557, Oct. 1965.

[43] D. V. Sarwate and N. R. Shanbhag, "High-speed architecture for Reed-Solomon decoders," *IEEE Trans. VLSI Syst.*, vol. 9, no. 5, pp. 641-655, Oct. 2001.

[44] K. Seth, *et. al.*, "Ultra folded high-speed architectures for Reed-Solomon decoders," *Proc. IEEE Intl. Conf. on VLSI Design*, 2006.

[45] G. D. Forney, Jr., "Generalized minimum distance decoding," *IEEE Trans. on Info. Theory*, vol. 12, no. 2, pp. 125-131, Apr. 1966.

[46] D. Chase, "A Class of algorithms for decoding block codes with channel measurement information," *IEEE Trans. on Info. Theory*, vol. 18, pp. 170-182, Jan. 1972.

[47] R. Koetter and A. Vardy, "Algebraic soft-decision decoding of Reed-Solomon codes," *IEEE Trans. on Info. Theory*, vol. 49, no. 11, pp. 2809-2825, Nov. 2003.

[48] F. Parvaresh and A. Vardy, "Multiplicity assignments for algebraic soft-decoding of Reed-Solomon codes," *Proc. IEEE Intl. Symp. Info. Theory*, pp. 205, Yokohama, Japan, Jul. 2003.

[49] M. El-Khamy and R. J. McEliece, "Interpolation multiplicity assignment algorithms for algebraic soft-decision decoding of Reed-Solomon codes," *AMS-DIMACS volume on Algebraic Coding Theory and Info. Theory*, vol. 68, 2005.

[50] N. Ratnakar and R. Köetter, "Exponential error bounds for algebraic soft-decision decoding of Reed-Solomon codes," *IEEE Trans. on Info. Theory*, vol. 51 pp. 3899-3917, Nov. 2005.

[51] J. Bellorado and A. Kavcic, "Low-complexity soft-decoding algorithms for Reed-Solomon codes-part I: an algebraic soft-in hard-out Chase decoder," *IEEE Trans. on Info. Theory*, vol. 56, pp. 945-959, Mar. 2010.

[52] J. Jiang and K. Narayanan, "Algebraic soft-decision decoding of Reed-Solomon codes using bit-level soft information," *IEEE Trans. on Info. Theory*, vol. 54, no. 9, pp. 3907-3928, Sep. 2008.

[53] M. Sudan, "Decoding of Reed-Solomon codes beyond the error correction bound," *Journal of Complexity*, vol. 12, pp. 180-193, 1997.

[54] V. Guruswami and M. Sudan, "Improved decoding of Reed-Solomon and algebraic-geometric codes," *IEEE Trans. on Info. Theory*, vol. 45, pp. 1755-1764, Sep. 1999.

[55] J. Zhu, X. Zhang and Z. Wang, "Backward interpolation for algebraic soft-decision Reed-Solomon decoding," *IEEE Trans. on VLSI Syst.*, vol. 17, no. 11, pp. 1602-1615, Nov. 2009.

[56] W. J. Gross, *et. al.*, "A VLSI architecture for interpolation in soft-decision decoding of Reed-Solomon codes," *Proc. IEEE Workshop on Signal Processing Systems*, pp. 39-44, San Diego, Oct. 2002.

[57] R. Köetter and A. Vardy, "A complexity reducing transformation in algebraic list decoding of Reed-Solomon codes," *Proc. Info. Theory Workshop*, Paris, France, Mar. 2003.

[58] J. Zhu and X. Zhang, "High-speed re-encoder design for algebraic soft-decision Reed-Solomon decoding," *Proc. IEEE Intl. Symp. on Circuits and Syst.*, pp. 465-468, Paris, France, May 2010.

[59] J. Ma, A. Vardy and Z. Wang, "Reencoder design for soft-decision decoding of an (255, 239) Reed-Solomon code," *Proc. IEEE Intl. Symp. Circuits and Syst.*, pp. 3550-3553, Island of Kos, Greece, May 2006.

[60] X. Zhang, J. Zhu and W. Zhang, "Efficient re-encoder architectures for algebraic soft-decision Reed-Solomon decoding," *IEEE Trans. on Circuits and Syst.-II*, vol. 59, no. 3, pp. 163-167, Mar. 2012.

[61] R. Kötter, "Fast generalized minimum-distance decoding of algebraic-geometry and Reed-Solomon codes," *IEEE Trans. Info. Theory*, vol. 42, no. 3, pp. 721-737, May 1996.

[62] R. Köetter, *On Algebraic Decoding of Algebraic-Geometric and Cyclic Codes*, Ph.D. thesis, Dept. of Elec. Engr., Linköping University, Linköping, Sweden, 1996.

[63] R. R. Nielson, *List Decoder of Linear Block Codes*, Ph.D. thesis, Technical University of Denmark, Denmark, 2001.

[64] Z. Wang and J. Ma, "High-speed interpolation architecture for soft-decision decoding of Reed-Solomon codes," *IEEE Trans. VLSI Syst.*, vol. 14, no. 9, pp. 937-950, Sep. 2006.

[65] X. Zhang and J. Zhu, "Efficient interpolration architecture for soft-decision Reed-Solomon decoding by applying slow-down," *Proc. IEEE Workshop Signal Processing Syst.*, pp. 19-24, Washington D. C., Oct. 2008.

[66] J. Zhu, X. Zhang and Z. Wang, "Combined interpolation architecture for soft-decision decoding of Reed-Solomon codes," *Proc. IEEE Intl. Conf. Computer Design*, pp. 526-531, Lake Tahoe, CA, Oct. 2008.

[67] X. Zhang and J. Zhu, "High-throughput interpolation architecture for algebraic soft-decision Reed-Solomon decoding," *IEEE Trans. Circuits and Syst.-I*, vol. 57, no. 3, pp. 581-591, Mar. 2010.

[68] X. Zhang and J. Zhu, "Algebraic soft-decision decoder architectures for long Reed-Solomon codes," *IEEE Trans. on Circuits and Syst.-II*, vol. 57. no. 10, pp. 787-792, Oct. 2010.

[69] K. Lee and M. O'Sullivan, "An interpolation algorithm using Gröbner bases for soft-decision decoding of Reed-Solomon codes," *Proc. IEEE Intl. Symp. Info. Theory*, pp. 2032-2036, Seattle, Washington, Jul. 2006.

[70] X. Zhang, "Reduced complexity interpolation architecture for soft-decision Reed-Solomon decoding," *IEEE Trans. on VLSI Syst.*, vol. 14, no. 10, pp. 1156-1161, Oct. 2006.

[71] A. Ahmed, R. Koetter and N. Shanbhag, "VLSI architecture for soft-decision decoding of Reed-Solomon codes," *Proc. IEEE Intl. Conf. Commun.*, pp. 2584-2590, Paris, France, Jun. 2004.

[72] H.O'Keeffe and P. Fitzpatrick, "Gröbner basis solutions of constrained interpolation problems," *Linear Algebra and its Applications*, vol. 351-352, pp. 533-551, 2002.

[73] X. Zhang and J. Zhu, "Low-complexity interpolation architecture for soft-decision Reed-Solomon decoding," *Proc. IEEE Intl. Symp. Circuits and Syst.*, pp. 1413-1416, New Orlean, Louisiana, May 2007.

[74] J. Zhu and X. Zhang, "Efficient VLSI architecture for soft-decision decoding of Reed-Solomon codes" *IEEE Trans. on VLSI Syst.*, no. 55. vol. 10, pp. 3050-3062, Nov. 2008.

[75] R. M. Roth and G. Ruckenstein, "Efficient decoding of Reed-Solomon codes beyond half the minimum distance," *IEEE Trans. Info. Theory*, vol. 46, no. 1, pp. 246-257, Jan. 2000.

[76] X. Zhang, "Further exploring the strength of prediction in the factorization of soft-decision Reed-Solomon decoding," *IEEE Trans. on VLSI Syst.*, vol. 15, no. 7, pp. 811-820, Jul. 2007.

[77] X. Zhang, "Partial parallel factorization in soft-decision Reed-Solomon decoding," *Proc. of ACM Great Lakes Symp. on VLSI*, pp. 272-277, Philadelphia, PA, Apr. 2006.

[78] J. Zhu and X. Zhang, "Factorization-free low-complexity Chase soft-decision decoding of Reed-Solomon codes," *Proc. IEEE Intl. Symp. on Circuits and Syst.*, pp. 2677-2680, Taiwan, May 2009.

[79] X. Zhang, Y. Wu and J. Zhu, "A novel polynomial selection scheme for low-complexity Chase algebraic soft-decision Reed-Solomon decoding," *Proc. IEEE Intl. Symp. on Circuits and Syst.*, pp 2689-2692, Brazil, May 2011.

[80] X. Zhang, Y. Wu, J. Zhu, and Y. Zheng, "Novel polynomial selection and interpolation for low-complexity Chase algebraic soft-decision Reed-Solomon decoding," *IEEE Trans. on VLSI Syst.*, vol. 20, no. 7, pp. 1318-1322, Jul. 2012.

[81] X. Zhang and Y. Zheng, "Efficient codeword recovery architecture for low-complexity Chase Reed-Solomon decoding," *Proc. Info. Theory and Applications Workshop*, San Diego, Feb. 2011.

[82] X. Zhang and Y. Zheng, "Systematically re-encoded algebraic soft-decision Reed-Solomon decoder," *IEEE Trans. on Circuits and Syst.-II*, vol. 59, no. 6, pp. 376-380, Jun. 2012.

[83] X. Zhang and Y. Zheng, "Generalized backward interpolation for algebraic soft-decision decoding of Reed-Solomon codes," *IEEE Trans. on Commun.*, vol. 61, no. 1, pp. 13-23, Jan. 2013.

[84] J. Zhu and X. Zhang, "Efficient generalized minimum-distance decoder of Reed-Solomon codes," *Proc. IEEE Intl. Conf. on Acoustics, Speech and Signal Processing*, pp. 1502-1505, Dallas, TX, Mar. 2010.

[85] J. Zhu and X. Zhang, "Efficient generalized minimum-distance decoders of Reed-Solomon codes," *Springer Journ. of Signal Processing Syst.*, vol. 66, no. 3, pp. 245-257, Mar. 2012.

[86] Y. Wu, "Fast Chase decoding algorithms and architectures for Reed-Solomon codes, *IEEE Trans. Info. Theory*, vol. 58, no. 1, pp. 109-129, Jan. 2012.

[87] X. Zhang, Y. Zheng, and Y. Wu, "A Chase-type Köetter-Vardy algorithm for soft-decision Reed-Solomon decoding," *Proc. Intl. Conf. on Computing, Networking and Commun.*, pp. 466-470, Maui, HI, Feb. 2012.

[88] A. Hocquenghem, *Codes Correcteurs d'Erreurs*, Chiffres, vol. 2, pp. 147-156, 1959.

[89] R. C. Bose, and D. K. Ray-Chaudhuri, "On a class of error correcting binary group codes," *Info. and Control*, vol. 3, pp. 68-79, Mar. 1960.

[90] K. K. Parhi, "Eliminating the fanout bottleneck in parallel long BCH encoders," *IEEE. Trans. on Circuits and Syst.-I*, vol. 51, No. 3, pp. 2611-2615, Mar.2004.

[91] X. Zhang and K. K. Parhi, "High-speed architectures for parallel long BCH encoders," *IEEE Trans. on VLSI Syst.*, vol. 13, no. 7, pp. 872-877, Jul. 2005.

[92] S. Paul, F. Cai, X. Zhang and S. Bhunia, "Reliability-driven ECC allocation for multiple bit error resilience in processor cache," *IEEE Trans. on Computers*, vol. 60, no. 1, pp. 20-34, Jan. 2011.

[93] X. Zhang, F. Cai, and M. P. Anantram, "Low-energy and low-latency error correction for phase change memory," *Proc. IEEE Intl. Symp. on Circuits and Syst.*, pp. 1236-1239, Beijing, China, May 2013.

[94] J. Justesen, K. J. Larsen, and L. A. Pedersen, "Error correcting coding for OTN", *IEEE Commun. Magazine*, vol. 48, no. 9, pp. 70-75, Sep. 2010.

[95] N. Kamiya, "An algebraic soft-decision decoding algorithms for BCH codes," *IEEE Trans. Info. Theory*, vol. 47, no. 1, pp. 45-58, Jan. 2001.

[96] Y. Wu, "Fast Chase decoding algorithms and architectures for Reed-Solomon codes," *IEEE Trans. Info. Theory*, vol. 58, no. 1, pp. 109-129, Jan. 2012.

[97] X. Zhang, J. Zhu and Y. Wu, "Efficient one-pass Chase soft-decision BCH decoder for multi-level cell NAND flash memory," *Proc. IEEE Intl. Midwest Symp. Circuits and Syst.*, Aug. 2011.

[98] X. Zhang, "An efficient interpolation-based BCH Chase decoder," *IEEE Trans. on Circuits and Syst.-II*, vol. 60, no. 4, pp. 212-216, Apr. 2013.

[99] R. G. Gallager, "Low desnity parity check codes," *IRE Trans. Info. Theory*, vol. 8, pp. 21-28, Jan, 1962.

[100] Y. Kou, S. Lin, and M. P. C. Fossorier, "Low-density parity-check codes based on finite geometries: a rediscovery and new results," *IEEE Trans. on Info. Theory*, vol. 47, no. 7, pp. 2711-2736, Jul. 2001.

[101] Z. Li, *et. al.*, "Efficient encoding of quasi-cyclic low-density parity-check codes," *IEEE Trans. on Commun.*, vol. 54, no. 1, pp. 71-81, Jan, 2006.

[102] X. Hu, E. Eleftheriou, and D. M. Arnold, "Regular and irregular progressive edge-growth Tanner graphs," *IEEE Trans. on Info. Theory*, vol. 51, no. 1, pp. 386-398, Jan. 2005.

[103] W. Ryan and S. Lin, *Channel Codes: Classical and Modern*, Cambridge University Press, 2009.

[104] M. Davey and D. J. MacKay, "Good error-correcting codes based on very sparse matrices, *IEEE Trans. Info. Theory*, vol. 45, pp. 399-431, Mar. 1999.

[105] J. Chen and M. P. C. Fossorier, "Density evolution for two improved BP-based decoding algorithms of LDPC codes," *IEEE Commu. Letters*, vol. 6, no. 5, pp. 208-210, May 2002.

[106] Y. Kou, S. Lin, and M. P. C. Fossorier, "Low density parity check codes based on finite geometries, a rediscovery and new result," *IEEE Trans. Info. Theory*, vol. 47, no. 2, pp. 619-637, Feb. 2001.

[107] J. Zhang and M. R. C. Fossorier, "A modified weighted bit-flipping decoding for low-density parity-check codes," *IEEE Commun. Lett.*, vol. 8, pp. 165-167, Mar. 2004.

[108] Q. Huang, *et. al.*, "Two reliability-based iterative majority-logic decoding algorithms for LDPC codes," *IEEE Trans. on Commun.*, vol. 57, no. 12, pp. 3597-3606, Dec. 2009.

[109] T. Richardson, "Error floors of LDPC codes," *Proc. 41st Annual Allerton Conf. on Commun. Control and Computing*, 2003.

[110] B. Vasic, S. K. Chilappagari, D. V. Nguyen, and S. K. Planjery, "Trapping set ontology," *Proc. 47th Annual Allerton Conf. on Commun., Control, and Computing*, Sep. 2009.

[111] Y. Han and W. E. Ryan, "Low-floor decoders for LDPC codes," *IEEE Trans. on Commun.*, vol. 57 , no. 6, pp.1663-1673, Jun. 2009.

[112] J. Kang, Q. Huang, S. Lin, and K. Abdel-Ghaffar, "An iterative decoding algorithm with backtracking to lower the error-floors of LDPC codes," *IEEE Trans. on Commun.*, vol. 59, no. 1, pp. 64-73, Jan. 2011.

[113] Z. Zhang, *et. al.*, "An efficient 10GBASE-T Ethernet LDPC decoder design with low error floors" *IEEE Journ. Solid-State Circuits*, vol. 45, no. 4, pp. 843-855, Apr. 2010.

[114] D. Declercq, B. Vasic, S.K. Planjery and E. Li, "Finite alphabet iterative decoders, part II: towards guaranteed error correction of LDPC codes via iterative decoder diversity", *IEEE Trans. on Commun.*, vol. 61, no. 10, pp. 4046-4057, Oct. 2013.

[115] F. Cai *et. al.*, "Finite alphabet iterative decoder for LDPC codes: optimization, architecture and analysis," *IEEE Trans. on Circuits and Syst.-I*, vol. 61, no. 5, pp. 1366-1375, May 2014.

[116] L. Liu and C.-J. Shi, "Sliced message passing: high throughput over-lapped decoding of high-rate low-density parity-check Codes," *IEEE Trans. on Circuits and Syst.-I*, vol. 55, no. 11, pp. 3697-3710, Dec. 2008.

[117] M. Mansour, N. Shanbhag, "A 640-Mb/s 2048-bit programmable LDPC decoder chip," *IEEE Journ. of Solid-State Circuits*, vol. 41, no. 3, pp. 684-698, 2006.

[118] J. Zhang and M. P. C. Fossorier, "Shuffled iterative decoding," *IEEE Trans. on Commun.*, vol. 53, no. 2, pp. 209-213, Feb. 2005.

[119] Y. Sun, G. Wang and J. Cavallaro, "Multi-layer parallel decoding algorithm and VLSI architecture for quasi-cyclic LDPC codes," *Proc. IEEE Intl. Symp. Circuits and Syst.*, pp. 1776-1779, Rio De Janeiro, Brazil, May 2011.

[120] X. Zhang, "High-speed multi-block-row layered decoding for quasi-cyclic LDPC codes," *Proc. IEEE Global Conf. on Signal and Info. Processing*, Feb. 2015.

[121] Z. Cui, Z. Wang, X. Zhang, Q. Jia, "Efficient decoder design for high-throughput LDPC decoding," *Proc. IEEE Asia Pacific Conf. on Circuits and Syst.*, pp. 1640-1643, Nov. 2008, Macro, China.

[122] D. E. Knuth, *The Art of Computer Programming*, 2nd ed., New York: Addison-Wesley.

[123] C.-L. Wey, M.-D. Shieh, and S.-Y. Lin, "Algorithm of finding the first two minimum values and their hardware implementation," *IEEE Trans on Circuits and Syst.-I*, vol. 55, no. 11, pp. 3430-3437, Dec. 2008.

[124] L. G. Amaru, M. Martina and G. Masera, "High-speed architectures for finding the first two maximum/minimum values," *IEEE Trans. on VLSI Syst.*, vol. 20, no. 12, pp. 2342-2346, Dec. 2012.

[125] A. Darabiha, A. C. Carusone, and F. R. Kschischang, "A bit-serial approximate Min-Sum LDPC decoder and FPGA implementation," *Proc. IEEE Intl. Symp. Circuits and Syst.*, May 2006.

[126] J. H. Lee and M. H. Sunwoo, "Bit-serial first two min values generator for LDPC decoding," *preprint*.

[127] D. Oh and K. K. Parhi, "Min-sum decoder architectures with reduced word length for LDPC codes," *IEEE Trans. on Circuits and Syst.-I*, vol. 17, no. 1, pp.105-115, Jan. 2010.

[128] A. Blad, O. Gustafsson, and L. Wanhammar, "An early decision decoding algorithm for LDPC codes using dynamic thresholds," *Proc. European Conf. Circuit Theory Design*, pp. 285-288, 2005.

[129] Z. Cui, L. Chen, and Z. Wang, "An efficient early stopping scheme for LDPC decoding", *Proc. 13th NASA Symp. on VLSI Design*, 2007.

[130] X. Zhang, F. Cai and R. Shi, "Low-power LDPC decoding based on iteration prediction," *Proc. IEEE Intl. Symp. on Circuits and Syst.*, pp. 3041-3044, Seoul, Korea, May 2012.

[131] M. Davey and D. J. MacKay, "Low density parity check codes over GF(q)," *IEEE Commun. Letter*, vol. 2, pp. 165-167, Jun. 1998.

[132] S. Pfletschinger *et. al.*, "Performance evaluation of non-binary LDPC codes on wireless channels," *Proc. ICT-Mobile Summit*, Santander, Spain, Jun. 2009.

[133] B. Zhou, et. al., "Array dispersions of matrices and constructions of quasi-cyclic LDPC codes over non-binary fields," *Proc. IEEE Intl. Symp. on Info. Theory*, pp. 1158-1162, Toronto, Canada, Jul. 2008.

[134] L. Barnault, D. Declercq, "Fast decoding algorithm for LDPC over $GF(2^q)$", *Proc. Info. Theory Workshop*, pp. 70-73, Paris, France, Mar. 2003.

[135] H. Wymeersch, H. Steendam and M. Moeneclaey, "Log-domain decoding of LDPC codes over $GF(q)$," *Proc. IEEE Intl. Conf. on Commun.*, pp. 772-776, Paris, France, Jun. 2004.

[136] C. Spagnol, E. Popovici and W. Marnane, "Hardware implementation of $GF(2^m)$ LDPC decoders," *IEEE Trans. on Circuits and Syst.-I*, vol 56, no. 12, pp. 2609-2620, Dec. 2009.

[137] D. Declercq, M. Fossorier, "Decoding algorithms for nonbinary LDPC codes over GF(q)", *IEEE Trans. on Commun.*, vol. 55, no. 4, pp. 633-643, Apr. 2007.

[138] V. Savin, "Min-Max decoding for non binary LDPC codes," *Proc. IEEE Intl. Symp. on Info. Theory*, pp. 960-964, Toronto, Canada, Jul. 2008.

[139] C.-Y. Chen, Q. Huang, C.-C. Chao, and S. Lin, "Two low-complexity reliability-based message-passing algorithms for decoding non-binary LDPC codes," *IEEE Trans. on Commu.*, vol. 58, no. 11, pp. 3140-3147, Nov. 2010.

[140] J. Lin, J. Sha, Z. Wang and L. Li, "Efficient decoder design for nonbinary quasicyclic LDPC codes", *IEEE Trans. on Circuits and Syst.-I*, vol. 57, no. 5, pp. 1071-1082, May. 2010.

[141] F. Cai and X. Zhang, "Efficient check node processing architectures for non-binary LDPC decoding using power representation." *Proc. IEEE Workshop on Signal Processing Syst.*, pp. 137-142. 2012.

[142] F. Cai and X. Zhang, "Effcient check node processing architectures for non-binary LDPC decoding using power representation," *Springer Journ. of Signal Processing Syst.*, vol. 76, no. 2, pp. 211-222, Aug. 2014.

[143] X. Zhang and F. Cai, "Efficient partial-parallel decoder architecture for quasi-cyclic non-binary LDPC codes," *IEEE Trans. on Circuits and Syst.-I*, vol. 58, no. 2, pp. 402-414, Feb. 2011.

[144] X. Zhang and F. Cai, "Reduced-complexity decoder architecture for non-binary LDPC codes," *IEEE Trans. on VLSI Syst.*, vol. 19, no. 7, pp. 1229-1238, Jul. 2011.

[145] X. Chen and C. Wang, "High-throughput efficient non-binary LDPC decoder based on the simplified Min-sum algorithm," *IEEE Trans. on Circuits and Syst.-I*, vol. 59, no. 11, pp. 2784-2794, Nov. 2012.

[146] E. Li, D. Declercq, and K. Gunnam, "Trellis-based extended min-sum algorithm for non-binary LDPC codes and its hardware structure," *IEEE Trans. on Commun.*, vol. 61, no. 7, pp. 2600-2611, Jul. 2013.

[147] X. Zhang, "Modified trellis-based min-max decoder for non-binary LDPC codes," *Proc. Intl. Conf. on Computing, Networking and Commun.*, San Diego, CA, Feb. 2015.

[148] F. Cai and X. Zhang, "Relaxed Min-max decoder architectures for non-binary LDPC code," *IEEE Trans. on VLSI Syst.*, vol. 21, no. 11, pp. 2010-2023, Nov. 2013.

[149] A. Voicila, *et. al.*, "Low-complexity decoding for non-binary LDPC codes in high order fields," *IEEE Trans. on Commun.*, vol. 58, no. 5, pp. 1365-1375, May 2010.

[150] E. Boutillon, and L. Conde-Canencia, "Simplified check node processing in nonbinary LDPC decoders," *Intl. Symp. on Turbo Codes and Iterative Info. Processing*, pp. 201-205, France, Sep. 2010.

[151] X. Zhang and F. Cai, "Reduced-complexity extended Min-sum check node processing for non-binary LDPC decoding," *Proc. IEEE Intl. Midwest Symp. on Circuits and Syst.*, pp. 737-740, Seattle, WA, Aug. 2010.

[152] X. Zhang, F. Cai, and S. Lin, "Low-complexity reliability-based message-passing decoder architectures for non-binary LDPC codes," *IEEE Trans. on VLSI Syst.*, vol. 20, no. 11, pp. 1938-1950. Nov. 2012.

[153] X. Zhang and F. Cai, "An efficient architecture for iterative soft reliability-based majority-logic non-binary LDPC decoding," *Proc. Asilomar Conf. on Signals, Syst., and Computers*, pp. 885-888, Pacific Grove, CA, Nov. 2011.

Index

A

Additive operations, 2
Additive white Gaussian noise (AWGN) channel, 79
Affine polynomial, 40, 41, 42
Algebraic soft-decision (ASD) decoding algorithms. *See also* Re-encoded algebraic soft-decision decoders erasure decoding, 87
 erasure decoding, 87
 factorization step, 77, 78
 interpolation step, 77, 78
 Kötter-Vardy (KV) ASD algorithm; *see* Kötter-Vardy (KV) ASD algorithm overview, 76
 pipelined, 88–89

B

Basis conversions, 114-117
Basis representations, 68
Belief propagation algorithms, 228
 binary LDPC coding, for, 271
 log domain, 275
Bellman-Ford algorithms, 28
Berlekamp's algorithm, 207, 208
Berlekamp-Massey algorithms (BMAs), 52, 54
 error locations, 59, 60
 error locator polynomials, 62, 67
 key equation solver (KES) step, 67

error magnitude computations, 67
 one-pass schemes, 144
 reformulated inversionless (riBM); *see* Reformulated inversionless Berlekamp-Massey algorithm (riBM)
 ultra-ultra folded inversionless Berlekamp-Massey algorithm (UiBM); *see* Ultra-ultra folded inversionless Berlekamp-Massey algorithm (UiBM)
 Berlekamp-Welch algorithms, 54
 key equation solver (KES) step, 67
Bit-flipping algorithms, 230–231
Bit-level generalized minimum distance (BGMD) decoder, 77–78, 79, 111, 156
Bit-serial multi-input sorter architectures, 255
Bose-Chaudhuri-Hocquenghem (BCH) algorithms, 1, 39
 binary, 56, 189–190, 204, 206, 208
 decoding, 189, 196–197, 208
 encoding, 189, 191–192, 195–196, 198, 200, 201
 error locator, 204, 207
 hard-decision; *see* Hard-decision BCH decoders
Bubble check schemes, 345

C

N

M

O